LONDON MATHEMATICAL SOCIE'

Managing Editor: Ian J. Leary,
Mathematical Sciences, University of Southampton, UK

57 Introduction to Banach algebras, operators and harmonic a
58 Computational algebraic geometry, HAL SCHENCK
59 Frobenius algebras and 2-D topological quantum field thec
60 Linear operators and linear systems, JONATHAN R. PART
61 An introduction to noncommutative Noetherian rings (2nd Edition), K. R. GOOD
R. B. WARFIELD, JR
62 Topics from one-dimensional dynamics, KAREN M. BRUCKS & HENK BRUIN
63 Singular points of plane curves, C. T. C. WALL
64 A short course on Banach space theory, N. L. CAROTHERS
65 Elements of the representation theory of associative algebras I, IBRAHIM ASSEM, DANIEL SIMSON & ANDRZEJ SKOWROŃSKI
66 An introduction to sieve methods and their applications, ALINA CARMEN COJOCARU & M. RAM MURTY
67 Elliptic functions, J. V. ARMITAGE & W. F. EBERLEIN
68 Hyperbolic geometry from a local viewpoint, LINDA KEEN & NIKOLA LAKIC
69 Lectures on Kähler geometry, ANDREI MOROIANU
70 Dependence logic, JOUKU VÄÄNÄNEN
71 Elements of the representation theory of associative algebras II, DANIEL SIMSON & ANDRZEJ SKOWROŃSKI
72 Elements of the representation theory of associative algebras III, DANIEL SIMSON & ANDRZEJ SKOWROŃSKI
73 Groups, graphs and trees, JOHN MEIER
74 Representation theorems in Hardy spaces, JAVAD MASHREGHI
75 An introduction to the theory of graph spectra, DRAGOŠ CVETKOVIĆ, PETER ROWLINSON & SLOBODAN SIMIĆ
76 Number theory in the spirit of Liouville, KENNETH S. WILLIAMS
77 Lectures on profinite topics in group theory, BENJAMIN KLOPSCH, NIKOLAY NIKOLOV & CHRISTOPHER VOLL
78 Clifford algebras: An introduction, D. J. H. GARLING
79 Introduction to compact Riemann surfaces and dessins d'enfants, ERNESTO GIRONDO & GABINO GONZÁLEZ-DIEZ
80 The Riemann hypothesis for function fields, MACHIEL VAN FRANKENHUIJSEN
81 Number theory, Fourier analysis and geometric discrepancy, GIANCARLO TRAVAGLINI
82 Finite geometry and combinatorial applications, SIMEON BALL
83 The geometry of celestial mechanics, HANSJÖRG GEIGES
84 Random graphs, geometry and asymptotic structure, MICHAEL KRIVELEVICH et al
85 Fourier analysis: Part I - Theory, ADRIAN CONSTANTIN
86 Dispersive partial differential equations, M. BURAK ERDOĞAN & NIKOLAOS TZIRAKIS
87 Riemann surfaces and algebraic curves, R. CAVALIERI & E. MILES
88 Groups, languages and automata, DEREK F. HOLT, SARAH REES & CLAAS E. RÖVER
89 Analysis on Polish spaces and an introduction to optimal transportation, D. J. H. GARLING
90 The homotopy theory of $(\infty, 1)$-categories, JULIA E. BERGNER
91 The block theory of finite group algebras I, M. LINCKELMANN
92 The block theory of finite group algebras II, M. LINCKELMANN
93 Semigroups of linear operators, D. APPLEBAUM
94 Introduction to approximate groups, M. C. H. TOINTON
95 Representations of finite groups of Lie type (2nd Edition), F. DIGNE & J. MICHEL
96 Tensor products of C*-algebras and operator spaces, G. PISIER

London Mathematical Society Student Texts 97

Topics in Cyclic Theory

DANIEL G. QUILLEN
University of Oxford

and

GORDON BLOWER
Lancaster University

CAMBRIDGE
UNIVERSITY PRESS

CAMBRIDGE
UNIVERSITY PRESS

University Printing House, Cambridge CB2 8BS, United Kingdom

One Liberty Plaza, 20th Floor, New York, NY 10006, USA

477 Williamstown Road, Port Melbourne, VIC 3207, Australia

314–321, 3rd Floor, Plot 3, Splendor Forum, Jasola District Centre, New Delhi – 110025, India

79 Anson Road, #06–04/06, Singapore 079906

Cambridge University Press is part of the University of Cambridge.

It furthers the University's mission by disseminating knowledge in the pursuit of education, learning, and research at the highest international levels of excellence.

www.cambridge.org
Information on this title: www.cambridge.org/9781108479615
DOI: 10.1017/9781108855846

First published 2020

Printed in the United Kingdom by TJ International Ltd. Padstow Cornwall

A catalogue record for this publication is available from the British Library.

ISBN 978-1-108-47961-5 Hardback
ISBN 978-1-108-79044-4 Paperback

To Daniel Quillen, *in memoriam*

Contents

Introduction

Daniel Gray Quillen gave graduate lectures at the Mathematical Institute in St Giles in Oxford as part of his duties as Waynflete Professor of Pure Mathematics. The audience consisted of postgraduate students, researchers and faculty members, mainly with interests in differential geometry, algebra and functional analysis. At the time, Yang–Mills theory and index theory for operators were pervasive topics in Oxford mathematics. The courses typically lasted for 16 lectures, and were intended as self-contained introductions to the research papers that Quillen was writing. Some of the material fed directly into the papers, whereas other ingredients included motivation and applications which do not feature in the published work. The latter is valuable to a newcomer to cyclic theory, and these notes emphasize the motivation and applications. Quillen kept a mathematical diary which records how he worked over topics to refine essential ideas and developed a profound understanding of diverse theories. The diaries show a remarkable breadth of interest, covering topics in several branches of mathematics. Quillen delivered lectures in traditional style, with calculations written in full detail, so the blackboard was soon filled with tensor products and commuting diagrams. As far as I am aware, he did not produce a collated set of lecture notes, and the record is primarily based upon my notes from the lectures.

The material in Chapters 3 and 4 of this volume is taken from the course 'Topics in K-theory and cyclic homology' which Quillen gave in Hilary term 1989, followed by 'Topics in K-theory and cyclic cohomology' in Michaelmas 1989. The Hilary 1989 course also covered material which I defer until Chapters 8 and 11, since it required more by way of prerequisites. The title of the present book reflects the emphasis of the courses on cyclic theory and its connection with other branches of mathematics. A special issue of the *Journal of K-Theory* **11** (3) (2013) describes the mathematical legacy of Quillen's work, particularly his contribution to K-theory. Segal [94] describes Quillen's professional career.

The original lectures were not always linearly ordered, as Quillen would revisit calculations and improve them, and digress into topics which were intended to provide motivation for forthcoming material. A significant simplification evidently emerged through his collaboration with Joachim Cuntz, and part of this work is produced here in Chapters 6 and 7, which is taken from the course 'Cyclic cohomology and Karoubi operators' given in Hilary term 1991, and updated in Trinity term 1992. The latter course also covered the material in Chapter 8, regarding connections, and Chapter 12 on Hodge decompositions.

Cyclic theory is an aspect of noncommutative geometry. Classical geometry uses commutative algebra, particularly to describe geometrical objects with additional structure such as differentiability or smoothness. Noncommutative geometry involves noncommutative algebras and seeks to describe geometrical objects which are possibly rough, or topologically complicated. In order to understand the basic definitions of cyclic theory, it is helpful to review some of the definitions of differential geometry that point towards the noncommutative theory.

These notes are intended as an introduction to some topics in cyclic theory and are assuredly not a systematic review. Apart from the historical interest in this material, Quillen's lecture presentations seem to me more accessible than the papers and emphasize the motivation of, and output from, cyclic theory. Some readers might be surprised at the apparent lack of generality, particularly that all the calculations are carried out over fields of characteristic zero, with $\mathbf{C}$ as the default choice of field. Generally, the presentation emphasizes analysis and differential geometry, rather than topology and homological algebra. Quillen motivated some of his results on noncommutative differential forms by comparing the theory with the algebraic approach to Kähler differential forms on algebraic varieties, as we discuss in Chapter 6. This may reflect the interests of Quillen's likely audience for the lectures, or his choice of contemporaneous reading material. His earliest researches were in the formal theory of partial differential equations, and he developed his interest in quantum mechanics and field theory. I have attempted to fill in some of the analytical details omitted from the lectures without making a meal of them, so the reader at least can appreciate what the analytical issues are and how they can be addressed. In so doing, I have used methods that can be readily understood from the viewpoint of a student; I have not introduced more difficult concepts such as metric tensor products, Lie group representations, diffusions on manifolds or Kasparov's KK theory.

There are some differences between Connes's cyclic theory and Quillen's which are worth highlighting. The Chern character is central to Connes's approach, and his B operator has a natural interpretation as a boundary operator. In these notes, the Chern character is mainly discussed for commutative algebras that arise in geometry, and the definition in Chapter 8 is taken from differential geometry, using connections. Connections are fundamental to the development of cyclic theory in this book. In Chapter 7, B is introduced via the Karoubi operator κ, and the properties of B emerge from some simple algebraic computations which do not reveal the geometrical motivation or interpretation. Instead, B and κ are used to turn graded complexes of cochains into modules over principal ideal domains or local rings. Quillen's approach to curvature and commutators fits neatly with the analytical theory of Helton and Howe, as mentioned in Chapters 3 and 9. Some of the results in Chapter 4 regarding algebraic approaches to dilations and extensions appear to be original contributions by Quillen that have the potential for further development. Chapters 6 and 7 emphasize both the similarities and differences between commutative differential forms and noncommutative differential forms. As motivation, we mention results about coordinate rings on algebraic varieties and Riemann surfaces in particular.

Quillen was an assiduous craftsman of mathematics, who refined manuscripts through several stages before publication. The present notes have not undergone such a process and should be read as an outline of his ideas, rather than the finished product. My main purpose in writing these notes is to make the material available to another generation of mathematicians in the hope that they can realize the potential for further development. Some of the topics are of current research interest and there is a legacy of resistant problems which deserve further consideration. For instance, projective modules over Banach algebras are still mysterious. The references do not cover more recent results on cyclic theory or operator K-theory. The first nine chapters should be accessible to any reader with a basic graduate-level knowledge of commutative algebra, differential geometry and functional analysis.

I am especially grateful to Jean Quillen for granting permission on behalf of the Quillen Estate to proceed with publication. Also, I am grateful to Glenys Luke and Roger Astley for encouragement with this project, to James Groves for inspecting the manuscript, and to Andrey Lazarev for helpful comments on an earlier draft. I record my belated but sincere thanks to Merton College Oxford, who supported the junior research fellowship that enabled me to attend Quillen's lectures. More recently, the University of Macau supported the writing of these notes through a Senior Visiting Fellowship.

Gordon Blower
Lancaster University

1
Background Results

The contents of this chapter are classical, in that they refer to the algebraic structures that are used to describe classical mechanics and often involve modules over commutative algebras as in [8]. Some of the results are used in subsequent chapters, while other aspects will be generalized to quantum mechanics and to noncommutative algebras in subsequent chapters. Advanced readers can proceed to Chapter 2.

1.1 Graded Algebras

A unital algebra over a field $\mathbf{k}$ can be regarded as triple $(A, m, 1)$ where $m\colon A \times A \to A$ is the associative multiplication and 1 is the multiplicative unit. The associativity law $(ab)c = a(bc)$ says that the following diagram commutes

$$\begin{array}{ccccccc} A \otimes A \otimes A & \longrightarrow & A \otimes A & & a \otimes b \otimes c & \mapsto & a \otimes bc \\ \downarrow & & \downarrow & & \downarrow & & \downarrow\ ; \\ A \otimes A & \to & A & & ab \otimes c & \to & abc \end{array} \tag{1.1.1}$$

the operation of scalar multiplication on the right gives a commuting diagram

$$\begin{array}{ccccccc} A \otimes A & \longrightarrow & A & & a \otimes \kappa 1 & \longrightarrow & \kappa a \\ \downarrow & & \downarrow & & \downarrow & & \downarrow ; \\ A \otimes \mathbf{k} & \to & A & & a \otimes \kappa & \to & a\kappa \end{array} \tag{1.1.2}$$

a similar diagram applies to multiplication from the left. Hence we identify $a \otimes \kappa b$ with $\kappa a \otimes b$, and as a default we work with $A \otimes_{\mathbf{k}} A$, the tensor products over $\mathbf{k}$.

An ideal I of A is a subalgebra such that $am \in I$ and $ma \in I$ for all $a \in A$ and $m \in I$. The square of I is $I^2 = \mathrm{span}_{\mathbf{k}}\{mp\colon m, p \in I\}$, which is also an ideal with $I^2 \subseteq I \subseteq A$, and we can proceed to form $I^3, I^4, \dots$ likewise.

Definition 1.1.1 (i) (*Graded algebra*) Let $\mathbf{k}$ be a field and A a unital algebra over $\mathbf{k}$. Suppose that $(A_n)_{n=0}^{\infty}$ is a sequence of $\mathbf{k}$-vector subspaces of A such that $A = \oplus_{n=0}^{\infty} A_n$, in the sense that each $a \in A$ has a unique expression as a finite sum $a = \sum_{j=0}^{\infty} a_j$ where $a_j \in A_j$ and the multiplication satisfies $A_m A_n \subseteq A_{m+n}$. We suppose that $A_0 = \mathbf{k}1$.

(ii) (*Principal ideal domain*) A commutative ring R with 1 is an integral domain if $xy = 0$ implies $x = 0$ or $y = 0$. Suppose further that every ideal I in R has the form $I = \{rx\colon r \in R\} = (x)$ for some $x \in R$. Then R is called a principal ideal domain (PID).

(iii) An ideal I in a ring R is called nilpotent if $I^n = 0$ for some $n \in \mathbf{N}$.

Example 1.1.2 (Polynomial algebra) (i) Let $\mathbf{k}$ be a field and t an indeterminate. Then the polynomial algebra $\mathbf{k}[t] = \{\sum_{j=0}^{m} a_j t^j\colon a_j \in \mathbf{k}, m = 0, 1, \dots\}$ with the usual addition and multiplication is a PID. Also with $A_m = \{a_m t^m\colon a_m \in \mathbf{k}\}$ we obtain a graded algebra. Then $\sum_{j=0}^{m} A_j$ is the space of polynomials of degree less than or equal to m, including the zero polynomial.

(ii) Let R be a unital algebra which is an integral domain, and let $t_1, \dots, t_n$ be commuting indeterminates. Then the algebra of polynomials $A = R[t_1, \dots, t_n]$ is a graded algebra where

$$A_m = \left\{ \sum a_{m_1,\dots,m_n} t_1^{m_1} \dots t_n^{m_n}\colon a_{m_1,\dots,m_n} \in A, \sum_{j=1}^{n} m_j = m \right\} \tag{1.1.3}$$

and where m is called the total degree, and the elements of A_m are called homogeneous components of degree m.

This R has derivatives ∂_j such that $\partial_j\colon A \to A$ is $\mathbf{k}$-linear and

$$\partial_j a_{m_1,\dots,m_n} t_1^{m_1} \dots t_n^{m_n} = m_j a_{m_1,\dots,m_n} t_1^{m_1} \dots t_j^{m_j-1} \dots t_n^{m_n}, \tag{1.1.4}$$

so Leibniz's rule holds

$$\partial_j(fg) = (\partial_j f)g + f(\partial_j g) \qquad (f, g \in A) \tag{1.1.5}$$

$\partial_j\colon A_m \subseteq A_{m-1}$ and $\partial_j \partial_k = \partial_k \partial_j$, and ∂_j is an operator of degree (-1). This is the fundamental example of a commutative graded algebra, and may be compared to the following one.

Example 1.1.3 (Formal power series) (i) For a given field $\mathbf{k}$, let $A = \mathbf{k}[[\hbar]]$ be the algebra of formal power series in $\hbar$. Then there is a derivation $\delta\colon \mathbf{k}[[\hbar]] \to \mathbf{k}[[\hbar]]$ given by formal differentiation

$$\delta\colon \sum_{n=0}^{\infty} k_n \hbar^n \mapsto \sum_{n=1}^{\infty} n k_n \hbar^{n-1}. \tag{1.1.6}$$

Let $I = \langle \hbar \rangle$ be the ideal generated by $\hbar$. Then $I^n = \langle \hbar^n \rangle$, and A/I^n is isomorphic to $\mathrm{span}\{1, \hbar, \dots, \hbar^{n-1}\}$ as a **k**-vector space. Then A/I^n is an Artinian algebra, which has a unique maximal ideal $(\hbar + I^n)$, which is nilpotent, since $(\hbar + I^n)^n = (0)$. Observe that $I^n I^m \subseteq I^{m+n}$ and $A \supset I \supset I^2 \supset \cdots$ is an infinite strictly decreasing sequence of ideals with $\cap_{n=1}^{\infty} I^n = \{0\}$.

There is a derivative $\partial\colon A \to A$ that satisfies $\partial(fg) = (\partial f)g + f\partial g$, and $\partial I^n \subset I^{n-1}$.

(ii) Now let $A[\hbar^{-1}] = \{\sum_{j=-n}^{\infty} a_j \hbar^j; n \in \mathbf{N}; a_j \in \mathbf{k}\}$ be the algebra of formal Laurent series with coefficient in **k** that have only finitely many nonzero terms. Then $A_n = \{\sum_{j=-n}^{\infty} a_j \hbar^j; a_j \in \mathbf{k}\}$ is the space of formal Laurent series that have a pole of order at most n at $\hbar = 0$, and A_n is a module over A.

Exercise 1.1.4 (Gradings by powers of an ideal) Let I be an ideal in a unital algebra A over a field **k** of characteristic zero. There is a natural filtration by powers of the ideal

$$A \supseteq I \supseteq I^2 \supseteq I^3 \supseteq \cdots$$

called the I-adic filtration. We also impose the condition $\cap_{n=1}^{\infty} I^n = \{0\}$.

(i) Show that there is a natural multiplication map

$$I^n/I^{n+1} \times I^m/I^{m+1} \to I^{n+m}/I^{n+m+1}. \tag{1.1.7}$$

(ii) Deduce that with $I^0 = A$, there is a graded algebra

$$g_I(A) = \bigoplus_{j=0}^{\infty} I^j/I^{j+1}. \tag{1.1.8}$$

(iii) Suppose that A is commutative and I is a maximal ideal. Show that $\mathbf{k} = A/I$ is a field, and $g_I(A)$ is a direct sum of **k**-vector spaces.

(iv) Suppose that A is commutative and I is a finitely generated maximal ideal. Show that the summands in $g_I(A)$ are finitely generated. (See [8] Theorem 11.22 for more on the structure of $g_I(A)$.)

(v) Show that I is an ideal of R with $I^2 = 0$, where

$$I = \left\{ \begin{bmatrix} 0 & B \\ 0 & 0 \end{bmatrix} : B \in M_n(\mathbf{C}) \right\} \quad R = \left\{ \begin{bmatrix} A & B \\ 0 & D \end{bmatrix} : A, B, D \in M_n(\mathbf{C}) \right\}.$$

1.2 Derivations

Let A and R be algebras over a field $\mathbf{k}$. A homomorphism is a map $\rho\colon A \to R$ such that

$$\rho(\lambda a + \mu b) = \lambda\rho(a) + \mu\rho(b), \tag{1.2.1}$$

$$\rho(ab) = \rho(a)\rho(b) \qquad (a,b \in A; \lambda, \mu \in \mathbf{k}). \tag{1.2.2}$$

If A and R are unital, and $\rho(1) = 1$, then ρ is said to be unital. Many of the algebras we consider are not unital.

The kernel (nullspace) $\{a\colon \rho(a) = 0\}$ of any homomorphism $\rho\colon A \to R$ is an ideal, and conversely, any ideal arises from the kernel of some homomorphism. Given any ideal I of A, the quotient space $R = A/I$ is an algebra with the usual multiplication and addition, and $\pi\colon A \to R\colon a \mapsto a + I$ is a homomorphism with kernel I; we call this the canonical (quotient) homomorphism, and write

$$0 \longrightarrow I \longrightarrow A \longrightarrow R \longrightarrow 0. \tag{1.2.3}$$

Let M be a bimodule over A. This means that M is a $\mathbf{k}$-vector space with operations $A \times M \to M$ and $M \times A \to M$ such that

$$\text{(i)} \quad a(m+n) = am + an; \qquad (m+n)a = ma + na; \tag{1.2.4}$$

$$\text{(ii)} \quad (a+b)m = am + bm; \qquad m(a+b) = ma + bm; \tag{1.2.5}$$

$$\text{(iii)} \quad a(bm) = (ab)m; \qquad (ma) = m(ab); \tag{1.2.6}$$

$$\text{and (iv)} \quad a(mb) = (am)b \qquad (a,b \in A; m,n \in M). \tag{1.2.7}$$

If A has a unit 1 we generally suppose that $1m = m1 = m$ for all $m \in M$.

In particular, A is a bimodule over A with the left and multiplication operations $(a,b) \mapsto ab$. Likewise, any ideal I of A is an A-bimodule. Given A-modules M and N, we write $\mathrm{Hom}_A(M,N)$ for the space of left A-module maps $\psi\colon M \to N$ such that $\psi(am + np) = a\psi(m) + b\psi(p)$ for all $a,b \in A$ and $m,p \in M$.

Definition 1.2.1 (Derivation) A derivation is a map $\delta\colon A \to M$ such that

$$\delta(\lambda a + \mu b) = \lambda\delta(a) + \mu\delta(b), \tag{1.2.8}$$

$$\delta(ab) = a\delta(b) + \delta(a)b \qquad (a,b \in A; \lambda, \mu \in \mathbf{k}). \tag{1.2.9}$$

We write the space of such derivations as $\mathrm{Der}_{\mathbf{k}}(A,M)$ to emphasize that $\mathbf{k}$ is a field of constants. Given any bimodule M over A, we can choose $m \in M$ and introduce the inner derivation $\delta_m\colon A \to M$ by

$$\delta_m(a) = [a,m] = am - ma. \tag{1.2.10}$$

It is of interest to determine the derivations that can be expressed in this form; see Section 2.6.

Example 1.2.2 (Derivations on polynomials) Let $\mathbf{k}[t]$ be the polynomial algebra over $\mathbf{k}$ with the derivation $\partial\colon \mathbf{k}[t] \to \mathbf{k}[t]$ given by the usual derivative $\partial(\sum_{j=0}^n a_j t^j) = \sum_{j=1}^n j a_j t^{j-1}$. Next let dt be the formal infinitesimal with $(dt)^2 = 0$, and introduce $\mathbf{k}[t,dt] = \mathbf{k}[t] \oplus \mathbf{k}[t]dt$ with the multiplication

$$\big(f_0(t)+f_1(t)dt\big)\big(g_0(t)+g_1(t)dt\big) = f_0(t)g_0(t) + (f_0(t)g_1(t) + f_1(t)g_0(t))dt,$$

so that $R = \mathbf{k}[t,dt]$ is a commutative algebra with ideal $I = \mathbf{k}[t]dt$ which has square zero in the sense that $I^2 = 0$, and we can identify $A = \mathbf{k}[t]$ with R/I. Also, let $d(f_0(t) + f_1(t)dt) = \partial f_0(t)dt$. Then R is an A-bimodule, and $d\colon A \to R$ gives a derivation.

Example 1.2.3 (Pseudo-differential operators) (i) Let A be a unital commutative ring, and let $\partial\colon A \to A$ be a derivation. Then the space of formal differential operators with coefficients in A is given by

$$DO_A = \left\{ \sum_{j=0}^n a_j \partial^j : n \in \mathbf{N}, a_j \in A \right\}$$

which forms a ring under the composition rule

$$\left(\sum_{j=0}^n a_j \partial^j\right) \circ \left(\sum_{k=0}^m b_j \partial^k\right) = \sum_{j,k,\ell: 0\le \ell \le j \le n, 0\le k\le m} \binom{j}{k} a_j (\partial^\ell b_k) \partial^{j+k-\ell}$$

so that A is a subring of DO_A. Now we introduce the integration operator ∂^{-1}, which is required to satisfy

$$\partial^{-1} \circ a = a\partial^{-1} - (\partial a)\partial^{-2} + (\partial^2 a)\partial^{-3} - \cdots,$$

and $\partial \circ \partial^{-1} = 1$, so $\partial \circ \partial^{-1} \circ a = a$ for all $a \in A$. As in [70], one can extend the composition rule to powers ∂^j with $j \in \mathbf{Z}$ via the generalized Leibniz rule. We introduce

$$\Psi DO_A^n = \left\{ \sum_{j=-\infty}^n a_j \partial^j : a_j \in A \right\} \qquad (n \in \mathbf{Z}).$$

Then $\Psi DO_A = DO_A \oplus \Psi DO_A^{-1}$ gives a graded algebra such that $\Psi DO_A^n \subseteq \Psi DO_A^{n+1}$ with $\partial \circ \Psi DO_A^j \subseteq \Psi DO_A^{j+1}$ and

$$\Psi DO_A^j \circ \Psi DO_A^k \subseteq \Psi DO_A^{j+k} \qquad (j,k \in \mathbf{Z}).$$

For later use, we note that the commutator $[X,Y] = X \circ Y - Y \circ X$ has the special property

$$[\Psi DO_A^j, \Psi DO_A^k] \subset \Psi DO_A^{j+k-1} \qquad (j,k \in \mathbf{Z}),$$

so we lose a derivative when taking the commutator. There is an exact sequence of algebras

$$0 \longrightarrow \Psi DO_A^{-1} \longrightarrow \Psi DO_A^0 \longrightarrow A \longrightarrow 0.$$

Mulase [76] considered this as a graded algebra.

(ii) (*Weyl algebra*) With $A = \mathbf{C}[z]$, the algebra of differential operators with polynomial coefficients $DO_A = \mathbf{C}\langle z, d/dz\rangle$ is known as the Weyl algebra; in Definition 9.2.1 we obtain this algebra via a more sophisticated construction. As in (i), we can extend DO_A to $\Psi DO_A = \mathbf{C}\langle z, d/dz, (d/dz)^{-1}\rangle$.

1.3 Commutators and Traces

Definition 1.3.1 (i) (*Commutators*) Let A be an associative algebra over $\mathbf{k}$, and M an A-bimodule. Then we write $[a,m] = am - ma$ for the commutator of $a \in A$ and $m \in M$, then let

$$[A,M] = \mathrm{span}_{\mathbf{k}}\{[a,m]; \quad a \in A, m \in M\}$$

for the commutator subspace, which is a $\mathbf{k}$-vector subspace of M. The commutator quotient space is $M/[A,M]$.

(ii) (*Traces*) In particular, A is an A-bimodule for the standard multiplication, so $[A,A]$ is the subspace of A spanned by the commutators. Then a trace $\tau: A \to \mathbf{k}$ is a $\mathbf{k}$-linear map such that $\tau([a,b]) = 0$ for all $a, b \in A$ equivalently, a trace is a linear function $\tau: A/[A,A] \to \mathbf{k}$.

Let I be an ideal in an associative algebra A, so that I is an A-bimodule. Then $[I,I]$ and $[I,A]$ are commutator subspaces of I. A trace on I is a linear functional $\tau: I \to \mathbf{k}$ such that $\tau|[I,I] = 0$. Such a trace may have the stronger property that $\tau|[I,M] = 0$. See Proposition 3.2.9(iv) for a significant example.

Example 1.3.2 (Quaternions) (i) We consider an example of a noncommutative algebra, namely the quaternions. We introduce Pauli's matrices

$$\sigma_0 = \begin{bmatrix} 1 & 0 \\ 0 & 1 \end{bmatrix}, \quad \sigma_1 = \begin{bmatrix} 1 & 0 \\ 0 & -1 \end{bmatrix}, \quad \sigma_2 = \begin{bmatrix} 0 & -i \\ i & 0 \end{bmatrix}, \quad \sigma_3 = \begin{bmatrix} 0 & 1 \\ 1 & 0 \end{bmatrix}, \tag{1.3.1}$$

which satisfy

$$\sigma_0^2 = \sigma_1^2 = \sigma_2^2 = \sigma_3^2 = I; \qquad \sigma_j\sigma_k = -i\sigma_\ell \tag{1.3.2}$$

for all cyclic permutations $(jk\ell)$ of (123). Note that $\{\sigma_0, \sigma_1, \sigma_2, \sigma_3, i\sigma_0, i\sigma_1, i\sigma_2, i\sigma_3\}$ gives the quaternion group of eight elements. The space $\mathbf{H} = \{a_0\sigma_0 + ia_1\sigma_1 + ia_2\sigma_2 + ia_3\sigma_3 : a_j \in \mathbf{R}\}$ gives the algebra of quaternions, a four-dimensional noncommutative division ring over $\mathbf{R}$. The elements with $a_0 = 0$ are called pure quaternions.

(ii) Let $\mathbf{R}^3$ have the standard unit vector basis $\{e_1, e_2, e_3\}$ and let $\mathbf{H} \to \mathbf{R} \times \mathbf{R}^3$ be the linear map $a_0e_0 + a_1e_1 + a_2e_2 + a_3e_3 \to (a_0, \mathbf{a})$ where $\mathbf{a} = (a_1, a_2, a_3)$. We write $(a_0, \mathbf{a})^* = (a_0, -\mathbf{a})$, and introduce the multiplication

$$(a_0, \mathbf{a})(b_0, \mathbf{b}) = (a_0b_0 - \mathbf{a} \cdot \mathbf{b}, a_0\mathbf{b} + b_0\mathbf{a} + \mathbf{a} \times \mathbf{b}) \tag{1.3.3}$$

where $\mathbf{a} \cdot \mathbf{b}$ is the usual scalar (dot) product of vectors and $\mathbf{a} \times \mathbf{b}$ is the usual vector (cross) product on $\mathbf{R}^3$. This $*$ gives an anti-automorphism of the real skew ring $\mathbf{H}$, such that

$$(q_1 + q_2)^* = q_1^* + q_2^*; \qquad (q_1q_2)^* = q_2^*q_1^*.$$

Then we define a norm by $\|(a_0, \mathbf{a})\| = \sqrt{a_0^2 + \mathbf{a} \cdot \mathbf{a}}$. By vector algebra, one checks that

$$\|(a_0, \mathbf{a})(b_0, \mathbf{b})\| = \|(a_0, \mathbf{a})\| \, \|(b_0, \mathbf{b})\|. \tag{1.3.4}$$

If $(a_0, \mathbf{a}) \in \mathbf{H}$ has $\|(a_0, \mathbf{a})\| = 1$, then we say that $(a_0, \mathbf{a})$ is a unit quaternion. One checks that $Sp(1) = \{(a_0, \mathbf{a}) \in \mathbf{H} : \|(a_0, \mathbf{a})\| = 1\}$ is a group.

Now observe that

$$a_0\sigma_0 + ia_1\sigma_1 + ia_2\sigma_2 + ia_3\sigma_3 = \begin{bmatrix} a_0 + ia_1 & a_2 + ia_3 \\ -a_2 + ia_3 & a_0 - ia_1 \end{bmatrix}$$

so that there is a bijective group homomorphism map between the group $Sp(1)$ of unit quaternions and the group $SU(2)$ of 2×2 unitary complex matrices that have determinant one; $Sp(1) \sim SU(2, \mathbf{C})$.

(iii) (*Rotations and spin*) Let $SO(3)$ be the group of all rotations of Euclidean space about the origin, namely $\{U \in M_3(\mathbf{R}) : U^tU = I, \det U = 1\}$ with matrix multiplication. There is an exact sequence of groups

$$0 \to \mathbf{Z}/(2) \to Sp(1) \to SO(3) \to 0, \tag{1.3.5}$$

arising as follows. For all $p \in Sp(1)$, there exists a real algebra automorphism of $\mathbf{H}$ given by conjugation $\alpha_p(q) = pqp^{-1}$. Note that $\alpha_p \circ \alpha_r = \alpha_{pr}$, so $p \mapsto \alpha_p$ is a group action; evidently the kernel is $\{\pm 1\}$. Conversely, given any real automorphism α of $\mathbf{H}$, α acts on the pure quaternions so that

$\alpha(e_1), \alpha(e_2)$ and $\alpha(e_3)$ are unit pure quaternions, satisfying $\alpha(e_1)^2 = \alpha(e_2)^2 = \alpha(e_3)^2 = -1$ with $\alpha(e_1)\alpha(e_2) = \alpha(e_3)$, hence they may be identified with unit vectors $\alpha(e_1), \alpha(e_2)$ and $\alpha(e_3) = \alpha(e_1) \times \alpha(e_2) \in \mathbf{R}^3$. We deduce that there exists $A \in SO(3)$ that takes the frame $\{e_1, e_2, e_3\}$ to $\{\alpha(e_1), \alpha(e_2), \alpha(e_3)\}$. One can check that A arises from conjugation by some $p \in Sp(1)$.

Let $M_n(\mathbf{H}) = \{[a_{j,k}]_{j,k=1}^n : a_{j,k} \in \mathbf{H}\}$, and $GL_n(\mathbf{H}) = \{A \in M_n(\mathbf{H}) : A$ invertible$\}$. Then one can extend the * operation to $M_n(\mathbf{H})$ by $[a_{jk}]^* = [a_{kj}^*]$ and consequently define the quaternionic symplectic group $Sp(n) = \{A \in GL_n(\mathbf{H}) : A^*A = I\}$.

Proposition 1.3.3 *Let A be a unital algebra which is a finite-dimensional vector space over the field $\mathbf{k}$ and suppose that the only nilpotent ideal of A is zero. Let τ be a trace on A.*

Then there exist indices $n_j \in \mathbf{N}$ for $j = 1, \dots, \ell$ and division rings D_j over $\mathbf{k}$ such that

$$A = \bigoplus_{j=1}^{\ell} M_{n_j}(D_j) \tag{1.3.6}$$

and linear functionals $\phi_j : D_j \to \mathbf{k}$ such that

$$\tau = \sum_{j=1}^{\ell} \phi_j \otimes \mathrm{trace}_{n_j} \tag{1.3.7}$$

where $\mathrm{trace}_{n_j} : M_{n_j}(\mathbf{k}) \to \mathbf{k}$ is the standard trace on $n_j \times n_j$ matrices.

Proof The algebra A is semisimple, and the Artin–Wedderburn theory applies to give the direct sum decomposition. We write $\tau = \sum_{j=1}^{\ell} \tau_j$, where $\tau_j : M_{n_j}(D_j) \to \mathbf{k}$ is a trace. Let e_{pq} be the usual system of matrix units, so e_{pq} has entry 1 in place pq, and zeros elsewhere. Note that

$$e_{pq} = e_{p1}e_{1q} - e_{1q}e_{p1} = [e_{p1}, e_{1q}] \qquad (p \neq q) \tag{1.3.8}$$

so $\tau_j(\alpha_{pq}e_{pq}) = 0$ for all $\alpha_{pq} \in D_j$; also

$$\begin{bmatrix} 0 & 1 \\ 1 & 0 \end{bmatrix} \begin{bmatrix} 1 & 0 \\ 0 & 0 \end{bmatrix} \begin{bmatrix} 0 & 1 \\ 1 & 0 \end{bmatrix} = \begin{bmatrix} 0 & 0 \\ 0 & 1 \end{bmatrix},$$

so one can deduce by considering suitable block matrices that $\tau_j(\alpha_{pp}e_{pp}) = \tau_j(\alpha_{pp}e_{11})$. We deduce that

$$\tau_j \left(\sum_{p,q=1}^{n_j} \alpha_{pq}e_{pq} \right) = \sum_{p=1}^{n_j} \tau_j(\alpha_{pp}e_{11}) \tag{1.3.9}$$

so that $\tau_j = \phi_j \otimes \mathrm{trace}_{n_j}$ for the **k**-linear map $\phi_j \colon D_j \to \mathbf{k}$ determined by $\phi_j(\alpha) = \tau_j(\alpha e_{11})$ for $\alpha \in D_j$. □

Remarks 1.3.4 Any finite-dimensional right A-module M is a direct sum of minimal A-modules, so $M = \oplus_{j=1}^{\ell} M_j$, where M_j is isomorphic to a minimal right ideal of A, namely a row of $M_{n_j}(D_j)$.

Complement 1.3.5 Let $\mathbf{k} = \mathbf{R}$. Then each D_j in Proposition 1.3.3 is either the real numbers **R**, the complex numbers **C** or the quaternions **H**. See [57] for a discussion of division rings over **R**.

1.4 Tensor Algebras

In this section, we introduce the tensor algebra, and then some related algebras with universal properties. Let **k** be a field of characteristic zero, which we later specialize to the complex numbers **C** or the real numbers **R**. Given a **k**-vector space V, the dual V^* is the space of all **k**-linear functionals $\phi \colon V \to \mathbf{k}$. Let $V^{\otimes 0} = \mathbf{k}$, and then let $V^{\otimes m} = V \oplus \cdots \otimes V$ be the m-fold tensor product of V with itself over **k**, so

$$\alpha v_1 \otimes v_2 \oplus \cdots \otimes v_n = v_1 \otimes \alpha v_2 \oplus \cdots \otimes v_n$$
$$= \cdots = v_1 \otimes v_2 \oplus \cdots \otimes \alpha v_n \ (\alpha \in \mathbf{k}, v_j \in V). \quad (1.4.1)$$

Often for convenience, we abbreviate $v_1 \otimes v_2 \oplus \cdots \otimes v_n$ by $(v_1, \ldots, v_n)$, where such elementary tensors span $V^{\otimes n}$. A multilinear functional on $V^{\times m}$ is equivalent to $\varphi \in (V^{\otimes m})^*$.

(i) *Tensor algebra*: As in Example 1.1.2 the tensor algebra over V generalizes the notion of the algebra of polynomials in several variables. Given a **k**-vector space V, the tensor algebra is $T(V) = \oplus_{n=0}^{\infty} V^{\otimes n}$ where $V^{\otimes 0} = \mathbf{k}$. The product is defined via concatenating elementary tensors

$$V^{\otimes n} \times V^{\otimes m} \to V^{\otimes(n+m)} : (a_1, \ldots, a_n) \cdot (b_1, \ldots, b_m) = (a_1, \ldots, a_n, b_1, \ldots, b_m) \quad (1.4.2)$$

and taking linear combinations. The product is not generally commutative, but there is a unit

$$1_{T(V)} = (1, 0, 0, \ldots). \quad (1.4.3)$$

The tensor algebra is characterized by the following universal property.

Proposition 1.4.1 *Let A be an algebra over $\mathbf{k}$ and $\phi\colon V \to A$ a $\mathbf{k}$-linear map. Then there exists a unique $\mathbf{k}$-algebra homomorphism $\Phi\colon T(V) \to A$ such that $\Phi|V = \phi$.*

Proof The basic idea is that in degree n, we can define $\Phi(a_1, \dots, a_n) = \phi(a_1)\phi(a_2)\dots\phi(a_n)$, and extend this by linearity. See Jacobson [57] for more details. □

(ii) *Symmetric algebra*: The symmetric algebra $S(V)$ is $T(V)/I$ where $I = (v \otimes w - w \otimes v; v, w \in V)$ is the ideal generated by all the antisymmetric tensors $v \otimes w - w \otimes v$ with $v, w \in V$. Hence $S(V)$ is a commutative algebra. Let $W = V^*$ be the dual space of linear maps $V \to \mathbf{k}$. Then $S(V)$ may be regarded as the algebra of polynomial functions on W.

Lemma 1.4.2 *Let $V = \mathbf{C}^n$. Then $S(V)$ is isomorphic to the algebra $\mathbf{C}[t_1, \dots, t_n]$.*

Proof Let V have basis $\{e_1, \dots, e_n\}$ and observe that $S(V)$ is spanned by 1 and products $e_{j_1} \dots e_{j_r}$ where $1 \leq j_1 \leq \dots \leq j_r \leq n$ are integers. We introduce a multi index $\alpha = (\alpha_1, \dots, \alpha_n)$ such that

$$e_{j_1} \dots e_{j_r} = e_1^{\alpha_1} \dots e_n^{\alpha_n} \tag{1.4.4}$$

and $|\alpha| = \sum_{j=1}^n \alpha_j = r$ is the degree of α. Then the isomorphism $S(V) \to \mathbf{C}[t_1, \dots, t_n]$ is given by

$$e_1^{\alpha_1} \dots e_n^{\alpha_n} \mapsto t_1^{\alpha_1} \dots t_n^{\alpha_n}. \tag{1.4.5}$$

□

(iii) *Clifford algebra:* Let V be an n-dimensional vector space over $\mathbf{k}$ of characteristic zero. A quadratic form on V is a map $q\colon V \to \mathbf{k}$ such that $q(tv) = t^2 q(v)$ for all $t \in \mathbf{k}$ and $v \in V$, and

$$\beta(v, w) = 2^{-1}\big(q(v) + q(w) - q(v - w)\big) \tag{1.4.6}$$

is a bilinear form $\beta\colon V \times V \to \mathbf{k}$. Evidently $\beta(v, w) = \beta(w, v)$, so β is symmetric and determines q as $q(v) = \beta(v, v)$. Then the Clifford algebra associated with q is

$$C_q = T(V)/(v^2 - q(v); v \in V), \tag{1.4.7}$$

which has the following universal property. Given any algebra A over $\mathbf{k}$ and $\mathbf{k}$-linear map $\phi\colon V \to A$ such that $q(v)1_A = \phi(v)^2$, there exists an algebra homomorphism $\Phi\colon T(V) \to A$ extending ϕ such that

$$\Phi(v^2) = \Phi(v)^2 = \phi(v)^2 = q(v)\Phi(1_k), \tag{1.4.8}$$

so all elements of the form $v^2 - q(v)$ are in the kernel; hence we can factor $\Phi = \tilde{\Phi} \circ \pi$, where $\pi : T(V) \to C_q$ is the canonical homomorphism, and $\tilde{\Phi} : C_q \to A$ is a homomorphism such that $\phi(v) = \tilde{\Phi} \circ \pi(v)$ for all $v \in V$. It follows easily from the universal property that C_q is unique.s

We can obtain a more explicit description of C_q in specific cases.

Proposition 1.4.3 (i) *There exists a unique Clifford algebra C_q for any such q.*

(ii) *There is an algebra automorphism $\varepsilon : C_q \to C_q$ given by $\varepsilon(v) = -v$ for $v \in V$.*

(iii) *For V of dimension n, C_q has dimension 2^n.*

Proof Given a vector space V of dimension n over $\mathbf{k}$, one can introduce a Clifford algebra as follows. Let $\{e_1, \dots, e_n\}$ be any basis for V and q be any quadratic form on V, and introduce β as above. Then let

$$A = \mathrm{span}_{\mathbf{k}}\{e_1^{i_1} e_2^{i_2} \dots e_n^{i_n} : i_1, \dots, i_n \in \{0, 1\}\} \tag{1.4.9}$$

where we identify $e_1^0 e_2^0 \dots e_n^0$ with $1 \in \mathbf{k}$. Then A has dimension 2^n as a $\mathbf{k}$-vector space. We multiply elements of A using associativity and the identities

$$e_j e_j = 2^{-1}\beta(e_j, e_j); \quad e_m e_j = -e_j e_m - 2\beta(e_j, e_m) \qquad (j < m) \tag{1.4.10}$$

to reduce the products to sums of monomials in A. By induction on n, one proves that the product $A \times A \to A$ is well-defined and satisfies $q(v) = v^2$. For each $a \in A$, let $\lambda_a : A \to A : b \mapsto ab$ for all $b \in A$. The map $a \mapsto \lambda_a$ gives a representation of A as linear transformations on A, so λ_a may be expressed as a matrix in $M_{2^n}(\mathbf{k})$. □

We have not yet assumed that q or β is non-degenerate, so the squares could all be zero; indeed, this is a significant case in geometry which we consider next.

Lemma 1.4.4 *The Clifford algebra with $q = 0$ is the Grassmann algebra with product $(v, w) \mapsto v \wedge w$ so that $v \wedge w = -w \wedge v$.*

Proof In the setting of (iii) let $q = 0$, which leads to a case which is important in geometry. Then C_0 has $v^2 = 0$ and $vw = -wv$, so

$$C_0 = \bigwedge^{*} V = \mathbf{k} \oplus V \oplus \wedge^2 V \oplus \dots \oplus \wedge^n V \oplus \cdots \tag{1.4.11}$$

is the Grassmann algebra over V. In this context, one often writes the product as $(v, w) \mapsto v \wedge w$, and one calls $\wedge^k V$ the kth exterior power of V. Suppose that V is finite dimensional, and let $\{e_1, \dots, e_n\}$ be a basis for V, and σ a permutation of $\{1, \dots, n\}$. Then one checks that

$$e_j^2 = 0, \quad e_j e_k = -e_k e_j \tag{1.4.12}$$

hence $e_{\sigma(1)} \dots e_{\sigma(n)} = \mathrm{sign}(\sigma) e_1 \dots e_n$, so the summand $\wedge^n V$ has dimension one. Generally, the subspace $\wedge^m V$ is spanned by $e_{i_1} \dots e_{i_m}$ where $i_1 < i_2 < \dots < i_m$ with all $i_j \in \{1, \dots, n\}$, so $\wedge^m V$ has dimension $\binom{n}{m}$, and C_0 has dimension 2^n. The highest exterior power of V is $\Lambda(V) = \wedge^n V$, all higher powers being zero. For a given permutation $(\sigma_1, \sigma_2, \dots \sigma_n)$, let N be the number of transpositions of adjacent terms needed to restore the natural order $(1, 2, \dots, n)$; then let $\mathrm{sgn}(\sigma) = (-1)^N$. We have

$$e_{\sigma(1)} e_{\sigma(2)} \dots e_{\sigma(n)} = \mathrm{sgn}\,(\sigma) e_1 e_2 \dots e_n \tag{1.4.13}$$

for all permutations σ. □

Let $T \colon V \to V$ be a linear transformation, and consider the matrix of T with respect to the basis (e_j). Using the definition of the determinant of this matrix, one checks that

$$T(e_1) \dots T(e_n) = \det(T) e_1 \dots e_n. \tag{1.4.14}$$

Exercise 1.4.5 (Determinants) Let V and W be $\mathbf{k}$-vector spaces of the same finite dimension, and $A \colon V \to W$ a linear transformation.

(i) Show that there is a $\mathbf{k}$-algebra homomorphism $\Gamma(A) \colon T(V) \to T(W)$ such that $\Gamma(A)|V = A$.

(ii) Show that there is a $\mathbf{k}$-algebra homomorphism $\Lambda(A) \colon \wedge(V) \to \wedge(W)$ such that $\Lambda(A)|V = A$.

(iii) Let $\Lambda(V)$ be the highest exterior power of V, and $\Lambda(W)$ be the highest exterior power of W. Show that the restriction $\Lambda(A) \colon \Lambda(V) \to \Lambda(W)$ gives the determinant of A.

Example 1.4.6 (Grassmann algebra) On the polynomial algebra $A = \mathbf{k}[t_1 \dots, t_n]$, introduce the usual partial derivatives ∂_j by $\partial_j f = \frac{\partial f}{\partial t_j}$. Then with $e_j = dt_j$ we let $V = \mathrm{span}_{\mathbf{k}}\{dt_1, \dots, dt_n\}$ and form the Grassmann algebra $\wedge V$; then let

$$R = \mathbf{k}[t_1, \dots, t_n] \otimes \wedge V,$$

which is an algebra and a left A-module. There is a $\mathbf{k}$-linear map $d \colon A \to A \otimes \wedge^1 V$ given by

$$df = \sum_{j=1}^{n} (\partial_j f)\, dt_j$$

such that $d(fg) = fdg + gdf$. We extend d to $d: A \otimes \wedge V \to A \otimes \wedge V$ by $d(f \otimes \omega) = (df) \wedge \omega$, so that d is a derivation on R as a left A-module. One proves that $d^2 = 0$.

1.5 Real Clifford Algebras

In this section, we take $\mathbf{k} = \mathbf{R}$, and consider the Clifford algebras that arise when the quadratic form q is negative definite. The case of the quaternions provides the motivating example.

Example 1.5.1 (Clifford matrices) Suppose that $\mathbf{k} = \mathbf{R}$, and the quadratic form on V satisfies $q(v) < 0$ for all $v \in V \setminus \{0\}$. Then $q(v) = -\|v\|^2$ for some norm $\|\cdot\|$ on V, and β is given by the corresponding inner product $\langle u, v\rangle = 2^{-1}(\|u\|^2 + \|v\|^2 - \|u - v\|^2)$, so one can use the Gram–Schmidt process to produce a basis $\{e_j\}$ of V such that

$$-\beta(e_j, e_m) = \langle e_j, e_m\rangle = \delta_{jm}, \tag{1.5.1}$$

which is known as an orthonormal basis for V. (The Gram–Schmidt process requires us to take square roots of positive numbers, hence we have restricted attention to $\mathbf{k} = \mathbf{R}$.)

In particular, let σ_j be the 2×2 Pauli matrices of Example 1.3.2. Then the 4×4 matrices, made up of 2×2 blocks

$$\gamma_0 = \begin{bmatrix} 0 & \sigma_0 \\ \sigma_0 & 0 \end{bmatrix}, \qquad \gamma_j = \begin{bmatrix} 0 & \sigma_j \\ -\sigma_j & 0 \end{bmatrix} \tag{1.5.2}$$

satisfy

$$\gamma_0^2 = I, \quad \gamma_1^2 = \gamma_2^2 = \gamma_3^2 = -I \tag{1.5.3}$$

and anti-commute so that

$$\gamma_j\gamma_k = -\gamma_k\gamma_j \qquad (j \neq k). \tag{1.5.4}$$

Generally, let $(e_j)_{j=1}^n$ be a family of $p \times p$ matrices, say with $p = 2^n$, which satisfy the Clifford relations

$$e_\ell e_m + e_m e_\ell = -2\delta_{m\ell} I_p; \tag{1.5.5}$$

then

$$\left(\sum_{m=1}^n a_m e_m\right)^2 = -\left(\sum_{m=1}^n a_m^2\right) I_p. \tag{1.5.6}$$

We write $v = \sum_{m=1}^n a_m e_m$, and $q(v) = -\sum_{m=1}^n a_m^2$.

Corollary 1.5.2 *For each $n = 1, 2, \ldots$, there exists a real Clifford algebra Cl_n of dimension 2^n with basis satisfying (1.5.5), which is unique up to automorphism.*

Proof This follows from the preceding discussion and Proposition 1.4.3. □

Example 1.5.3 (i) We have $Cl_0 = \mathbf{R}$, $Cl_1 = \mathbf{C}$ and $Cl_2 = \mathbf{H}$.

(ii) *(Spin groups)* On Cl_n, there is an anti-homomorphism $x \mapsto \bar{x}$ such that $(\lambda x + \mu y)^- = \lambda\bar{x} + \mu\bar{y}$ and $\overline{xy} = \bar{y}\bar{x}$. One can define this by taking

$$e_{i_1} \ldots e_{i_k} \mapsto (-1)^k e_{i_k} \ldots e_{i_1} \tag{1.5.7}$$

for $i_1 < \cdots < i_k$ and extending linearly. We observe that $\varepsilon(\bar{x}) = \overline{\varepsilon(x)}$, and that both ε and $^-$ are linear involutions. On V, we have $\alpha(v) = \bar{v} = -v$, so $\alpha(v)v = \bar{v}v = -v^2 = \|v\|^2$. We proceed to construct some groups which have natural representations on this subspace. The set $\{v \in V : \|v\| = 1\}$ is simply the sphere S^{n-1} in $\mathbf{R}$ of unit radius and centre 0, and the orthogonal group $O(n) = \{U \in M_n(\mathbf{R}) : UU^* = I\}$ is the group of linear transformations of V that stabilize S^{n-1} as a set. The discussion follows the approach taken to the quaternions in Example 1.3.2.

First, the set of invertible elements of the algebra Cl_n forms a multiplicative group, denoted $Cl_n^\times$. This has a natural representation on Cl_n by conjugation, so $u \mapsto \rho_u$, where $\rho_n(a) = \varepsilon(u)au^{-1}$ for $a \in Cl_n$. Next we introduce the Clifford group Γ_n, which is the subgroup of $Cl_n^\times$ that stabilizes V as a subspace, namely

$$\Gamma_n = \{u \in Cl_n^\times : \varepsilon(u)vu^{-1} \in V, \forall v \in V\}. \tag{1.5.8}$$

The Clifford group has a subgroup

$$\mathrm{Spin}(n) = \{u \in \Gamma_n : \varepsilon(u) = u; \bar{u}u = 1\};$$

one checks readily that $\mathrm{Spin}(n)$ is a subgroup since ε is a homomorphism and $^-$ an anti-homomorphism. Clearly, ρ restricts to an action of $\mathrm{Spin}(n)$ on V; moreover, one shows that

$$\begin{aligned} \|\rho_u(v)\|^2 &= -\rho_u(v)\rho_u(v) \\ &= -uv\bar{u}uv\bar{u} = u\|v\|^2\bar{u} = \|v\|^2 \qquad (u \in \mathrm{Spin}(n), v \in V) \end{aligned} \tag{1.5.9}$$

so each ρ_u is a linear isometry on V, hence there is a group homomorphism $\mathrm{Spin}(n) \to O(n)$. In [9], it is shown that there is a short exact sequence of multiplicative matrix groups

$$\{I\} \to \{\pm I\} \to \mathrm{Spin}(n) \to SO(n) \to \{I\}. \tag{1.5.10}$$

1.6 Lie Bracket

Definition 1.6.1 (Lie algebra) Let $\mathbf{k}$ be a field of characteristic zero and A an associative algebra over $\mathbf{k}$. A Lie bracket on A is a map $[\, ,\,]: A \times A \to A$ that satisfies:

(i) $[\lambda f + \mu g, h] = \lambda[f,h] + \mu[g,h]$;

(ii) $[f,g] = -[f,g]$;

(iii) Jacobi's identity

$$\big[[f,g],h\big] + \big[[h,f],g\big] + \big[[g,h],f\big] = 0 \qquad (1.6.1)$$

for all $f,g,h \in A$ and $\lambda, \mu \in \mathbf{k}$, Given such a bracket, A is a Lie algebra. The centre of $(A,[\, ,\,])$ is $Z = \{f \in A\colon [f,g] = 0, \forall g \in A\}$.

Except in trivial cases, the algebra $(A,[\, ,\,])$ is not commutative and not associative.

Exercise 1.6.2 (Lie algebras) (i) Let $A = C^\infty(\mathbf{R})$. Verify that $[f,g] = fg' - f'g$ gives a Lie bracket.

(ii) For $V = \mathbf{R}^3$, let $[f,g] = f \times g$ be the usual cross product of vectors. Verify Jacobi's identity.

Let g be an n-dimensional Lie algebra over $\mathbf{k}$. Each $X \in g$ determines a linear transformation ad_X on g by $ad_X\colon Y \mapsto [X,Y]$. From Jacobi's identity, we have

$$ad_X([Y,Z]) = [ad_X(Y), Z] + [Y, ad_X(Z)], \qquad (1.6.2)$$

so ad_X is a derivation of $(g,[\, ,\,])$. Let trace be the trace on $End_{\mathbf{k}}(g)$, which we can obtain from the usual trace on $M_n(\mathbf{k})$ with respect to some basis of g.

Definition 1.6.3 (Killing form) The Killing form on g is the symmetric bilinear form $K\colon g \times g \to \mathbf{k}$ given by

$$K(X,Y) = \mathrm{trace}(ad_X ad_Y) \qquad (X,Y \in g), \qquad (1.6.3)$$

where $ad_X ad_Y$ is the composition of the endomorphisms, or equivalently the product of their matrix representations. Associated with K, there is a quadratic form

$$q(X) = K(X,X) = \mathrm{trace}((ad_X)^2).$$

Cartan introduced K to classify Lie algebras. When $\mathbf{k} = \mathbf{R}$ and g is isomorphic to $\mathbf{R}^n$, we can use Lagrange's method to reduce the quadratic form by change of basis to a diagonal quadratic form. This involves replacing the matrix K_0 that represents K as a bilinear form by $S^t K_0 S$, where S is a real and invertible matrix with transpose S^t. Then K and K_0 are said to be congruent.

Lemma 1.6.4 (Sylvester's law of inertia) *Let the rank of K be the dimension of the image of K; let the signature of K be the number of positive eigenvalues of K minus the number of negative eigenvalues of K, counted according to algebraic multiplicity. Then the rank and signature of K do not change under congruence.*

(1) Suppose that $\mathbf{k} = \mathbf{R}$ and that $K(X,Y) = 0$ for all X if and only if $Y = 0$. Then K is said to be non-degenerate, and the associated q has full rank.

Proposition 1.6.5 (Cartan's criterion) *A finite-dimensional real Lie algebra g has a non-degenerate Killing form, if and only if g is semisimple and can be expressed as a direct sum of simple Lie algebras.*

Example 1.6.6 (Lie algebras) The basic examples of real Lie algebras include the following.

(i) Let $gl(n, \mathbf{C})$ be the space $M_n(\mathbf{C})$ of $n \times n$ complex matrices with the usual product and the Lie bracket $[A, B] = AB - BA$. Let $sl(n, \mathbf{C})$ be the subspace of $X \in M_n(\mathbf{C})$ such that $\mathrm{trace}(X) = 0$. Then $sl(n, \mathbf{C})$ is a semisimple Lie algebra.

(ii) For $p, q, r, s \in \mathbf{C}^n$ with $p = (p_j)_{j=1}^n$, etc., we introduce a bilinear form $\Phi \colon \mathbf{C}^{2n} \times \mathbf{C}^{2n} \to \mathbf{C}$ by

$$\Phi\left(\begin{bmatrix} q \\ p \end{bmatrix}, \begin{bmatrix} s \\ r \end{bmatrix}\right) = \sum_{j=1}^{n} \begin{vmatrix} q_j & s_j \\ p_j & r_j \end{vmatrix}, \tag{1.6.4}$$

where $|\,|$ denotes the determinant. Then we let $sp(n, \mathbf{C})$ be the space of $X \in M_{2n}(\mathbf{C})$ such that

$$\Phi\left(X\begin{bmatrix} q \\ p \end{bmatrix}, \begin{bmatrix} s \\ r \end{bmatrix}\right) + \Phi\left(\begin{bmatrix} q \\ p \end{bmatrix}, X\begin{bmatrix} s \\ r \end{bmatrix}\right) = 0. \tag{1.6.5}$$

One easily checks that for all $X, Y \in sp(n, \mathbf{C})$, the commutator $[X, Y]$ also satisfies $[X, Y] \in sp(n, \mathbf{C})$, so $sp(n, \mathbf{C})$ is also a Lie algebra.

(2) Suppose that $\mathbf{k} = \mathbf{R}$ and that $K(X, X) < 0$ for all $X \in g \setminus \{0\}$, so q is negative definite. Then g is said to be of compact type. If G is a compact, connected and finite-dimensional Lie group, then its Lie algebra is of compact type. We encountered examples such as the spin group $Sp(n)$ in Section 1.5.

Example 1.6.7 (Infinite-dimensional Lie algebras) There are also infinite-dimensional examples of spaces with Lie brackets.

(3) Let M be an n-dimensional differentiable manifold, and $T(M)$ the space of smooth vector fields, or tangent fields, so that locally $F, G \in T(M)$ have the form

$$F = \sum_{j=1}^{n} f_j(x)\frac{\partial}{\partial x_j}, \quad G = \sum_{j=1}^{n} g_j(x)\frac{\partial}{\partial x_j} \tag{1.6.6}$$

where $x \in M$ and $f_j \in C^\infty(M); \mathbf{R})$. Then $T(M)$ has a Lie bracket

$$[F, G] = \sum_{i=1}^{n}\left(\sum_{j=1}^{n}\left(f_j(x)\frac{\partial g_i}{\partial x_j} - g_j(x)\frac{\partial f_i}{\partial x_j}\right)\right)\frac{\partial}{\partial x_i}. \tag{1.6.7}$$

The Lie bracket is intended to be consistent with the usual composition rule for differential operators. Given $F = \sum_{j=1}^{n} f_j(x)\frac{\partial}{\partial x_j}$ and a smooth scalar function h, we write $D_F h = \sum_{j=1}^{n} f_j(x)\frac{\partial h}{\partial x_j}$, for the derivative of h in the direction of the vector field $(f_j(x))_{j=1}^{n}$ at x, and one checks that $D_{[F,G]}h = D_F(D_G h) - D_G(D_F h)$, so

$$D_{[F,G]} = D_F D_G - D_F D_G. \tag{1.6.8}$$

There is a linear map $C^\infty(M) \to T(M)\colon f \mapsto X_f$, where $X_f = \sum_{j=1}^{n} \frac{\partial f}{\partial x_j} dx_j$, which is associated with the gradient of f.

Example 1.6.8 (Divergence free fields) On $T(M)$, we define the divergence by

$$\operatorname{div} F = \sum_{j=1}^{n} \frac{\partial f_j}{\partial x_j}. \tag{1.6.9}$$

Let $T(M)_0 = \{F \in T(M)\colon \operatorname{div} F = 0\}$. By a simple calculation, one checks that for all $F, G \in T(M)_0$, the commutator $[F, G]$ is also in $T(M)_0$, so $T(M)_0$ is a Lie subalgebra of $T(M)$. In the special case of $M = \mathbf{R}^3$, divergence free vector fields are important in fluid mechanics.

With $V = \operatorname{span}\{dx_j; j = 1, \dots, n\}$ we can form the skew-symmetric algebra $\wedge V$ and then the algebra $C^\infty(M) \otimes \wedge^* V$; in particular, we can let

$$T^*M = \left\{\sum_{j=1}^{n} f_j dx_j\colon f_j \in C^\infty(M)\right\}, \tag{1.6.10}$$

and generally let $\Omega^k M$ be the space of k-forms with coefficients in $C^\infty(M)$. Then we define $d\colon C^\infty(M) \to \Omega^1 M$ by $df = \sum_{j=1}^{n} \frac{\partial f}{\partial x_j} dx_j$, and $d(fg) = g(df) + f dg$. Note the close analogy between $df \in \Omega^1 M$ and the gradient $X_f \in T(M)$. Also, there is $d\colon \Omega^1 \to \Omega^2$, given by

$$d\sum_{j=1}^{n} f_j dx_j = \sum_{j=1}^{n}(df_j)dx_j = \sum_{1\le j<k\le n}\left(\frac{\partial f_j}{\partial x_k} - \frac{\partial f_k}{\partial x_j}\right)dx_j dx_k. \tag{1.6.11}$$

This satisfies $d^2 = 0$.

Exercise 1.6.9 (Circulant identity) Let A be a unital algebra, made into a Lie algebra for $[a,b] = ab - ba$, and X a non-empty subset of A; let I be an ideal in A. Let $[U,V] = \operatorname{span}\{[u,v]: u \in U, v \in V\}$ for subsets $U, V \subseteq A$.

(i) Prove the circulant identity

$$[y, x_1 \dots x_n] = \sum_{j=1}^{n} [x_{j+1} \dots x_n y x_1 \dots x_{j-1}, x_j].$$

(ii) Show that

$$[A, I^m] \subseteq [I^{m-j}, I^j] \subseteq [I^{m-1}, I].$$

(iii) Suppose that $I = AXA = \operatorname{span}\{axb: a,b \in A; x \in X\}$ is the ideal of A generated by X. Deduce that

$$[I^m, I] \subseteq [I^m AX, A] + [AI^m A, X] + [XAI^m, A],$$

and

$$[I^m, I] \subseteq [I^{m+1}, A].$$

Exercise 1.6.10 (Commutator ideal) For noncommutative algebras, it can be useful to replace the notion of the I-adic filtration by the commutator filtration; see [51, 52].

(i) Let J be an ideal in a **k**-algebra A_0 that contains the commutator subspace $[A_0, A_0]$. Show that $C_0 = A_0/J$ is a commutative algebra.

(ii) Let A_1 be the ideal in A_0 that is generated by $[A_0, A_0]$, and let $C_1 = A/A_1$. Show that there is a natural surjective quotient map $C_1 \to C_0$. This A_1 is the commutator ideal.

(iii) Let A_2 be the ideal generated by A_1^2 and the commutators $[A_0, A_1]$. Then let A_3 be the ideal generated by A_1^3 and the commutators $[A_0, A_2]$, etc. Show that there is a decreasing filtration

$$A_0 \supseteq A_1 \supseteq A_2 \supseteq A_3 \supseteq \cdots$$

of ideals of A_0 such that A_j/A_{j+1} is commutative.

(iv) Show that $A_i A_j \subseteq A_{i+j}$ and $[A_i, A_j] \subseteq A_{i+j+1}$. The key step is to prove by induction that $A_1 A_j \subseteq A_{j+1}$.

1.7 The Poisson Bracket

Definition 1.7.1 (Poisson bracket) Let **k** be a field of characteristic zero and A an associative algebra over **k**. A Poisson bracket on A is a map $\{\cdot,\cdot\}: A \times A \to A$ that satisfies:

(i) $\{\lambda f + \mu g, h\} = \lambda\{f,h\} + \mu\{g,h\}$;

(ii) $\{f,g\} = -\{f,g\}$;
(iii) Jacobi's identity

$$\{\{f,g\},h\} + \{\{h,f\},g\} + \{\{g,h\},f\} = 0; \tag{1.7.1}$$

(iv) $\{fg,h\} = f\{g,h\} + g\{f,h\}$; for all $f,g,h \in A$ and $\lambda,\mu \in k$.

The axioms (i), (ii) and (iii) show that $\{\cdot,\cdot\}$ defines a Lie bracket on A. The axioms (i) and (iv) show that $f \mapsto \{f,g\}$ defines a derivation on A. So there are many examples in geometry and mechanics. The following is the fundamental one from classical mechanics.

Example 1.7.2 (Phase space) Let $V = \mathbf{C} \otimes \mathbf{R}^{2n}$ and form the symmetric algebra $S(V)$; we write the coordinates of $\mathbf{R}^{2n}$ in the traditional style $(q_1,\dots,q_n;p_1,\dots,p_n)$ and take $S(V) = \mathbf{C}[q_1,\dots,q_n;p_1,\dots,p_n]$. Then define

$$\{f,g\} = \sum_{j=1}^{n}\left(\frac{\partial f}{\partial q_j}\frac{\partial g}{\partial p_j} - \frac{\partial f}{\partial p_j}\frac{\partial g}{\partial q_j}\right) \qquad (f,g \in S(V)). \tag{1.7.2}$$

Then $\{\cdot,\cdot\}$ defines a Poisson bracket on $S(V)$ over $\mathbf{C}$. Traditionally in Hamiltonian mechanics, $\mathbf{R}^{2n}$ is called a phase space, with $q_1,\dots,q_n$ the generalized coordinates and $p_1,\dots,p_n$ the generalized momenta. See [37].

We can extend this to the Poisson algebra, namely the space of formal power series

$$A = \mathbf{C}[[q_1,\dots,q_n;p_1,\dots,p_n]]$$

with the bracket $\{\cdot,\cdot\}$ to make it into a Lie algebra. The centre consists of the constant functions; quotienting out by the constants we get the space of Hamiltonian vector fields

$$\sum_{j=1}^{n}\left(\frac{\partial f}{\partial q_j}\frac{\partial}{\partial p_j} - \frac{\partial f}{\partial p_j}\frac{\partial}{\partial q_j}\right). \tag{1.7.3}$$

Suppose further that all coordinates depend upon time t and there is a Hamiltonian function $H\colon \mathbf{R} \times \mathbf{R}^{2n} \to \mathbf{R}$. Then

$$\sum_{j=1}^{n}\left(\frac{\partial H}{\partial q_j}\frac{\partial}{\partial p_j} - \frac{\partial H}{\partial p_j}\frac{\partial}{\partial q_j}\right)$$

gives a Hamiltonian vector field associated with H and the canonical equations are

$$\frac{dq_j}{dt} = \frac{\partial H}{\partial p_j}, \quad \frac{dp_j}{dt} = -\frac{\partial H}{\partial q_j}. \tag{1.7.4}$$

Given this system of equation, any function $\Phi : \mathbf{R} \times \mathbf{R}^{2n} \to \mathbf{R}$ of the form $\Phi(t, q_1, \dots, q_n, p_1, \dots, p_n)$ satisfies

$$\begin{aligned} \frac{d\Phi}{dt} &= \frac{\partial \Phi}{\partial t} + \sum_{j=1}^{n} \left(\frac{\partial \Phi}{\partial q_j} \frac{dq_j}{dt} + \frac{\partial \Phi}{\partial p_j} \frac{dp_j}{dt} \right) \\ &= \frac{\partial \Phi}{\partial t} + \sum_{j=1}^{n} \left(\frac{\partial \Phi}{\partial q_j} \frac{\partial H}{\partial p_j} - \frac{\partial \Phi}{\partial p_j} \frac{\partial H}{\partial q_j} \right) \\ &= \frac{\partial \Phi}{\partial t} + \{\Phi, H\}; \end{aligned} \tag{1.7.5}$$

so the Hamiltonian generates the motion through phase space with respect to time.

Example 1.7.3 (Partition function) With potential function V we take the Hamiltonian to be

$$H(q, p) = \frac{1}{2} \sum_{j=1}^{n} p_j^2 + V(q), \tag{1.7.6}$$

so by (1.7.5) H is itself a constant of the motion. Suppose that V is bounded below, and grows at infinity, so

$$V(q) \geq c_1 \|q\|^2 - c_2 \qquad (q \in \mathbf{R}^n)$$

for some $c_1, \dots, c_2 > 0$, so V is of polynomial growth. Then the Liouville measure

$$e^{-\beta H(q,p)} dq_1 \cdots dq_n dp_1 \cdots dp_n \tag{1.7.7}$$

is invariant under the flow generated by the canonical equations of motion, so the classical partition function

$$Z(\beta) = \int_{\mathbf{R}^{2n}} e^{-\beta H(q,p)} dq_1 \cdots dq_n dp_1 \cdots dp_n \qquad (\beta > 0) \tag{1.7.8}$$

is constant with respect to time. The partition function reduces to an integral over $\mathbf{R}^n$

$$Z(\beta) = \left(\frac{\sqrt{2\pi}}{\beta} \right)^n \int_{\mathbf{R}^n} e^{-\beta V(q)} dq_1 \cdots dq_n.$$

Canonical forms 1.7.4 (i) On $\mathbf{R}^{2n}$, the canonical 1 form is $\omega = \sum_{j=1}^{n} p_j dq_j$. The motivation for this choice is given by the calculus of variations.

(ii) The canonical 2 form is $\omega_2 = \sum_{j=1}^{n} dq_j \wedge dp_j$, which determines the Poisson bracket as follows. As in the preceding Section 1.6.5, $f, g \in C^\infty$ are associated with vector fields

$$X_f = \sum_{j=1}^{n} \left(\frac{\partial f}{\partial q_j} dq_j + \frac{\partial f}{\partial p_j} dp_j \right), \qquad X_g = \sum_{j=1}^{n} \left(\frac{\partial g}{\partial q_j} dq_j + \frac{\partial g}{\partial p_j} dp_j \right); \tag{1.7.9}$$

since $dq_j \wedge dp_j = -dp_j \wedge dq_j$, we obtain

$$\omega_2(X_f, X_g) = \sum_{j=1}^{n} \left(\frac{\partial f}{\partial q_j} \frac{\partial g}{\partial p_j} - \frac{\partial f}{\partial p_j} \frac{\partial g}{\partial q_j} \right) = \{f, g\}, \tag{1.7.10}$$

and we recover the Poisson bracket. One can take the existence of a non-degenerate 2 form as the starting point and deduce the existence of the Poisson bracket. An even-dimensional manifold with a non-degenerate 2 form satisfying various other axioms is called a symplectic manifold.

Example 1.7.5 (A cyclic cocycle from Jacobians) In particular, we consider $A = C_c^\infty(\mathbf{R}^2; \mathbf{C})$ with the above Poisson bracket and consider

$$\varphi(f, g, h) = \int_{\mathbf{R}^2} f\{g, h\} dq dp. \tag{1.7.11}$$

Suppose that the original variables (q, p) are written as differentiable functions of new variables (u, v) so that $q = q(u, v)$ and $p = p(u, v)$ under a bijective correspondence. Then

$$\begin{aligned}\varphi(f, g, h) &= \int_{\mathbf{R}^2} f \frac{\partial(g, h)}{\partial(q, p)} dq dp \\ &= \int_{\mathbf{R}^2} f \frac{\partial(g, h)}{\partial(q, p)} \frac{\partial(q, p)}{\partial(u, v)} du dv \\ &= \int_{\mathbf{R}^2} f \frac{\partial(g, h)}{\partial(u, v)} du dv \end{aligned} \tag{1.7.12}$$

by the Jacobian change of variables formula. This expression appears in multivariable calculus and in operator theory.

Also $\varphi \colon A^3 \to \mathbf{C}$ is linear in each variable and we introduce $b\varphi \colon A^4 \to \mathbf{C}$ as the following linear expression, in which we multiply successive pairs of variables and introduce a sign according to the position of the product in a cyclical fashion,

$$\begin{aligned} b\varphi(f_0, f_1, f_2, f_3) &= \varphi(f_0 f_1, f_2, f_3) - \varphi(f_0, f_1 f_2, f_3) \\ &\quad + \varphi(f_0, f_1, f_2 f_3) - \varphi(f_3 f_0, f_1, f_2). \end{aligned} \tag{1.7.13}$$

We now show that

$$b\varphi(f_0, f_1, f_2, f_3) = 0. \tag{1.7.14}$$

Using axiom (iv), we rearrange the left-hand side as

$$\begin{aligned}
&\int f_0 f_1\{f_2, f_3\} - \int f_0\{f_1 f_2, f_3\} + \int f_0\{f_1, f_2 f_3\} - \int f_3 f_0\{f_1, f_2\} \\
&= \int f_0 f_1\{f_2, f_3\} - \int f_0 f_1\{f_2, f_3\} - \int f_0 f_2\{f_1, f_3\} \\
&\quad + \int f_0\{f_1, f_2\} f_3 + \int f_0\{f_1, f_3\} f_2 - \int f_3 f_0\{f_1, f_2\} \\
&= 0.
\end{aligned} \tag{1.7.15}$$

1.8 Extensions of Algebras via Modules

Let A be an algebra over $\mathbf{k}$ and M a left A-module. Then we form the $\mathbf{k}$-vector space $R = A \oplus M$ and endow it with the multiplication

$$(a,m) \cdot (b,n) = (ab, an + mb) \tag{1.8.1}$$

so that R becomes an algebra over $\mathbf{k}$. We identify M with the subspace $\{(0,m) \colon m \in M\}$ of R and observe that $(0,m) \cdot (b,n) = (0,mb)$ and $(b,n) \cdot (0,m) = (0,bm)$, so M is an ideal of R. Furthermore, M has the trivial multiplication $(0,m) \cdot (0,n) = (0,0)$ for all $m,n \in M$, so $M^2 = 0$. We can realize M as the kernel of the canonical homomorphism $\pi : R \to A$ $(a,m) \mapsto a$, so we have an exact sequence

$$0 \longrightarrow M \longrightarrow R \longrightarrow A \longrightarrow 0. \tag{1.8.2}$$

We call this the extension of A by the square-zero ideal M. Conversely, given such an exact sequence with $\pi : R \to R/M$, the canonical homomorphism, a linear lifting is a $\mathbf{k}$-linear map $\rho \colon A \to R$ such that $\pi \circ \rho = Id$. Then $e = \rho \circ \pi$ is a linear map on R such that $e^2 = e$, so $R = A \oplus M$, and we can write $\rho(a) = (a, \delta(a))$ for some $\mathbf{k}$-linear $\delta \colon A \to M$. The extension is called trivial if ρ is a homomorphism.

Proposition 1.8.1 *Let M be an A-bimodule and $\delta \colon A \to M$ a derivation. Then there exists a trivial extension of A with M a square-zero ideal, a homomorphism $\rho \colon A \to E$ such that the homomorphism $\pi : E \to A$ satisfies $\pi \circ \rho = Id_A$. In particular, any inner derivation of the form $\delta_m(a) = am - ma$ for $a \in A$ and $m \in M$ gives a trivial extension.*

Proof We write

$$E = \left\{ \begin{bmatrix} a & m \\ 0 & a \end{bmatrix} : a \in A; m \in M \right\} \tag{1.8.3}$$

so E is an algebra and consider the homomorphism

$$\pi : E \to A : \quad \begin{bmatrix} a & m \\ 0 & a \end{bmatrix} \mapsto a \tag{1.8.4}$$

to have nullspace M and

$$\rho : A \to E : a \mapsto \begin{bmatrix} a & \delta(a) \\ 0 & a \end{bmatrix} \tag{1.8.5}$$

which is a homomorphism with $\pi \circ \rho = Id$ and which is an algebra isomorphism of A to a subalgebra of E. Indeed, one checks that ρ is a homomorphism if and only if δ is a derivation. The map $e : E \to E$ given by $\rho \circ \pi = e$ is an algebra homomorphism and satisfies $e^2 = e$ and $e(E)$ is isomorphic to A as an A-bimodule. We can regard E as having the multiplication of A deformed by the derivation δ. □

Definition 1.8.2 (i) *(Homomorphism modulo an ideal)* Let A and R be algebras over $\mathbf{k}$, and I an ideal of R with canonical quotient homomorphism $\pi : R \to R/I$. We say that a $\mathbf{k}$-linear map $\rho : A \to R$ is a homomorphism modulo I if the map $\pi \circ \rho : A \to R \to R/I$ is a homomorphism; equivalently, $-\rho(a_0 a_1) + \rho(a_0)\rho(a_1) \in I$ for all $a_0, a_1 \in R$.

(ii) *(Curvature)* Let $\rho : A \to R$ be a $\mathbf{k}$-linear map. Then the curvature of ρ is $\omega : A^2 \to R$

$$\omega(a_0, a_1) = \rho(a_0 a_1) - \rho(a_0)\rho(a_0). \tag{1.8.6}$$

So ρ is a homomorphism if and only if its curvature is zero, so that $\omega = 0$. Also, ρ is a homomorphism modulo I if and only if the range of ω is in I.

In Sections 5.4.3 and 8.1.3 we see how this relates to curvature from differential geometry. In operator theory, the expression $\omega(a_0, a_1)$ is sometimes called the semi-commutator of $\rho(a_0)$ and $\rho(a_1)$, and the following result is used for Toeplitz operators, as in Section 3.6.

Lemma 1.8.3 *Let A and R be algebras over $\mathbf{k}$, and let $\rho : A \to R$ be a homomorphism, modulo the ideal I of R. Then ρ induces a natural map $A/[A, A] \to (R/I)/[R/I, R/I]$, with dual mapping from the traces on (R/I) to the traces on A.*

Proof We have

$$\begin{aligned}\rho([a,b]) &= \rho(ab) - \rho(a)\rho(b) + \rho(b)\rho(a) - \rho(ba) + \rho(a)\rho(b) - \rho(b)\rho(a) \\ &= \omega(a,b) - \omega(b,a) + [\rho(a),\rho(b)] \qquad (1.8.7)\end{aligned}$$

so the composition of ρ with the canonical quotient homomorphism π gives $\pi \circ \rho\colon A \to R \to R/I$ such that $\pi \circ \rho[a,b] = [\pi \circ \rho(a), \pi \circ \rho(b)]$, so $\pi \circ \rho[A,A] \subseteq [R/I, R/I]$. The dual map takes a trace $\phi \in (R/I)'$ to $\phi \circ \pi \circ \rho \in A'$, where $\phi \circ \pi \circ \rho | [A,A] = 0$ since $\phi | [R/I, R/I] = 0$. □

Example 1.8.4 Let $A = C^\infty(\mathbf{R}^n, \mathbf{R})$ and $R = A[[\hbar]]$ be the algebra of formal power series in the indeterminate $\hbar$ with coefficients in A with the standard multiplication. Let $\rho\colon A \to R$ have the form

$$\rho(a) = a + \sum_{j=1}^{\infty} \frac{\hbar^j}{j!} \delta_j(a) \qquad (1.8.8)$$

where $\delta_j\colon A \to A$ is $\mathbf{k}$-linear. Let $I = (\hbar)$, so $I^2 = (\hbar^2)$, etc. Then ρ is a homomorphism modulo I^2, provided that $\rho(fg) - \rho(f)\rho(g) \in I^2$, so by comparing coefficients of $\hbar$, we have

$$\delta_1(fg) = \delta_1(f)g + f\delta_1(g);$$

that is, $\delta_1 : A \to A$ is a derivation. Given this, ρ is a homomorphism modulo I^3, provided that

$$f\delta_2(g) - \delta_2(fg) + \delta_2(f)g = \delta_1(f)\delta_1(g). \qquad (1.8.9)$$

To satisfy the first of these conditions, we can choose a smooth vector field $v\colon \mathbf{R}^n \to \mathbf{R}^n$, and $\nabla\colon C^\infty(\mathbf{R}^n, \mathbf{R}) \to C^\infty(\mathbf{R}^n, \mathbf{R}^n)$ the usual gradient operator on $\mathbf{R}^n$, so $\nabla f = (\frac{\partial f}{\partial x_1}, \dots, \frac{\partial f}{\partial x_n})$. Then

$$\delta_1(f) = \sum_{j=1}^{n} v_j(x) \frac{\partial f}{\partial x_j} \qquad (1.8.10)$$

is a derivation on A.

1.9 Deformations of the Standard Product

Let A be a commutative algebra over $\mathbf{k}$, and for indeterminate $\hbar$, let $\mathbf{k}[[\hbar]]$ be the algebra of formal power series in $\hbar$. Then supposing $\hbar a = a\hbar$, we can introduce the $\mathbf{k}$-vector space $A[[\hbar]]$ of formal power series in $\hbar$ with coefficients from A and made it into an $\mathbf{k}[[\hbar]]$-bimodule with the obvious operations. Likewise, given any $\mathbf{k}$-bilinear map $P\colon A\times A\to A$, we can extend it to a $\mathbf{k}[[\hbar]]$-bimodule map $P\colon A[[\hbar]]\times A[[\hbar]]\to A[[\hbar]]$. The space $A[[\hbar]]$ has an obvious multiplication arising from the formal product of power series in $\hbar$ with coefficients in $\hbar$, and we write fg for this product of $f,g\in[[\hbar]]$. There are other multiplications one could define on $A[[\hbar]]$, and they satisfy the following uniqueness theorem, as in [67].

Proposition 1.9.1 *Suppose that $*$ is an associative multiplication on $A[[\hbar]]$ such that*

$$f\star g = fg + \sum_{n=1}^{\infty}\hbar^n P_n(f,g) \tag{1.9.1}$$

where $P_n\colon A\times A\to A$ are $\mathbf{k}$-bilinear maps. Then

$$\{f,g\} = P_1(f,g) - P_1(g,f) \tag{1.9.2}$$

defines a Poisson bracket $\{\cdot,\cdot\}\colon A\times A\to A$.

Proof The associative rule gives

$$(a\star b)\star c = a\star(b\star c) \qquad (a,b,c\in A) \tag{1.9.3}$$

so when we expand up to terms in $\hbar^2$, we obtain

$$(ab+\hbar P_1(a,b)+\hbar^2P_2(a,b))\star c = a\star(bc+\hbar P_1(b,c)+\hbar^2P_2(b,c)) \tag{1.9.4}$$

so

$$\begin{aligned}abc&+\hbar P_1(a,b)c+\hbar^2P_2(a,b)c+\hbar P_1(ab,c)+\hbar^2P_1(P_1(a,b),c)+\hbar^2P_2(ab,c)\\&=abc+\hbar aP_1(b,c)+\hbar^2aP_2(b,c)+\hbar P_1(a,bc)+\hbar^2P_1(a,P_1(b,c))\\&\quad+\hbar^2P_2(a,bc)+O(\hbar^3)\end{aligned} \tag{1.9.5}$$

where the coefficients of $\hbar$ are

$$P_1(a,b)c+P_1(ab,c) = aP_1(b,c)+P_1(a,bc) \tag{1.9.6}$$

while the coefficients of $\hbar^2$ are

$$\begin{aligned}P_2(a,b)c+P_1(P_1(a,b),c)+P_2(ab,c) &= aP_1(b,c)+P_1(a,P_1(b,c))\\&\quad+P_2(a,bc).\end{aligned} \tag{1.9.7}$$

Now we verify the axioms of the Poisson bracket in turn.

(i) P_1 is bilinear by hypothesis.

(ii) Clearly the bracket is anti-symmetric.

(iii) We expand the bracket in terms of P_1, obtaining

$$\begin{aligned}\{ab,c\} - a\{b,c\} - b\{a,c\} &= P_1(ab,c) - P_1(c,ab) - aP_1(b,c) \\ &\quad + aP_1(c,b) - bP_1(a,c) + bP_1(c,a) \\ &= 0, \end{aligned} \tag{1.9.8}$$

after some reduction.

(iv) We take one triple product and express it in terms of P_1 and then P_2, obtaining

$$\begin{aligned}\{\{a,b\},c\} &= \{P_1(a,b) - P_1(b,a)\},c\} \\ &= P_1(P_1(a,b),c) - P_1(P_1(b,a),c) - P_1(c,P_1(a,b)) + P_1(c,P_1(b,a)) \\ &= -P_2(a,b)c + aP_2(b,c) - P_2(ab,c) + P_2(a,bc) \\ &\quad + P_2(b,a)c - bP_2(a,c) + P_2(ab,c) - P_2(a,bc); \end{aligned} \tag{1.9.9}$$

similar expansions are obtained for the other brackets by cyclically permuting a,b,c. The required identity follows by cancelling the terms. □

This is a uniqueness theorem and does not of itself indicate that the bracket $\{,\}$ is useful or even that it is non-zero. However, we can take $A = C^\infty(\mathbf{R}^{2n};\mathbf{R})$ with the Poisson bracket of Hamiltonian mechanics, introduce $A[[\hbar]]$ and ask: Does there exist an associative $*$ product on $A[[\hbar]]$ such that

$$f \star g = fg + \hbar\{f,g\} + O(\hbar^2)? \tag{1.9.10}$$

If so, can we classify all such $\star$ products? In Section 6.5, we consider the Fedosov product from [35], which gives algebras that to a first-order approximation reproduce the dynamical systems of classical mechanics.

2
Cyclic Cocycles and Basic Operators

Cyclic theory involves graded vector spaces which are linked by differentials to form chain complexes and hence produce homology groups by standard constructions from homological algebra, which we review here. The special feature of cyclic theory is the role of cyclic permutation operators on these chain complexes. In this chapter, we introduce the basic operators of cyclic theory and the notion of a cyclic cocycle for an associative noncommutative algebra. We also introduce one of the motivating examples of a cyclic cocycle on a compact differentiable manifold. The other notion is that of homomorphism modulo, an ideal of an algebra, which leads to an algebraic notion of curvature. One of the fundamental aspects of cyclic theory is the pairing with K-theory. For a commutative algebra A, we introduce the group $K_0(A)$ for finitely generated projective A modules.

2.1 The Chain Complex

Definition 2.1.1 (Chain complex) (i) A chain complex is a sequence of **k**-vector spaces K_j, linked by **k**-linear maps $d_j\colon K_j \to K_{j-1}$ such that $d_{j-1}d_j = 0$, which we write:

$$\cdots \xrightarrow{d_{j+2}} K_{j+1} \xrightarrow{d_{j+1}} K_j \xrightarrow{d_j} K_{j-1} \xrightarrow{d_{j-1}} \cdots \tag{2.1.1}$$

We abbreviate this by $(K,d) = (K_j,d_j)$, where d has degree (-1), and we write $d^2 = 0$ as shorthand, even when d^2 does not make sense as a composition of operators. If $\sigma_j \in K_j$ has $d_j\sigma_j = 0$, then we say that σ_j is a cycle. If $\beta_j \in K_j$ has the form $\beta_j = d_{j+1}\omega_{j+1}$ for some $\omega_{j+1} \in K_{j+1}$, then we say that β_j is a boundary. By hypothesis, every boundary is a cycle. By analogy with the case of simplicial homology, we call d a differential or boundary map.

Let $Z_j = \mathrm{Ker}(d_j)$ and $B_j = \mathrm{Im}(d_{j+1})$, so that $B_j \subseteq Z_j$, and define the homology group $H_j = H_j(M_j) = Z_j/B_j$ to be the quotient vector space. When $Z_j = B_j$, the sequence is said to be exact at K_j; otherwise, H_j measures the discrepancy of the complex from exactness. A short exact sequence consists of

$$0 \longrightarrow K_2 \overset{d_2}{\longrightarrow} K_1 \overset{d_1}{\longrightarrow} K_0 \longrightarrow 0 \tag{2.1.2}$$

with $d_1 d_2 = 0$, d_2 injective, d_1 surjective and $\ker(d_1) = \mathrm{im}(d_2)$. We think of K_2 as the kernel of d_1 and K_0 as the cokernel of d_1.

(ii) Let $(M,d) = (M_j, d_j)_{j=0}^N$ and $(M',d') = (M'_j, d'_j)_{j=0}^N$ be complexes, and suppose that $\alpha_j \colon M'_j \to M_j$ are $\mathbf{k}$-linear maps such that $d_j \alpha_j = \alpha_{j-1} d'_j$ so the maps round each basic cell

$$\begin{array}{ccc} M'_j & \overset{d'_j}{\longrightarrow} & M'_{j-1} \\ \alpha_j \downarrow & & \downarrow \alpha_{j-1} \\ M_j & \underset{d_j}{\longrightarrow} & M_{j-1} \end{array} \tag{2.1.3}$$

commute. Then we say that $\alpha \colon M' \to M$ is a map of complexes. Evidently $\alpha_j \colon \mathrm{im}(d'_{j+1}) \to \mathrm{im}(d_{j+1})$ and $\alpha_j \colon \ker(d'_j) \to \ker(d_j)$, so there is an induced map on the homology groups $\tilde{\alpha}_j \colon H'_j \to H_j$. If $\tilde{\alpha}_j$ is an isomorphism for all j, then we say that α is a quasi-isomorphism.

(iii) Let $(M,d) = (M_j, d_j)_{j=0}^N$, $(M',d') = (M'_j, d'_j)_{j=0}^N$ and $(M'',d'') = (M''_j, d''_j)_{j=0}^N$ be complexes, with $\alpha \colon M' \to M$ and $\beta \colon M \to M''$ maps of complexes such that $\beta_j \alpha_j = 0$ for all $j = 0, \ldots, N$. Then we write

$$0 \longrightarrow M' \overset{\alpha}{\longrightarrow} M \overset{\beta}{\longrightarrow} M'' \longrightarrow 0. \tag{2.1.4}$$

In particular, if all of the sequences

$$0 \longrightarrow M'_j \overset{\alpha_j}{\longrightarrow} M_j \overset{\beta_j}{\longrightarrow} M''_j \longrightarrow 0 \tag{2.1.5}$$

are short exact, then we say that the map of complexes is exact.

Lemma 2.1.2 *Let* (2.1.5) *be a short exact sequence of complexes. Then there are connecting homomorphisms $\delta_{j+1} \colon H_{j+1}(M''_{j+1}) \to H_j(M'_j)$ such that*

$$H_{j+1}(M''_{j+1}) \overset{\delta_{j+1}}{\longrightarrow} H_j(M'_j) \overset{\tilde{\alpha}_j}{\longrightarrow} H_j(M_j) \overset{\tilde{\beta}_j}{\longrightarrow} H_j(M''_j) \overset{\delta_j}{\longrightarrow} H_{j-1}(M'_j) \tag{2.1.6}$$

is a long exact sequence of the homology groups.

Proof This is a standard result of algebraic topology. Consider the following diagram, in which the downward maps are differentials d.

$$\begin{array}{ccccc} M'_{j+1} & \overset{\alpha_{j+1}}{\longrightarrow} & M_{j+1} & \overset{\beta_{j+1}}{\longrightarrow} & M''_{j+1} \\ \downarrow & & \downarrow & & \downarrow \\ M'_j & \overset{\alpha_j}{\longrightarrow} & M_j & \overset{\beta_j}{\longrightarrow} & M''_j \\ \downarrow & & \downarrow & & \downarrow \\ M'_{j-1} & \overset{\alpha_{j-1}}{\longrightarrow} & M_{j-1} & \overset{\beta_{j-1}}{\longrightarrow} & M''_{j-1} \end{array} \tag{2.1.7}$$

Given $x \in M''_{j+1}$ such that $dx = 0$, we need to trace back to $z \in M'_j$ such that $dz = 0$. See [33, p. 638] for discussion. □

Definition 2.1.3 (Cochain complex) Given a chain complex, namely a sequence of $\mathbf{k}$-vector spaces K_j that are linked by $\mathbf{k}$-linear maps $d_j\colon K_j \to K_{j-1}$ such that $d_{j-1}d_j = 0$, we introduce $C^j = \mathrm{Hom}_{\mathbf{k}}(K_j, \mathbf{k})$ and call the elements of C^j cochains. We introduce the dual map $d_j\colon C^j \to C^{j+1}$ so that for a suitable linear pairing

$$\langle c_j, d_{j+1}\omega_{j+1}\rangle = \langle d_j c_j, \omega_{j+1}\rangle \qquad (\omega_{j+1} \in K_{j+1}, c_j \in C^j) \tag{2.1.8}$$

we introduce the following array, called a cochain complex

$$\cdots \overset{d_{j-1}}{\longrightarrow} C^j \overset{d_j}{\longrightarrow} C^{j+1} \overset{d_{j+1}}{\longrightarrow} C^{j+2} \overset{d_{j+2}}{\longrightarrow} \cdots \tag{2.1.9}$$

where $d_{j+1}d_j = 0$. We abbreviate this as $(C,d) = (C^j, d_j)$, where d has degree $(+1)$, and we write $d^2 = 0$ as shorthand, even when d^2 does not make sense as a composition of operators. If $\sigma_j \in C^j$ has $d_j\sigma_j = 0$, then we say that σ_j is closed. If $\beta_j \in C^j$ has the form $\beta_j = d_{j-1}\omega_{j-1}$ for some $\omega_{j-1} \in C^{j-1}$, then we say that β_j is exact, as in the particular case of de Rham cohomology. By construction, every exact form is closed. Let $Z_j = \mathrm{Ker}(d_j)$ and $B_j = \mathrm{Im}(d_{j-1})$, so that $B_j \subseteq Z_j$, and we define the cohomology group $H^j = H^j(C^j) = Z_j/B_j$ to be the quotient vector space.

Exercise 2.1.4 (Vector calculus) Let $\Omega^1 = \{\sum_{j=1}^n f_j(x)dx_j\colon f_j \in C^\infty(\mathbf{R}^n;\mathbf{C})\}$ and let $d_0\colon C^\infty(\mathbf{R}^n;\mathbf{C}) \to \Omega^1$ be the differential $df = \sum_{j=1}^n \frac{\partial f}{\partial x_j}dx_j$. Then we let Ω^p be the space of differential p-forms with coefficients in $C^\infty(\mathbf{R}^n;\mathbf{C})$ which is spanned by $\omega = f_0 df_1 \wedge \cdots \wedge df_p$ where $f_j \in C^\infty(\mathbf{R}^n;\mathbf{C})$. There is a pairing between differential p-forms and simplices of dimension p via integration $\langle \omega, \sigma\rangle = \int_\sigma \omega$. As in Lemma 1.4.4 and Example 1.4.6, $\Omega^{n+1} = 0$. Then we define $d_p\colon \Omega^p \to \Omega^{p+1}$ by $d\omega = df_0 \wedge df_1 \wedge \cdots \wedge df_p$. Then

$d_{p+1}d_p = 0$, so here d is a differential of degree $(+1)$. So we can introduce a cochain complex

$$0 \longrightarrow \mathbf{C} \longrightarrow C^\infty(\mathbf{R}^n;\mathbf{C}) \longrightarrow \Omega^1 \longrightarrow \cdots \longrightarrow \Omega^n \longrightarrow 0. \qquad (2.1.10)$$

(i) Let $n = 3$, and show by vector calculus that the chain complex (2.1.10) is exact. Exactness at Ω^1 is a standard result of vector calculus which states that a vector field is irrotational if and only if it is the gradient field of some function. Thus $\nabla \times (\nabla f) = 0$ for all smooth functions f, and for all smooth vector fields $\mathbf{F}$ such that $\nabla \times \mathbf{F} = 0$, there exists some smooth scalar field f such that $\mathbf{F} = \nabla f$. Likewise, show that

$$\eta = F_1 dx_2 dx_3 + F_2 dx_3 dx_1 + F_3 dx_1 dx_2$$

has $d\eta = \mathrm{div}(\mathbf{F}) dx_1 dx_2 dx_3$ where $\mathbf{F} = (F_1, F_2, F_3)$, and that $\mathrm{div}(\mathbf{F}) = 0$ if and only if $\mathbf{F} = \nabla \times \mathbf{G}$ for some smooth vector field $\mathbf{G}$. See [71, p. 558] and [97].

(ii) One can show that (2.1.10) is exact for all $n \geq 1$. The key point is that for $\omega \in \Omega^p$ such that $d\omega = 0$, there exists $\sigma \in \Omega^{p-1}$ such that $d\sigma = \omega$ by Poincaré's Lemma [97, p. 145]. Hence the cohomology of this cochain complex is trivial.

(iii) We can introduce $C^\infty(S^1, \mathbf{C})$ and Ω^1 over the circle S^1. The differential form $\omega = d\theta$ has $d\omega = 0$, but ω is not an exact differential. In fact, we have an exact sequence

$$0 \longrightarrow \mathbf{C} \longrightarrow C^\infty(S^1;\mathbf{C}) \longrightarrow \Omega^1 \longrightarrow \mathbf{C} \longrightarrow 0, \qquad (2.1.11)$$

so $H^1(\Omega^\bullet) = \mathbf{C}$.

Exercise 2.1.5 (Exactness by formal integration) (i) Let $\mathbf{k}$ be a field of characteristic zero, and on the polynomial algebra $A = \mathbf{k}[x_1, \dots, x_n]$ introduce the differential $df = \sum_{j=1}^n \frac{\partial f}{\partial x_j} dx_j$ and the formal integral $\int x_j^p dx_j = x_j^{p+1}/(p+1)$. Prove that

$$0 \longrightarrow \mathbf{k} \longrightarrow A \overset{d}{\longrightarrow} \Omega^1 \overset{d}{\longrightarrow} \cdots \overset{d}{\longrightarrow} \Omega^n \longrightarrow 0 \qquad (2.1.12)$$

is an exact sequence. One approach, as in [48], is to use induction on n and eliminate variables one at a time by integration. In applications [98], it is important that one can use formal integration, without introducing the analytical definition of the indefinite integral.

(ii) Show that $H^0(\mathbf{k}^n) = H_{2n}(\mathbf{k}^n) = \mathbf{k}$, and that all the other cohomology and homology groups are zero. See [48, p. 53].

(iii) Extend these results to the algebra $\mathbf{k}[[x_1, \dots, x_n]]$ of formal power series.

2.2 The λ and b Operators

Let A be an associative algebra over $\mathbf{k}$ which is generally non-unital and noncommutative. In this context, the qualifier 'non-' means not necessarily. Let $[a,b] = ab - ba$ be the commutator of $a, b \in A$ and $[A,A] = \text{span}\{[a,b] : a, b \in A\}$ the commutator subspace. A trace on A is a linear functional $\tau : A \to \mathbf{k}$ such that

$$\tau(ab) = \tau(ba) \qquad (a, b \in A); \tag{2.2.1}$$

equivalently, $\tau \in (A/[A,A])^*$. See Proposition 1.3.3.

Elements of $A^{\otimes n}$ will often be called chains, whereas elements of $(A^{\otimes n})^*$ will be called cochains. The latter appear more directly in geometry, as in the next section.

Consider the cyclic permutation $(1, 2, \ldots, p) \mapsto (p, 1, 2, \ldots, p-1)$ on p symbols. The set of powers of this permutation gives the cyclic group of order p. Associated to this is the linear operator $\lambda : (A^{\otimes p})^* \to (A^{\otimes p})^*$, namely the cyclic permutation with sign, so $\varphi \in (A^{\otimes p})^*$ has

$$\lambda\varphi(a_1, \ldots, a_p) = (-1)^{p-1}\varphi(a_p, a_1, \ldots, a_{p-1}); \tag{2.2.2}$$

clearly $\lambda^p = id$. Taking linear combinations, we define $b : (A^{\otimes p})^* \to (A^{\otimes p+1})^*$ by

$$\begin{aligned} b\varphi(a_0, a_1, \ldots, a_p) = &\sum_{i=0}^{p-1}(-1)^i \varphi(a_0, \ldots, a_i a_{i+1}, \ldots, a_p) \\ &+ (-1)^p \varphi(a_p a_0, a_1, \ldots, a_{p-1}). \end{aligned} \tag{2.2.3}$$

The final summand is called the cross-over term and it warns us of curious features in calculations (compare 'crossover music').

Definition 2.2.1 (Cyclic cocycle) A cyclic n-cocycle on A is a cochain $\varphi \in (A^{\otimes n+1})^*$ such that $b\varphi = 0$ and $\lambda\varphi = \varphi$.

In particular, a cyclic 0-cocycle on A is precisely a trace on A, since

$$b\varphi(a_0, a_1) = \varphi(a_0 a_1) - \varphi(a_1 a_0). \tag{2.2.4}$$

2.3 Cyclic Cocycles on a Manifold

Let V be a smooth differentiable manifold and $A = C^\infty(V; \mathbf{C})$ the space of infinitely differentiable complex functions on V, with topology defined by a suitable family of seminorms. The multiplication is defined pointwise, so A is commutative and $[A,A] = \{0\}$. A continuous trace on A is a distribution on V, in the sense of Schwartz; namely a linear functional that is continuous with respect to the family of seminorms that determine the topology on V. Examples include $f \mapsto f(x_0)$ for $x_0 \in V$, which is given by a point mass measure at x_0. Let $\wedge^n A$ be the subspace of $A^{\otimes n}$ consisting of alternating tensors; see Lemma 1.4.4. Then an n-dimensional current γ on A is a continuous linear functional on the space of n-forms over V. Let $\varphi_\gamma : \wedge^n A \to \mathbf{C}$ be

$$\omega \mapsto \int_\gamma \omega \tag{2.3.1}$$

and an n-dimensional current, which we write as an integral, and associate it with

$$\varphi_\gamma(a_0, \dots, a_n) = \int_\gamma a_0 da_1 \dots da_n. \tag{2.3.2}$$

A current is said to be closed if $\int_\gamma d\omega = 0$ for all $\omega \in \wedge^{n-1} A$. The following result relates to Example 1.7.5.

Proposition 2.3.1 *Let φ_γ be a closed n-current. Then*
(i) $b\varphi_\gamma = 0$;
(ii) $\lambda\varphi_\gamma = \varphi_\gamma$.
So every closed n-current defines a cyclic cocycle on A.

Proof (i) For notational convenience we take $p = 3$ and look at

$$\begin{aligned} b\varphi(a_0, \dots, a_3) &= \int_\gamma a_0 a_1 da_2 da_3 - \int_\gamma a_0 d(a_1 a_2) da_3 \\ &\quad + \int_\gamma a_0 da_1 d(a_2 a_3) - \int_\gamma a_3 a_0 da_1 da_2 \\ &= \int_\gamma a_0 a_1 da_2 da_3 - \int_\gamma a_0 a_1 da_2 da_3 - \int_\gamma a_0 da_1\, a_2 da_3 \\ &\quad + \int_\gamma a_0 da_1\, a_2 da_3 + \int_\gamma a_0 a_1 da_2\, da_3 - \int_\gamma a_3 a_0 da_1 da_2 \\ &= 0. \end{aligned} \tag{2.3.3}$$

(ii) Note that

$$\begin{aligned} \lambda\varphi_\gamma(a_0, \dots, a_p) &= (-1)^p \varphi_\gamma(a_p, a_0, \dots, a_{p-1}) \\ &= (-1)^p \int_\gamma a_p da_0 \dots da_{p-1} \end{aligned} \tag{2.3.4}$$

where

$$a_p da_0 \dots da_{p-1} = d\big(a_0 da_1 \dots da_{p-1} da_p\big) - (-1)^{p-1} a_0 da_1 \dots da_p, \tag{2.3.5}$$

so when the current is closed

$$(-1)^p \int_\gamma a_p da_0 \dots da_{p-1} = \int_\gamma a_0 da_1 \dots da_p, \tag{2.3.6}$$

hence we have cyclic permutations with sign

$$\lambda\varphi_\gamma(a_0, \dots, a_p) = \varphi_\gamma(a_0, \dots, a_p). \tag{2.3.7}$$

□

Corollary 2.3.2 *Let V be a compact C^∞ differentiable manifold of dimension n which is orientable. Then V has a canonical cyclic n-cocycle*

$$\varphi(a_0, \dots, a_n) = \int_V a_0 da_1 \dots da_n. \tag{2.3.8}$$

Proof The space of alternating tensors of order n has dimension one. Suppose that V admits a continuous exterior differentiable form of degree n that is nowhere zero, and choose the form that has a particular orientation. Then the integral gives a cyclic cocycle by Proposition 2.3.1. □

Remarks 2.3.3 The term *cyclic theory* refers to the special status of the operator λ. We consider the noncommutative algebra $C^\infty(V) \otimes M_N$, which we can identify with the algebra of smooth functions $f\colon V \to M_N(\mathbf{C})$. We can extend $\varphi\colon A^{\otimes(n+1)} \to \mathbf{C}$ to $\varphi_{(N)}\colon (A \otimes M_N)^{\otimes(n+1)} \to \mathbf{C}$ by

$$\varphi_{(N)}(a_0 \otimes T_0, \dots, a_n \otimes T_n) = \int_V a_0 da_1 \dots da_n \mathrm{trace}(T_0 \dots T_n), \tag{2.3.9}$$

where trace : $M_N \to \mathbf{C}$ is the usual matrix trace. Note that

$$\varphi_{(N)}(a_{\sigma(0)} \otimes T_{\sigma(0)}, \dots, a_{\sigma(n)} \otimes T_{\sigma n} = \varepsilon(\sigma)\varphi_{(N)}(a_0 \otimes T_0, \dots, a_n \otimes T_n) \tag{2.3.10}$$

for any cyclic permutation σ of $\{0, \dots, n\}$ with sign $\varepsilon(\sigma)$. Whereas φ is symmetric up to the change of sign with respect to all permutations, the extension $\varphi_{(N)}$ to a noncommutative algebra is symmetric up to the change of sign only for cyclic permutations. This is the origin of the term cyclic theory. A cyclic cocycle is not the same as an element of $\wedge(A)^*$.

2.4 Double Complexes

Let $C_{n,m}$ be a doubly indexed array of $\mathbf{k}$-vector spaces which are connected by $\mathbf{k}$-linear maps such that the vertical differentials satisfy $d^v_{j+1,p}d^v_{j+1,p} = 0$, so the columns give chain complexes and the horizontal differentials satisfy $d^h_{j,p+1}d^h_{j,p} = 0$, so the rows give chain complexes. We also impose the condition that the maps round each basic cell

$$\begin{array}{ccc} C_{j,p+1} & \overset{d^h_{j+1,p}}{\longrightarrow} & C_{j+1,p+1} \\ d^v_{j,p} \uparrow & & \uparrow d^v_{j,p+1} \\ C_{j,p} & \underset{d^h_{j,p}}{\longrightarrow} & C_{j,p+1} \end{array} \tag{2.4.1}$$

anti-commute so

$$d^h_{j,p+1}d^v_{j,p} + d^v_{j+1,p}d^h_{j,p} = 0. \tag{2.4.2}$$

We sometimes add operators that have the same domain $C_{j,k}$ but different codomains. So $d = d^h + d^v$ is to be interpreted with this in mind. Under these conditions, $d^2 = 0$, since for example

$$\begin{aligned} &(d^v_{1,1} + d^h_{1,1})d^v_{1,0} + (d^v_{2,0} + d^h_{2,0})d^h_{1,0} \\ &= d^v_{1,1}d^v_{1,0} + d^h_{2,0}d^h_{1,0} + (d^h_{1,1}d^v_{1,0} + d^v_{2,0}d^h_{1,0}) \\ &= 0 \end{aligned} \tag{2.4.3}$$

and likewise at other entries in the array. Hence the array is a double complex, and we can form sums over the leading diagonals

$$C_k = \oplus_{j=-\infty}^{\infty} C_{k-j,j} \tag{2.4.4}$$

so that $d^h_{k-j,j} + d^v_{k-j,j}\colon C_{k-j,j} \to C_{k-j,j+1} + C_{k+1-j,j}$ gives the total differential:

$$\begin{array}{cccccc} C_{0,2} & \overset{d^h_{0,2}}{\longrightarrow} & C_{1,2} & \overset{d^h_{1,2}}{\longrightarrow} & C_{2,2} & \overset{d^h_{2,2}}{\longrightarrow} \\ d^v_{0,1} \uparrow & & d^v_{1,1} \uparrow & & d^v_{2,1} \uparrow & \\ C_{0,1} & \overset{d^h_{0,1}}{\longrightarrow} & C_{1,1} & \overset{d^h_{1,1}}{\longrightarrow} & C_{2,1} & \overset{d^h_{2,1}}{\longrightarrow} \\ d^v_{0,0} \uparrow & & d^v_{1,0} \uparrow & & d^v_{2,0} \uparrow & \\ C_{0,0} & \overset{d^h_{0,0}}{\longrightarrow} & C_{1,0} & \overset{d^h_{1,0}}{\longrightarrow} & C_{2,0} & \overset{d^h_{2,0}}{\longrightarrow} \end{array} \tag{2.4.5}$$

In examples, we often suppress the subscripts and superscripts on the differentials.

2.5 The b' and N Operators

Let V be a **k**-vector space and let the V-valued cochains of order p be

$$C^p(A, V) = \mathrm{Hom}_{\mathbf{k}}(A^{\otimes p}, V) = \{f\colon A^p \to V \quad \text{multilinear}\} \tag{2.5.1}$$

where we identify multilinear maps on A^p with linear maps on $A^{\otimes p}$. Let

$$C(A, V) = \oplus_{p=0}^{\infty} C^p(A, V) \tag{2.5.2}$$

be the space of sequences of cochains which are eventually zero. The basic operations on $C(A, V)$ are λ and b from Section 2.3, extended in the obvious way to cochains with values in V, and $b'\colon C^p(A, V) \to C^{p+1}(A, V)$ given by

$$b'f(a_0, \dots, a_p) = \sum_{i=0}^{p-1} (-1)^i f(a_0, \dots, a_i a_{i+1}, \dots, a_p); \tag{2.5.3}$$

and $N\colon C^p(A, V) \to C^p(A, V)$ by

$$Nf(a_1, \dots, a_p) = \sum_{i=0}^{p-1} \lambda^i f(a_1, \dots, a_p). \tag{2.5.4}$$

Note that b and b' differ only in the cross-over term, which is missing from b'. Clearly $\lambda N = N$, so N produces cyclically symmetric cochains.

Proposition 2.5.1 *The operators on $C(A, V)$ satisfy the basic identities*

$$b^2 = 0, \quad (b')^2 = 0, \quad (1 - \lambda)b = b'(1 - \lambda), \quad Nb' = bN \tag{2.5.5}$$

so that there is a double complex:

$$\begin{array}{ccccccc}
b\uparrow & & -b'\uparrow & & b\uparrow & & \\
C^{p+1}(A,V) & \xrightarrow{1-\lambda} & C^{p+1}(A,V) & \xrightarrow{N} & C^{p+1}A,C) & \xrightarrow{1-\lambda} & \\
b\uparrow & & -b'\uparrow & & b\uparrow & & \\
C^{p}(A,V) & \xrightarrow{1-\lambda} & C^{p}(A,V) & \xrightarrow{N} & C^{p}A,C) & \xrightarrow{1-\lambda} & \\
b\uparrow & & -b'\uparrow & & b\uparrow & & \\
C^{p-1}(A,V) & \xrightarrow{1-\lambda} & C^{p-1}(A,V) & \xrightarrow{N} & C^{p-1}A,C) & \xrightarrow{1-\lambda} &
\end{array} \tag{2.5.6}$$

Proof This follows by direct calculation. □

Now let R be another **k**-algebra and consider the (unsigned) cup product

$$C^p(A, R) \otimes C^q(A, R) \to C^{p+q}(A, R) \tag{2.5.7}$$

given by taking the product of $f \in C^p(A, R)$ and $g \in C^q(A, R)$ in R,

$$(f \cdot g)(a_1, \dots, a_{p+q}) = f(a_1, \dots, a_p) g(a_{p+1}, \dots, a_{p+q}). \tag{2.5.8}$$

Example 2.5.2 (b' and curvature) Let $\delta = -b'$. Then δ satisfies the following variant of Leibniz's rule

$$\delta(f \cdot g) = (\delta f) \cdot g + (-1)^p f \cdot (\delta g). \tag{2.5.9}$$

For $g, f \in C^1(A, R)$, we have

$$\delta f(a_0, a_1) = -b' f(a_0, a_1) = -f(a_0 a_1), \tag{2.5.10}$$

so

$$\begin{aligned}(\delta f) \cdot g(a_0, a_1, a_2) - f \cdot (\delta g)(a_0, a_1, a_2) &= -f(a_0 a_1) g(a_2) + f(a_0) g(a_1 a_2) \\ &= \delta(f \cdot g)(a_0, a_1, a_2). \end{aligned} \tag{2.5.11}$$

In later discussion, the following example will be important. For $f \in C^1(A, R)$, $\delta f + f \cdot f = 0$ if and only if $f : A \to R$ is an algebra homomorphism. Indeed

$$(\delta f + f \cdot f)(a_0, a_1) = -f(a_0 a_1) + f(a_0) f(a_1). \tag{2.5.12}$$

2.6 Hochschild Cohomology

First we review Hochschild cohomology for a unital associative algebra A over $\mathbf{C}$. We write $\mathbf{C} = \mathbf{C}1$ and $\bar{A} = A/\mathbf{C}$. An A-bimodule M as in Section 1.2 is equivalent to a left $A \otimes A^{op}$ module M with $(a \otimes c, m) \mapsto amc$. Let P be the resolution

$$\xrightarrow{b'} A \otimes \bar{A}^{\otimes 2} \otimes A \xrightarrow{b'} A \otimes \bar{A} \otimes A \xrightarrow{b'} A \otimes A \xrightarrow{b'} A \xrightarrow{b'} 0 \tag{2.6.1}$$

with

$$C^n(A, M) = \mathrm{Hom}((\bar{A})^{\otimes n}, M) = \mathrm{Hom}_{A \otimes A^{op}}(A \otimes (\bar{A})^{\otimes n} \otimes A, M) \tag{2.6.2}$$

consisting of multilinear $f : A^{\times n} \to M$ such that $f(a_1, \dots, a_n) = 0$ if $a_j = 1$ for some j. The differential is $\delta : C^n(A, M) \to C^{n+1}(A, M)$

$$\begin{aligned}(\delta f)(a_0, \dots, a_n) &= a_0 f(a_1, \dots, a_n) + \sum_{i=0}^{n-1} (-1)^{i+1} f(a_0, \dots, a_i a_{i+1}, \dots, a_n) \\ &\quad + (-1)^{n+1} f(a_0, \dots, a_{n-1}) a_n. \end{aligned} \tag{2.6.3}$$

Now we interpret $H^j(A, M)$ for $j = 0, 1, 2$.

(0) First, $C^0(A, M) = M$ and $\delta_m \in C^1(A, M)$ is the inner derivation $\delta_m(a) = am - ma$. It follows that

$$H^0(A,M) = \mathrm{Ker}(\delta\colon C^0(A,M) \to C^1(A,M))$$
$$= \{m \in M\colon am - ma = 0, \forall a \in A\} \qquad (2.6.4)$$

is the centre of M as an A-bimodule.

(1) Next we have

$$Z^1(A,M) = \mathrm{Ker}(\delta\colon C^1(A,M) \to C^2(A,M))$$
$$= \{f\colon \bar{A} \to M\colon \delta f(a_1,a_2) = 0\} \qquad (2.6.5)$$

is the set of all derivations $D\colon A \to M$. Hence

$$H^1(A,M) = \{\text{Derivations}\colon A \to M\}/\{\text{Inner derivations}\colon A \to M\}. \quad (2.6.6)$$

Computing $H^1(A,M)$ can be difficult, even for specific examples. We consider a significant example in Proposition 4.7.7, and in Section 6.4 we consider conditions that ensure that $H^1(A,M) = 0$ for all M. We recall from Proposition 1.8.1 that any derivation D and linear lifting map gives a trivial extension of A by a square-zero ideal M. Next we consider non-trivial extensions.

(2) As in Section 1.8, we consider an algebra extension

$$0 \longrightarrow M \longrightarrow E \overset{\pi}{\longrightarrow} A \longrightarrow 0 \qquad (2.6.7)$$

with $M^2 = 0$. In particular, we can choose a linear lifting map $\rho\colon A \to E$ with $\pi \circ \rho = Id$ and $\rho(a) = (a, \psi(a))$, so $E = A \oplus M$ as $\mathbf{k}$-vector spaces. Then we consider $\varphi\colon A^2 \to M$ and $(a,x)\cdot(b,y) = (ab, ay + xb + \varphi(a,b))$. Then the associative law holds for this multiplication, provided that $\varphi\colon A^2 \to M$ satisfies $\delta\varphi(a_0,a_1,a_2) = 0$, or more explicitly

$$a_0\varphi(a_1,a_2) - \varphi(a_0a_1,a_2) + \varphi(a_0,a_1a_2) - \varphi(a_0,a_1)a_2 = 0. \qquad (2.6.8)$$

Then

$$\omega(a_0,a_1) = \rho(a_0a_1) - \rho(a_0)\rho(a_1) = (0, \psi(a_0a_1) - a_0\psi(a_1) - \psi(a_0)a_1) \qquad (2.6.9)$$

where $\sigma(a_0,a_1) = \psi(a_0a_1) - a_0\psi(a_1) - \psi(a_0)a_1$ satisfies $\sigma = -\delta\psi$ and $\delta\sigma(a_0,a_1,a_2) = 0$; so σ is a coboundary. A Hochschild 2-cocycle $\varphi\colon A \times A \to M$ can be identified with an algebra extension with $M^2 = 0$ and a linear lifting map $\rho\colon A \to E$ such that $\pi \circ \rho = id$. We call such an extension with square zero. Changing the lifting map by a linear map alters φ by a coboundary. Hence $H^2(A,M)$ is the set of isomorphism classes of square-zero algebra extensions of A by M.

(3) The $H^n(A,M)$ with $n \geq 3$ are harder to interpret. Loday [68, p. 43] describes $H^3(A,M)$ in terms of crossed bimodules.

In this discussion, we have said little about M. In Section 2.9, we consider some specific classes of modules for specific A.

2.7 Vector Traces

Let V be a **k**-vector space. Let τ be a vector trace; that is $\tau\colon R \to V$ is a **k**-linear function such that $\tau|[R,R] = 0$, so τ induces a linear map $R/[R,R] \to V$. Let $C^p_\lambda(A,V)$ be the space of $f \in \mathrm{Hom}_{\mathbf{k}}(A^{\otimes p}, V)$ such that $\lambda f = f$; we call this the cyclic cochains of degree p. Define $\mathrm{tr}_\tau\colon C^p(A,R) \to C^{p-1}_\lambda(A,V)$ by $\mathrm{tr}_\tau f = N\tau f$, or more explicitly

$$\mathrm{tr}_\tau f(a_1,\dots,a_p) = \sum_{i=1}^{p-1}(-1)^{i(p-1)}\tau f(a_{i+1},\dots,a_p,a_1,\dots,a_i). \qquad (2.7.1)$$

In algebraic topology, there is an operation called the cup product defined on singular cohomology and a coboundary operator which is a derivation for this multiplication; see [42, p. 196]. The following result is in this spirit.

Proposition 2.7.1 (i) *With differential δ and the cup product $\cdot$, the cochains $C(A,R)$ form a differential graded algebra.*

(ii) *Let $r = degree(f)$ and $s = degree(g)$. Then*

$$\mathrm{tr}_\tau(f\cdot g) = (-1)^{rs}\mathrm{tr}_\tau(g\cdot f), \qquad (2.7.2)$$

$$\mathrm{tr}_\tau(\delta f) = -b\,\mathrm{tr}_\tau(f). \qquad (2.7.3)$$

Proof (ii) Note that $\tau(g\cdot f) = \lambda^p\tau(f\cdot g)$, so

$$N\tau(g\cdot f) = N\lambda^p\tau(f\cdot g). \qquad (2.7.4)$$

For $r+s=p$, we have $(-1)^{i(p-1)} = (-1)^{i(r-1)(s-1)}(-1)^{rs}$. □

Definition 2.7.2 (Homomorphism modulo an ideal) We say that a **k**-linear map $\rho\colon A \to R$ is a homomorphism modulo I if the map $\pi\circ\rho\colon A \to R \to R/I$ is a homomorphism; equivalently, $-\rho(a_0a_1) + \rho(a_0)\rho(a_1) \in I$ for all $a_0, a_1 \in R$. Alternatively, one can say that $\omega = \delta\rho + \rho\cdot\rho$ has $\omega \in C^2(A,I)$.

Write $I^2 = \mathrm{span}\{rs\colon r,s \in R\}$. Then we have a descending sequence of ideals of R, $R \supseteq I \supseteq I^2 \supseteq I^3 \supseteq \dots$. Let $\tau\colon I^n \to V$ be a **k**-linear map such that $\tau|[R,I^n] = \{0\}$, and define

$$\mathrm{tr}_\tau(f) = N\tau(f) \in C^{p-1}_\lambda(A,V) \qquad \big(f \in C^p_\lambda(A,I^n)\big); \qquad (2.7.5)$$

this gives a map $\mathrm{tr}_\tau\colon C(A,I^n) \to C_\lambda(A,V)$.

Proposition 2.7.3 (i) *$C(A, I^n)$ is an ideal in the differential graded algebra $(C(A, R), \cdot, \delta)$ which is closed under δ.*

(ii) *For $f \in C^r(A, I^n)$ and $g \in C^s(A, R)$,*

$$\mathrm{tr}_\tau(f \cdot g) = (-1)^{rs}\mathrm{tr}_\tau(g \cdot f). \tag{2.7.6}$$

Proof This is largely a formal consequence of Proposition 2.7.1. □

2.8 Bianchi's Identity

Let A and R be algebras over $\mathbf{k}$, and recall that a $\mathbf{k}$-linear map $\theta \colon A \to R$ is an (algebra) homomorphism if $\theta(ab) = \theta(a)\theta(b)$ for all $a, b, \in A$. Now let I be an ideal in R, so R/I is also an algebra; let $\pi : R \to R/I$ be the canonical quotient homomorphism as in Section 1.2.

Definition 2.8.1 (Curvature) The curvature of $\rho \in C^1(A, R)$ is

$$\omega = \delta\rho + \rho \cdot \rho \in C^2(A, R). \tag{2.8.1}$$

Let $\omega^n \in C^{2n}(A, R)$ be $\omega^n(a_1, a_2, \dots, a_{2n}) = \omega(a_1, a_2)\omega(a_3, a_4) \cdots \omega(a_{2n-1}, a_{2n})$, which has n factors of curvatures in the order of paired indices.

Proposition 2.8.2 *Let ρ be a homomorphism modulo I^n. Then*

(i) $\delta(\omega^n) = [\omega^n, \rho]$;
(ii) $\mathrm{tr}_\tau(\omega^n)$ *is a cyclic $2n - 1$ cocycle on A.*

Proof (i) With $\omega = \delta\rho + \rho \cdot \rho$, we have

$$\begin{aligned}\delta\omega &= (\delta\rho) \cdot \rho - \rho \cdot (\delta\rho) \\ &= (\delta\rho + \rho \cdot \rho) \cdot \rho - \rho \cdot (\delta\rho + \rho \cdot \rho) \\ &= \omega \cdot \rho - \rho \cdot \omega. \end{aligned} \tag{2.8.2}$$

Suppressing the $\cdot$ in the cup product, we use the derivation rule to obtain

$$\begin{aligned}\delta(\omega^n) &= \sum_{i=1}^{n} \omega^{i-1}\delta(\omega)\omega^{n-i} \\ &= \sum_{i=1}^{n} \omega^{i-1}(\omega\rho - \rho\omega))\omega^{n-i} \\ &= [\omega^n, \rho]. \end{aligned} \tag{2.8.3}$$

(ii) Then

$$b\mathrm{tr}_\tau(\omega^n) = -\mathrm{tr}_\tau(\delta(\omega^n)) = \mathrm{tr}_\tau(\rho\omega^n - \omega^n\rho) = 0, \tag{2.8.4}$$

since $\omega^n \in C^{2n}(A, I^n)$ and τ is a trace on I^n. Hence $\mathrm{tr}_\tau(\omega^n)$ is a $2n-1$ cocycle on the algebra A. Also,

$$\mathrm{tr}_\tau(\omega^n)(a_1, \ldots, a_{2n}) = \tau(\omega(a_1, a_2) \cdots \omega(a_{2n-1}, a_{2n})) \\ + \text{cyclic permutations with sign} \qquad (2.8.5)$$

is a cyclic $(2n-1)$-cocycle. □

2.9 Projective Modules

Let A be a ring, which is unital but not necessarily commutative, and consider left modules over A. The basic idea is that projective modules over A share many of the properties of vector spaces over fields. Mainly we formulate definitions for arbitrary rings and then give specific results for commutative unital algebras.

Definition 2.9.1 (1) (*Free module*) Let $A^n = \oplus_{j=1}^n A$ be the direct sum of A as left A-module. Then a left A-module F is free and finitely generated if F is isomorphic to A^n as a left A-module.

(2) (*Projective*) A left A-module P is said to be projective if for all left A-modules M and all surjective left A-module maps $\alpha : M \to P$, there exists a left A-module map $\beta : P \to M$ such that $\alpha \circ \beta = id$.

(3) (*Finitely generated*) A left A-module P is finitely generated if there exist $n \in \mathbf{N}$ and a surjective A-module map $A^n \to P$.

Lemma 2.9.2 *A left A-module P is finitely generated and projective if and only if there exist n and a left A-module P' such that $A^n \cong P \oplus P'$ as left A-modules.*

Such a P' is called a complementary module, and both P and P' are the images of a $\mathbf{k}$-linear projection on A^n that is also an A-module map. Proposition 4.6 of [33] shows that projectivity of P is equivalent to the condition that $R \to S \to 0$ exact implies $\mathrm{Hom}_A(P, R) \to \mathrm{Hom}_A(P, S) \to 0$ exact. This means that any A-module map $\psi : P \to S$ lifts to an A-module map $\tilde{\psi} : P \to R$, as in:

$$\begin{array}{ccc} \tilde{\psi} & & R \\ & \nearrow & \downarrow \\ P & \rightarrow & S \\ & \psi & \downarrow \\ & & 0 \end{array}$$

The map $\tilde{\psi}$ is called a lifting of A.

Example 2.9.3 (Projective modules) (i) Let V be a finite-dimensional $\mathbf{k}$-vector space. Then V is projective and finitely generated as a $\mathbf{k}$-module. This is essentially the rank-plus-nullity theorem of linear algebra; indeed, for any finite-dimensional W and surjective $\mathbf{k}$-linear map $\pi : W \to V$, one can select the bases of V and W so that $W \sim V \oplus \ker(\pi)$ and π may be represented as the block matrix

$$\pi \sim \begin{bmatrix} I & 0 \end{bmatrix}. \tag{2.9.1}$$

(ii) We can take $H = \mathbf{C}^n$ and write the free module as $A^n = A \otimes H$. Note that for all A-modules N, we have $N \otimes_A A^n \cong N^n$. For any finite-dimensional vector space V, the left A-module $A \otimes V$ is projective and finitely generated. In Section 4.4, we discuss $A \otimes W$ as a projective module for any vector space W. We pursue this idea in later sections.

Lemma 2.9.4 *For a finitely generated module M over a principal ideal domain A, let the torsion submodule of M be*

$$T = \{m \in M : xm = 0 \quad \text{for some} \quad x \in A \setminus \{0\}\}. \tag{2.9.2}$$

Then the following conditions are equivalent:

(1) M is free;
(2) M is projective;
(3) M is torsion free, that is, $T = \{0\}$.

Proof By [47, p. 125], T is a submodule of M, and there exists a finitely generated free submodule F of M such that $M \cong T \oplus F$. All the listed conditions are equivalent to $T = 0$. □

(iii) Lemma 2.9.4 applies to the principal ideal domains $A = \mathbf{C}[X]$ and $A = \mathbf{Z}$. See [47].

(iv) A commutative and unital ring R is said to be a local ring if R has only one maximal ideal. A finitely generated module over a local ring is projective if and only if it is free; by [64]. For example, $R = \mathbf{C}[X]/(X^2) = \{a1 + bX + (X^2) : a, b \in \mathbf{C}\}$ is a local ring, where $(X + (X^2))$ is the maximal ideal.

(v) Let $R = \mathbf{C}(X)$ be the field of rational complex functions, which contains the polynomial algebra $\mathbf{C}[X]$ as a subalgebra. Indeed, $\mathbf{C}[X]$ is the subalgebra of $f \in R$ such that f has no poles in $\mathbf{C}$, only possible poles at ∞. Then by Lemma 2.9.4, all finitely generated modules over R and all torsion-free finitely generated modules over $\mathbf{C}[X]$ are projective. More generally, let S be a finite subset of the Riemann sphere $\mathbf{C} \cup \{\infty\}$ and R_S the algebra of rational functions on $\mathbf{C} \cup \{\infty\}$ with possible poles in S, where we regard ∞ as a pole of a monic non-constant polynomial.

Proposition 2.9.5 *All finitely generated projective modules over R_S are free.*

Proof Let T be the multiplicatively closed subset of $\mathbf{C}[X]$ that is generated by 1, $(X-a)$ for $a \in S$, and let the ring of fractions be

$$T^{-1}\mathbf{C}[X] = \{p(X)/q(X) \colon p(X) \in \mathbf{C}[X]; q(X) \in T\}$$

as in [8, p. 36]. There is a canonical homomorphism $\varphi \colon \mathbf{C}[X] \to T^{-1}\mathbf{C}[X] : p(X) \mapsto p(X)$, and any ideal I in $T^{-1}\mathbf{C}[X]$ is pulled back to a contractive ideal $\varphi^{-1}(I)$ in $\mathbf{C}[X]$, which is a principal ideal domain, so $\varphi^{-1}(I) = (p(X))$ for some $p(X) \in \mathbf{C}[X]$. Hence $T^{-1}\mathbf{C}[X]$ is also a PID, which we can identify with R_S. By Lemma 2.9.4, all finitely generated projective modules over R_S are free. □

(vi) Let M be a finitely generated projective module over the polynomial ring $A = \mathbf{C}[X_1, \dots, X_n]$. Then M is free, by the Quillen–Suslin theorem. This is much more difficult to prove than the preceding examples; see [64].

Definition 2.9.6 (Projective resolution) (1) A module M has a finite projective resolution if there exists an exact sequence

$$0 \longrightarrow P_n \longrightarrow \cdots \longrightarrow P_1 \longrightarrow P_0 \longrightarrow M \longrightarrow 0 \qquad (2.9.3)$$

where the P_j are projective A-modules. The projective dimension of M is the minimum value of n for which such a sequence exists. Thus M is projective if and only if it has projective dimension zero. In Section 6.4, we consider modules of projective dimension 0 and 1. (In [55], the authors show how the projective dimension is related to other notations of dimension for modules over commutative algebras related to algebraic varieties.)

(2) Let $\mathrm{Proj}(A)$ be the set of isomorphism classes $[P]$ of finitely generated projective left A-modules P, and introduce the commutative addition rule $[P] + [Q] = [P \oplus Q]$. On $\mathrm{Proj}(A)^2$ we introduce an equivalence relation $\sim$ by

$$([P],[Q]) \sim ([P'],[Q']) \Leftrightarrow [P \oplus Q' \oplus V] = [Q \oplus P' \oplus V] \qquad (2.9.4)$$

for some projective finitely generated left A-module V.

Definition 2.9.7 (K_0) Let $K_0(A)$ be $\mathrm{Proj}(A)^{\times 2}/\sim$, with

$$([P],[Q]) + ([P'],[Q']) = ([P \oplus P'],[Q \oplus Q']) \qquad (\mathrm{mod} \sim). \qquad (2.9.5)$$

Note that

$$([P],[Q]) + ([P'],[Q']) \sim ([P'],[Q']) + ([P],[Q])$$

and

$$([P],[Q]) + ([Q],[P]) \sim ([P \oplus Q],[Q \oplus P]) \sim (0,0),$$

so $K_0(A)$ is an abelian group. The map $\mathrm{Proj}(A) \to K_0(A)\colon [P] \mapsto ([P],0)/\sim$ is a homomorphism of additive semigroups, but is not necessarily injective.

Definition 2.9.8 (Augmented algebra) Let A be a commutative algebra over $\mathbf{k}$, not necessarily with a unit. Then the augmented algebra of A is $A^+ = \{(f,c)\colon f \in A, c \in \mathbf{k}\}$ with the multiplication

$$((f,c),(g,d)) \mapsto (f,c)\cdot(g,d) = (fg + df + cg, cd).$$

We regard A^+ as A with a unit adjoined, and think of A^+ as an abbreviation for $A \oplus \mathbf{k}1$. (In the literature, the notation $\tilde{A}$ is also used for augmented algebra, but we use $\bar{A} = a/\mathbf{k}1$ and seek to avoid confusion between $\tilde{A}$ and $\bar{A}$.)

There is a surjective homomorphism $\psi\colon A^+ \to A$ such that $A^+/\mathbf{k}$ is isomorphic to A, and $\iota\colon \mathbf{k} \to A^+\colon\ c \mapsto (0,c)$ is a surjection onto the nullspace of ψ, so that A^+ has a unit $(0,1)$, and A^+ contains A as an ideal, so $A^+ = A \oplus \mathbf{k}$, and $[A^+,A^+] = [A,A]$; hence $A^+/[A^+,A^+] = (A/[A,A]) \oplus \mathbf{k}$.

This ψ induces a surjective homomorphism $\psi^*\colon K_0(A^+) \to K_0(\mathbf{k})$, while ι induces $\iota^*\colon K_0(\mathbf{k}) \to K_0(A^+)$ such that $\psi^* \circ \iota^* = id\colon K_0(\mathbf{k}) \to K_0(\mathbf{k})$. Thus we can write $K_0(A^+) = \tilde{K}_0(A) \oplus K_0(\mathbf{k})$ for some group $\tilde{K}_0(A)$. Observe that if A already has a unit, then $\tilde{K}_0(A) = K_0(A)$, so $\tilde{K}_0$ gives us nothing new; however, $\tilde{K}_0$ extends the definition of K_0 from unital to non-unital algebras.

Exercise 2.9.9 (Grothendieck groups) (1) Let A be a commutative algebra and M an A-module; let $[M]$ be the isomorphism class. Let $\mathcal{F}_A$ be the free abelian group that is generated by the classes of finitely generated A-modules. Let $\mathcal{E}_A$ be the subgroup of $\mathcal{F}_A$ that is generated by $[M_2] - [M_1] - [M_3]$, where

$$0 \longrightarrow M_1 \longrightarrow M_2 \longrightarrow M_3 \longrightarrow 0$$

is an exact sequence of A-modules. Then the Grothendieck group of A is the abelian group $\mathcal{G}_A = \mathcal{F}_A/\mathcal{E}_A$.

(2) Suppose that all finitely generated A-modules have finite projective dimension. Show how one can compute $\mathcal{G}_A$ using the projective resolutions. See [64, p. 210] for details.

We now give a description of $K_0(A)$ which emphasizes the analogy with linear algebra. Let A be a unital algebra over the field $\mathbf{k}$ of characteristic zero and $M_n(A) = M_n(\mathbf{k}) \otimes_{\mathbf{k}} A$ be the algebra of $n \times n$ matrices with entries in A, with the usual matrix multiplication. We introduce $\iota_n\colon M_n(A) \to M_{n+1}(A)$ via the non-unital homomorphism

$$T \mapsto \begin{bmatrix} T & 0 \\ 0 & 0 \end{bmatrix}. \tag{2.9.6}$$

We write $M_\infty(A) = \cup_{n=1}^{\infty} M_n(A)$ for the algebra of matrices with entries in A that have only finitely many non-zero entries. Now $M_\infty(A)$ is non-unital, so we form the augmented algebra $M_\infty(A)^+ = M_\infty(A) + \mathbf{k}1$ which is obtained by adjoining a unit; then $M_\infty(A)$ is a subalgebra of $M_\infty(A)^+$.

Let $\mathrm{Idem}_n(A) = \{e \in M_n(A)\colon e^2 = e\}$ be the set of idempotents in $M_n(A)$, and $\mathrm{Idem}(A) = \cup_{n=1}^{\infty}\mathrm{Idem}_n(A)$. Under ι_n, the identity of $M_n(A)$ is mapped to an idempotent $e = \iota_n(1)$, where e is not the identity of $M_{n+1}(A)$.

Let $GL_n(A)$ be the group of invertible $n \times n$ matrices with values in A under multiplication, and $\iota\colon GL_n(A) \to GL_{n+1}(A)$ the group homomorphism (different from (2.9.6))

$$T \mapsto \begin{bmatrix} T & 0 \\ 0 & 1 \end{bmatrix}. \tag{2.9.7}$$

Then $GL(A) = \cup_{n=1}^{\infty} GL_n(A)$ is a subset of $M_\infty(A)^+$ and a multiplicative group. One operates on $\mathrm{Idem}(A)$ with $GL(A)$ via conjugation, so $e \mapsto geg^{-1}$.

Lemma 2.9.10 *Proj(A) is canonically isomorphic to the orbit space of* Idem(A) *under the action of $GL(A)$ by conjugation. The addition rule corresponds to*

$$[P] + [Q] \mapsto \begin{bmatrix} [P] & 0 \\ 0 & [Q] \end{bmatrix}. \tag{2.9.8}$$

Proof Each $p \in \mathrm{Idem}_n(A)$ is associated with the projective left A-module $A^n p$, where $A^n = A^n p \oplus A^n(1-p)$ is a free left A-module. We check that if such projective left A-modules are isomorphic, then the isomorphism can be implemented by multiplying on the right by an invertible matrix with entries from A. Let $p \in \mathrm{Idem}_n(A)$ and $q \in \mathrm{Idem}_m(A)$, and suppose that $A^n p$ and $A^m q$ are isomorphic as left A-modules. Then we can introduce $a \in M_{n\times m}(A)$ and $b \in M_{m\times n}(A)$ such that $ab = p$ and $ba = q$; we also impose the conditions $a = pa$; $a(1-q) = 0$ and likewise $bp = b$ and $b(1-p) = 0$; then the isomorphism $S\colon A^n p \oplus A^n(1-p) \to A^m(1-q) \oplus A^m q$ is represented by right multiplication by

$$T \mapsto TS\colon [A^n p \oplus A^n(1-p)] \to [A^n(1-q) \oplus A^n q] \quad S = \begin{bmatrix} 1-p & a \\ b & 1-q \end{bmatrix} \tag{2.9.9}$$

such that $S^2 = Id$, so S is invertible and conjugation by S implements an isomorphism on $\mathrm{Idem}_{n+m}(A)$ such that

$$S\begin{bmatrix} p & 0 \\ 0 & 0 \end{bmatrix} S = \begin{bmatrix} 0 & 0 \\ 0 & q \end{bmatrix}. \tag{2.9.10}$$

We can also think of S as an intertwining operator

$$S\begin{bmatrix} p & 0 \\ 0 & 0 \end{bmatrix} = \begin{bmatrix} 0 & 0 \\ 0 & q \end{bmatrix} S. \tag{2.9.11}$$

□

Let tr be the standard trace on M_n. Note that tr gives an isomorphism of $M_n(\mathbf{C})/[M_n(\mathbf{C}), M_n(\mathbf{C})]$ with $\mathbf{C}$ by Proposition 1.3.3. Given an A-bimodule P, we can regard P as a left $A \otimes A^{op}$ module with operation $(a \otimes b)\colon p \mapsto apb$. Then $[A, P] = \{ap - pa\colon a \in A, p \in P\}$ is the space of commutators. We let $P_\natural = P/[A, P]$ be the commutator quotient space, and write $\natural\colon P \to P_\natural$ for the quotient map. We obtain the first glimpse of the Chern character, which involves the vector trace Tr taking values in the vector space $V = A/[A, A]$.

Proposition 2.9.11 (Chern character) *There is a natural map* $Tr = \natural(tr \otimes 1)\colon K_0(A) \to A_\natural$.

Proof Let e be an idempotent in $M_n(A) = M_n \otimes A$, and $g \in GL_n(A)$ so that $f = geg^{-1}$ is also an idempotent. We check that $Tr(e) = Tr(f)$, so that $Tr(e)$ depends only on the isomorphism class of e. Consider $h = eg^{-1}$, so

$$(\mathrm{tr} \otimes 1)(e) = (\mathrm{tr} \otimes 1)(hg) = \sum_{ij} h_{ij} g_{ji} \tag{2.9.12}$$

while

$$(\mathrm{tr} \otimes 1)(f) = (\mathrm{tr} \otimes 1)(gh) = \sum_{ij} g_{ji} h_{ij}, \tag{2.9.13}$$

and so

$$(\mathrm{tr} \otimes 1)(e) - (\mathrm{tr} \otimes 1)(f) = \sum_{ij} [h_{ij}, g_{ij}] \in [A, A]. \tag{2.9.14}$$

By considering (2.3.3), we obtain the map $K_0(A) \to A/[A, A]$. □

Proposition 2.9.12 *Let A be a commutative and unital algebra. Then $K_0(A)$ is a commutative and unital ring.*

Proof Let P and Q be finitely generated projective A-modules; then $P \otimes Q$ is also a finitely generated and projective A-module, as we now check. Since P is an A-module, and A is commutative, we can regard P as a right A-module by $pa = ap$ so we form $P \otimes_A Q$ in which $pa \otimes q = p \otimes aq$. Then we introduce complementary A-modules such that $A^n = P \oplus P'$ and $A^m = Q \oplus Q'$, then observe that $P \otimes Q$ is a direct summand of $A^{nm} = A^n \otimes A^m$, and $P \otimes Q$ is isomorphic to $Q \otimes P$. By [33, p. 573], there is a natural isomorphism

$$\mathrm{Hom}_A(P, \mathrm{Hom}_A(Q, R)) \sim \mathrm{Hom}_A(P \otimes_A Q, R)$$

for all A-modules R, and we deduce that $P \otimes_A Q$ is projective as a left A-module. Furthermore, $P \otimes_A A$ may be identified with P, so tensor multiplication on A-modules by A acts as a multiplicative unit.

If R is also a finitely generated projective left A-module, then $R \otimes (P \oplus Q)$ is isomorphic to $(R \otimes P) \oplus (R \otimes Q)$. Hence we can introduce binary operations $\oplus$ and $\otimes$ on $\mathrm{Proj}(A)$ such that $\otimes$ is distributive over $\oplus$. □

Whereas some proofs in this section easily extend to noncommutative algebras, the proof of Proposition 2.9.12 soon runs into problems that can only be remedied by reformulating the statement.

Exercise 2.9.13 (Ranks of modules) (1) Let $p \in \mathrm{Idem}(\mathbf{k})$, so p is an $n \times n$ idempotent matrix p which represents a linear projection on $\mathbf{k}^n$ for some n. Then $Tr(p) = \mathrm{rank}\,(p)$, since $\mathbf{k}_\natural = \mathbf{k}$ and $\mathrm{tr}(p) = \mathrm{trace}(p) = \mathrm{rank}(p)$. Then $Tr(p)$ takes values in $\mathbf{Z}_+$.

(2) For X, a compact metric space, let $A = C(X; \mathbf{R})$. Then $p \in \mathrm{Idem}(A)$ is represented by a matrix with entries in A such that $p(x)$ is an idempotent matrix for each $x \in X$. (One can say that $p(x)$ is a projection, but it may not be self-adjoint.) Hence $Tr(p) = \mathrm{rank}(p(x))$ gives the rank of the vector bundle $A^n p$, hence is a continuous function with values in $\mathbf{Z}_+$. When X is connected, the rank is constant on X.

(3) Let T be a real $m \times m$ matrix of rank one. Then $P = T^*T/\mathrm{trace}\,(T^*T)$ is a self-adjoint projection such that $\ker(P) = \ker(T)$. Likewise $Q = TT^*/\mathrm{trace}\,(TT^*)$ is a self-adjoint projection such that $\ker(Q) = \ker(T^*)$, so $\mathrm{im}(Q) = \mathrm{im}(T)$.

(4) Deduce that T may be expressed as $T = \|T\| \xi \otimes \eta$, where ξ, η are unit vectors. Discuss the matrix $\|T\|^{-1} \eta \otimes \xi$.

Exercise 2.9.14 ($K_0(PID)$) Let A be a principal ideal domain. Use Lemma 2.9.4 and Exercise 2.9.13 to show that $K_0(A) \cong \mathbf{Z}$.

Definition 2.9.15 (Vector bundle) Let X be a compact metric space. Then the trivial real vector bundle of dimension n is the metric space $X \times \mathbf{R}^n$ with the projection $p : X \times \mathbf{R}^n \to X$: $(x, v) \mapsto x$. Generally we say that a metric space E is a continuous real vector bundle over X, when there exist:

(1) a continuous surjection $p\colon E \to X$ such that the fibre $E_x = \{v \in E\colon p(v) = x\}$ is a real finite-dimensional vector space for all $x \in X$;
(2) for all $x \in X$ there exist an open $U \subset X$ such that $x \in X$, a finite-dimensional real vector space V and a homeomorphism $\varphi\colon \{v \in E\colon p(v) \in U\} \to U \times V$, such that
(3) $\varphi|E_x$ is a linear homeomorphism from E_x to V.

Then the dimension of E_x as a real vector space is a locally constant function on X, hence bounded and constant on each connected component, and $\sup_{x\in X} \dim_{\mathbf{R}} E_x$ is called the dimension of E as a vector bundle (which is possibly smaller than the dimension of E as a metric space). Let $\mathrm{Vect}(X)$ be the set of isomorphism classes $[E]$ of continuous real vector bundles E over X; also let $\mathrm{Vect}_n(X)$ be the set of isomorphism classes $[E]$ of continuous, n-dimensional, real vector bundles E over X. A bundle E is trivial if $[E]$ contains $X \times \mathbf{R}^n$.

The direct sum of vector spaces leads to a direct sum operation on (the fibres of) vector bundles, and we define $[E] \oplus [F] = [E \oplus F]$ for $[E], [F] \in \mathrm{Vect}(X)$. Now let E be a locally trivial finite-dimensional real vector bundle over a compact metric space X, so $p\colon E \to X$ is a continuous and surjective map and $V_x = \{f \in E\colon p(f) = x\}$ is a family of finite-dimensional real vector spaces, such that the dimension is locally constant. The space of sections of E is

$$\Gamma(X, E) = \{s\colon X \to E\colon \text{continuous} \quad p \circ s = id\}, \tag{2.9.15}$$

and the space of sections has a natural addition structure.

Theorem 2.9.16 (Swan) (i) *Let $A = C(X; \mathbf{R})$. Then $\Gamma(X; E)$ is a finitely generated and projective A-module.*

(ii) *Up to isomorphism, every finitely generated and projective A module arises as the spaces of sections of some continuous finite-dimensional real vector bundle over X.*

Proof (i) The details are presented in [91, p. 34; 101], so the following sketches the basic ideas. Observe that A^n is the typical finitely generated free left A-module, and that we can identify $A^n \cong C(\mathbf{X}; \mathbf{R}^n)$ so that each $s \in C(\mathbf{X}; \mathbf{R}^n)$ gives a section of the trivial rank n vector bundle over X. Any n-dimensional vector bundle E over X has a complementary bundle E' such that $E \oplus E'$ is isomorphic to the trivial vector bundle $C(\mathbf{X}; \mathbf{R}^n)$. A pair of sections s of E and s' of E' corresponds to a pair P, P' of $C(\mathbf{X}; \mathbf{R})$ modules so that $P \oplus P' \cong C(\mathbf{X}; \mathbf{R}^n)$.

(ii) Note also that any finitely generated projective left A-module P is a direct summand of A^n. Using the identification of sections of A^n with functions $X \to \mathbf{R}^n$, one can produce a locally trivial vector bundle E with section s to represent P. □

Thus one can define with $A = C(X; \mathbf{C})$ an abelian K_0-group $K_0(C(X; \mathbf{C}))$. In Chapter 3 we consider how to compute this group in specific cases. By Example 6.2.4(iii) relating to $X = S^2$, we cannot replace projective by free in the statement of Theorem 2.9.19; local triviality is not the same as global triviality for vector bundles.

Exercise 2.9.17 (Picard group) (See 5.4.2 of [45], and 6.12 of [49].) Let A be a unital commutative algebra so $K_0(A)$ is a commutative unital ring by Proposition 2.9.12. Let the Picard group Pic (A) be the set of equivalence classes of finitely generated projective A-modules P that have a multiplicative inverse under $\otimes$. Show that Pic (A) defines an abelian group. See Example 8.2.5 for a geometrical example of this.

(i) Let $G(A) = \{u \in A \colon \exists v \in A, \quad uv = 1\}$ be the multiplicative group of units in A and let P be a finitely generated projective A bimodule. Show that $\phi \in \mathrm{Hom}_A(A, A)$ is determined by $\phi(1)$, and $A\phi(1) = A$ if and only if $\phi(1) \in G(A)$.

(ii) Show that there is an A-module map $A \to \mathrm{Hom}_A(P, P)$ given by $a \mapsto \lambda_a$, where $\lambda_a(p) = ap$ for all $a \in A, p \in P$. Describe the image of $G(A)$.

(iii) Let $\check{P} = \mathrm{Hom}_A(P, A)$, known as the dual module of P. Show that $\check{P}$ is an A-module and that there is an A-module map $\pi : \check{P} \otimes_A P \to \mathrm{Hom}_A(P, P)$, given by

$$\pi\left(\sum_j \check{p}_j \otimes p_j\right) : r \mapsto \sum_j \check{p}_j(r) p_j.$$

(iv) State a condition on P under which $\check{P} \otimes_A P$ is a projective module that is isomorphic to A via the maps π and λ. (See lemma 3.17, p. 118 of [64].) In this circumstance, one writes $P = \mathcal{L}$, since $\mathcal{L}$ has rank one and is some sort of line bundle, and $\mathcal{L}^{-1} = \check{\mathcal{L}} = \mathrm{Hom}_A(\mathcal{L}; A)$, since

$$\mathrm{Hom}_A(\mathcal{L}; A) \otimes_A \mathcal{L} \sim A,$$

and A is the multiplicative unit of Pic (A).

(v) Find Pic (A) where $A = C(\mathbf{X}; \mathbf{R})$, and X is a connected and compact Hausdorff space. In [42, 5.4.2], Pic (A) is identified with a group of line bundles on X. See also [63] page 35.

Definition 2.9.18 (Group algebra) Let G be a group, and introduce the complex group algebra $\mathbf{C}[G] = \{\sum_{g\in G} a_g g : a_g \in \mathbf{C}\}$ with multiplication

$$\Big(\sum_{g\in G} a_g g\Big)\Big(\sum_{h\in G} b_h h\Big) = \sum_{k\in G}\Big(\sum_{h\in G} a_{kh^{-1}} b_h\Big) k. \tag{2.9.16}$$

Proposition 2.9.19 (Maschke's theorem) *Let G be a finite group, and P a submodule of a finitely generated $\mathbf{C}[G]$-module M. Then P is projective.*

Proof Since M is a finite-dimensional $\mathbf{C}$-vector space, we can choose a basis $\{p_1,\dots,p_k\}$ of P and extend to a basis $\{p_1,\dots,p_k,m_{k+1},\dots,m_n\}$ of M. Then we select a linear map $e\colon M \to M$ such that $e(p_j) = p_j$ and $e(m_\ell) = 0$ for all j,ℓ; evidently e is a projection with range P. Now let $e^G : M \to M$ be the linear map

$$e^G = \frac{1}{\sharp G}\sum_{g\in G} g e g^{-1},$$

so that $e^G h = h e^G$ and range$(e^G) \subseteq$ range$(e) = P$ since P is a submodule, so $heh^{-1}e^G = hee^G h^{-1} = he^G h^{-1} = e^G$, hence by averaging over $h \in G$ we deduce that $e^G e^G = e^G$. Now $e^G M$ and $(1-e^G)M$ are complementary $\mathbf{C}[G]$-modules. Finally, $\dim e^G M = \operatorname{trace}(e^G) = \operatorname{trace}(e) = \dim P$, so $e^G M = P$, and P has a complementary module. □

Maschke's theorem does not necessarily hold if one replaces $\mathbf{C}$ by a field of characteristic p dividing the order of G. Also, examples in [59] show that one cannot necessarily replace G by an infinite group. Nevertheless, in Corollary 4.7.3, we obtain a result about Hilbert modules over $\mathbf{C}[\mathbf{Z}]$. In the representation theory of finite groups, complemented modules are sometimes called completely reducible.

2.10 Singular Homology

In the context of arbitrary compact topological spaces, the most appropriate topological theory is Čech cohomology. Rather than develop this theory, we prefer to impose special assumptions on X so that we can work with the elementary theory of simplicial homology.

Let X be a compact topological space which is a simplicial complex.

(0) Let r be the number of path components of X. Then $H_0(X,\mathbf{Z}) \cong \mathbf{Z}^r$ is a free module.

(1) A loop is a continuous function $\ell\colon [0,1] \to X$ such that $\ell(0) = \ell(1)$. Suppose that $x_0 \in X$ and let $\pi_1(X,x_0)$ be the set of homotopy classes of loops in X that are based at x_0. Then $\pi_1(X,x_0)$ forms a group, namely the fundamental group for the space X with point x_0, as described in [50]. The derived group $\pi_1(X,x_0)'$ is the normal subgroup of $\pi_1(X,x_0)$ that is generated by the commutators $XYX^{-1}Y^{-1}$. Let $H_1(X,\mathbf{Z})$ be the group of singular 1-simplices in X. Any loop $\ell\colon [0,1] \to X$ with $\ell(0) = \ell(1) = x_0$ gives a singular 1-simplex.

Proposition 2.10.1 (i) *There is a natural homomorphism* $\chi : \pi_1(X,x_0) \to H_1(X,\mathbf{Z})$ *formed by taking the homotopy class of a loop* ℓ *to the homology class of the singular* 1*-simplex* ℓ.

(ii) *If* X *is path connected, then there is an exact sequence of groups*

$$\{1\} \longrightarrow \pi_1(X,x_0)' \longrightarrow \pi_1(X,x_0) \longrightarrow H_1(X,\mathbf{Z}) \longrightarrow \{1\}.$$

Proof See [92, Theorems 4.2.7 and 4.2.9]. □

Example 2.10.2 (Punctured sphere) (a) One can express the fundamental group of a connected simplicial complex with base point in terms of generators and relations. One can deduce that (i) the circle has fundamental group $\mathbf{Z}$; (ii) the torus has fundamental group $\mathbf{Z} \oplus \mathbf{Z}$; (iii) a figure of eight has fundamental group $\mathbf{F}_2$, the free group on two generators; see [92, Ex. 7.17].

(b) Let $S^2 = \mathbf{C} \cup \{0\}$ be the Riemann sphere. Then

(i) $S^2 \setminus \{0\}$ has fundamental group $\{1\}$;

(ii) $S^2 \setminus \{0,1\}$ has fundamental group $\mathbf{Z}$;

(iii) $X = S^2 \setminus \{0,1,\infty\}$ has fundamental group $\mathbf{F}_2$ and $H_1(X,\mathbf{Z}) = \mathbf{Z} \oplus \mathbf{Z}$.

Definition 2.10.3 (Smooth triangulation) Let $\tilde{K}$ be a simplicial complex, and K the point set associated with $\tilde{K}$, so K is a compact set. Given a C^∞ manifold X, a smooth triangulation consists of a homeomorphism $h\colon K \to X$ such that for each simplex $\tilde{\sigma}$ of $\tilde{K}$, $h|_\sigma : \sigma \to X$ has an extension h_σ to a neighbourhood U of σ in the plane of σ such that $h_\sigma : U \to X$ is a smooth submanifold.

By [97, p. 146], every smooth compact manifold has a smooth triangulation. In particular, for algebraic curves, one can produce smooth triangulations via Theorem 6.1.3; see [62] and [34]. The significance of this result for cyclic theory is that it provides a means for relating the homology theories in topological and smooth categories. For a smooth compact manifold M, we have $K_0(C(M;\mathbf{C})) = K_0(C^\infty(M;\mathbf{C}))$ by 8.2.6 of [68].

Exercise 2.10.4 (Euler characteristic) Let $\mathbf{X}$ be a compact topological space that has a triangulation. Then the simplicial homology groups $H_q(\mathbf{X};\mathbf{Z})$ are finitely generated abelian groups, namely finitely generated $\mathbf{Z}$-modules. Any finitely generated abelian group A may be written as $A = T \oplus F$ where T is the subgroup of torsion elements, and $F = A/T$ is a free abelian group isomorphic to $\mathbf{Z}^n$, so we define the rank of A to be n. See [47, p. 125]. In particular, we define the qth Betti number β_q to be the rank of $H_q(\mathbf{X};\mathbf{Z})$. Then the Euler characteristic is

$$\chi(\mathbf{X}) = \sum_q (-1)^q \beta_q, \tag{2.10.1}$$

and the genus g satisfies $2g = \beta_1$. For the triangulation with V vertices, E edges and F faces, we have $\chi = V - E + F$. Show that $\chi(S^2) = 2$, and that for $\mathbf{X}$ the complex torus of Example 6.1.5, we have $\chi(\mathbf{X}) = 0$. See [42, p. 120].

3
Algebras of Operators

Cyclic theory incorporates the index theory of elliptic operators as a particular application. In this chapter, we consider the simplest interesting case of index theory, namely Toeplitz operators on Hardy space over the circle. The chapter begins by introducing the basic facts about operators on Hilbert space that are required subsequently.

In Section 2.9, we imposed the condition that modules over algebras are finitely generated. To enable us to use such results, we consider Fredholm operators that have finite-dimensional kernels and cokernels, so that an index can be defined.

The basic idea is to consider an algebra R over $\mathbf{C}$ with an ideal I and a trace on I, namely a linear map $\tau: I/[R, I] \to \mathbf{C}$. Then we introduce an algebra A over $\mathbf{C}$ and a linear map $A \to R$ which is a homomorphism modulo I, so that the curvature ω takes values in I. This enables us to consider $\varphi(a_0, a_1) = \tau\omega(a_0, a_1)$ for all $a_0, a_1 \in A$. However, the algebras are infinite dimensional, so I and τ require careful definition.

3.1 The Gelfand Transform

(i) Let $\mathcal{A}$ be a complex, unital and commutative Banach algebra. Then $\mathcal{A}$ has a norm such that $\|S\|_{\mathcal{A}} \geq 0$ for all with $\|S\|_{\mathcal{A}} = 0$ only if $S = 0$;

$$\|S + T\|_{\mathcal{A}} \leq \|S\|_{\mathcal{A}} + \|T\|_{\mathcal{A}}; \qquad \|\lambda S\|_{\mathcal{A}} = |\lambda|\|S\|_{\mathcal{A}}; \tag{3.1.1}$$

$$\|ST\|_{\mathcal{A}} \leq \|S\|_{\mathcal{A}}\|T\|_{\mathcal{A}} \qquad (S, T \in \mathcal{A}, \lambda \in \mathbf{C}), \tag{3.1.2}$$

and $\mathcal{A}$ is complete for the metric $\|S - T\|$. The space of bounded linear functionals $\psi: \mathcal{A} \to \mathbf{C}$ is denoted $\mathcal{A}'$, and known as the dual space.

Exercise 3.1.1 (Connectedness) Let Y be a compact Hausdorff space, and $C(Y;\mathbf{C})$ the space of continuous functions $f\colon Y \to \mathbf{C}$ with pointwise multiplication and the norm $\|f\|_\infty = \sup\{|f(y)|\colon y \in Y\}$. Verify that $C(Y;\mathbf{C})$ is a unital and commutative Banach algebra.

(1) Show that the space Y is disconnected, if and only if $C(Y,\mathbf{C})$ has non-trivial idempotent elements $e^2 = e$, namely $e \in C(Y,\mathbf{C})$ such that $e(x) \in \{0,1\}$ for all $x \in Y$ corresponding to the indicator functions of connected components of Y.

(2) Let $G(C(Y;\mathbf{C}))$ be the set of $f \in C(Y;\mathbf{C})$ that have $gf = 1$ for some $g \in C(Y;\mathbf{C})$. By elementary results, if $f(y) \neq 0$ for all $y \in y$, then there exists $\delta > 0$ such that $|f(y)| \geq \delta$ for all $y \in Y$, and $1/f$ is continuous. Deduce that $G(C(Y;\mathbf{C})) = C(Y;\mathbf{C}^*)$ where $\mathbf{C}^* = \mathbf{C} \setminus \{0\}$.

(3) Let $f \in C(Y;\mathbf{C})$. Deduce that $\sigma = \{f(y)\colon y \in Y\}$ is a compact subset of $\mathbf{C}$, and $\lambda - f$ is invertible for all $\lambda \in \mathbf{C} \setminus \sigma$. Let the spectrum of f be the set of $\lambda \in \mathbf{C}$ such that f does not have a multiplicative inverse in $C(Y;\mathbf{C})$. Deduce that the spectrum of f is equal to the range of f.

There is a systematic way of mapping a complex, unital and commutative Banach algebra into a space of continuous functions, known as the Gelfand transform.

(ii) We introduce the unit ball of $\mathcal{A}$ by $B_{\mathcal{A}} = \{S \in \mathcal{A}\colon \|S\|_{\mathcal{A}} \leq 1\}$ and the unit ball of the dual space by $B_{\mathcal{A}'} = \{\phi \in \mathcal{A}'\colon |\phi(S)| \leq 1, \forall S \in B_{\mathcal{A}}\}$. We have a weak* topology on $B_{\mathcal{A}'}$ given by the basic open sets

$$\begin{aligned} &U(T_1,\dots,T_m; z_1,\dots,z_m; \varepsilon_1,\dots,\varepsilon_m) \\ &\quad = \{\phi \in B_{\mathcal{A}'}\colon |\phi(T_j) - z_j| < \varepsilon_j, j = 1\dots,m\} \end{aligned} \tag{3.1.3}$$

for all $m \in \mathbf{N}$, $T_j \in B_{\mathcal{A}}$, $z_j \in \mathbf{C}$ and $\varepsilon_j > 0$. One shows using Tychonov's theorem that $B_{\mathcal{A}'}$ with the weak* topology is compact and Hausdorff. Furthermore, if $\mathcal{A}$ is separable for the norm topology, then $B_{\mathcal{A}'}$ with the weak* topology is metrisable.

(iii) Let $G(\mathcal{A})$ be the set of invertible elements of $\mathcal{A}$. Then $I \in G(\mathcal{A})$, and $G(\mathcal{A})$ is a multiplicative group. By considering perturbation series, one can show that $G(\mathcal{A})$ is an open subset of $\mathcal{A}$ for the norm topology. For all $T \in \mathcal{A}$, the series $\exp(T) = 1 + T + T^2/2! + \cdots$ converges, and one checks that $\exp(S+T) = \exp(S)\exp(T)$ and $\exp(tS) \to 1$ as $t \to 0$. Hence $\exp(\mathcal{A}) = \{\exp(T)\colon T \in \mathcal{A}\}$ forms a connected subgroup of $G(\mathcal{A})$ that contains 1, and we can form the quotient group

$$G(\mathcal{A})/\exp(\mathcal{A}). \tag{3.1.4}$$

(iv) An ideal M of $\mathcal{A}$ is said to be maximal if (1) M is proper and (2) if M' is any proper ideal such that $M \subseteq M' \subset \mathcal{A}$, then $M = M'$. Suppose that M is a maximal ideal. Then all the elements of M are not invertible and the closure $\bar{M}$ is an ideal containing M such that all the elements of $\bar{M}$ are limits of elements in M, hence are not invertible. We deduce that $\bar{M}$ is proper, and hence $\bar{M} = M$. It follows that M is closed and that the quotient map $\phi\colon \mathcal{A} \to \mathcal{A}/M$ is a homomorphism and $\mathcal{A}/M$ is a field containing $\{\phi(\lambda I)\colon \lambda \in \mathbf{C}\}$.

Definition (Spectrum) Let $\mathcal{A}$ be a unital Banach algebra. Then the spectrum $\sigma(T)$ of $T \in \mathcal{A}$ consists of those $\lambda \in \mathbf{C}$ such that $\lambda I - T$ does not have an inverse in $\mathcal{A}$.

Theorem 3.1.2 (Gelfand) *The spectrum of $T \in \mathcal{A}$ is a compact and non-empty subset of* $\mathbf{C}$.

Proof This follows from Liouville's theorem. See [2]. □

Corollary 3.1.3 (Gelfand–Mazur) *Let $\mathcal{A}$ be a commutative and unital Banach algebra in which every non-zero element has an inverse. Then $\mathcal{A}$ is isomorphic to* $\mathbf{C}$.

(v) *(Multiplicative linear functional) A linear functional $\phi\colon \mathcal{A} \to \mathbf{C}$ is multiplicative if $\phi(ST) = \phi(S)\phi(T)$ for all $S, T \in \mathcal{A}$ and $\phi(I) = 1$. By (iv), all multiplicative linear functionals are continuous, and one can easily check that $|\phi(T)| \leq \|T\|_{\mathcal{A}}$ for all $T \in \mathcal{A}$. We write*

$$X = \{\phi \in B_{\mathcal{A}'}\colon \phi(ST) = \phi(S)\phi(T), \forall S, T \in \mathcal{A}\} \tag{3.1.5}$$

*which is a closed subspace of $B_{\mathcal{A}'}$ for the weak * topology, hence compact and Hausdorff. Furthermore, if $\mathcal{A}$ is separable for the norm topology, then X with the weak* topology is metrisable.*

The kernel of ϕ is $\ker\phi = \{S \in \mathcal{A}\colon \phi(S) = 0\}$. One can show that $\ker\phi$ is a maximal ideal in $\mathcal{A}$ such that $\mathcal{A}/\ker\phi$ is isomorphic to $\mathbf{C}$ and there is a bijective correspondence between multiplicative linear functionals and maximal ideals

$$\mathrm{Hom}_{\mathbf{C}}(\mathcal{A}; \mathbf{C}) \leftrightarrow X\colon \phi \leftrightarrow \ker\phi. \tag{3.1.6}$$

We associate with each $T \in \mathcal{A}$ the continuous function $\hat{T}\colon X \to \mathbf{C}$ by $\hat{T}(\phi) = \phi(T)$ on the maximal ideal space X. This map $T \mapsto \hat{T}$ is known as the Gelfand transform.

Theorem 3.1.4 (Gelfand) *The maximal ideal space of a unital complex commutative Banach algebra is a non-empty compact Hausdorff space X.*

(i) *Then map $T \mapsto \hat{T}$ is an algebra homomorphism $\mathcal{A} \to C(X, \mathbf{C})$ such that $\widehat{ST} = \hat{T}\hat{S}$;*

(ii) $\|\hat{T}\|_\infty \leq \|T\|_{\mathcal{A}}$;
(iii) *If* $\hat{T}(\phi) \neq 0$ *for all* $\phi \in X$, *then* T *is invertible.*

Proof (iii) Suppose that T is not invertible, and let $(T) = \{ST : S \in \mathcal{A}\}$. Then (T) is an ideal of $\mathcal{A}$ that contains T, but does not contain I. So by Zorn's Lemma, (T) is contained in some maximal ideal M; then $\phi\colon \mathcal{A} \to \mathcal{A}/M$ is a multiplicative linear functional such that $\phi(T) = 0$. By Corollary 3.1.3, $\mathcal{A}/M$ is isomorphic to $\mathbf{C}$, so $\phi \in X$, and $\hat{T}(\phi) = 0$. □

Corollary 3.1.5 *The Gelfand transform gives a group homomorphism*

$$G(\mathcal{A})/\exp(\mathcal{A}) \to G(C(X;\mathbf{C}))/\exp(C(X;\mathbf{C})). \tag{3.1.7}$$

By the Arens–Royden theorem, this is actually a group isomorphism. Proving surjectivity needs the theory of several complex variables, as in [2].

Example 3.1.6 (exp on continuous functions) We return to the example of $C(X;\mathbf{C})$, as in Exercise 3.1.1.

Let $\mathbf{C}^* = \mathbf{C} \setminus \{0\}$. Then we observe that $G(C(X,\mathbf{C})) = C(X,\mathbf{C}^*)$. Also, $\exp(C(X;\mathbf{C}))$ is the connected component of $C(X,\mathbf{C}^*)$ that contains the identity. Using this result, it is easy to see that $C(X;\mathbf{C}^*)/\exp(C(X;\mathbf{C}))$ is the group of homotopy classes of continuous functions $X \to \mathbf{C}^*$. Hence the cohomology is

$$H^1(X,\mathbf{Z}) \cong G(C(X;\mathbf{C}))/\exp(C(X;\mathbf{C})). \tag{3.1.8}$$

Let $\mathcal{A}$ be a commutative and unital Banach algebra such that the maximal ideal space X is homeomorphic to a simplicial complex. Then one can prove the following.

(0) $H^0(X,\mathbf{Z})$ is isomorphic to the group generated by the idempotents in $\mathcal{A}$;
(1) $H^1(X,\mathbf{Z})$ is isomorphic to $G(\mathcal{A})/\exp(\mathcal{A})$;
(2) $H^2(X,\mathbf{Z})$ is isomorphic to the Picard group Pic $(\mathcal{A})$ of Exercise 2.9.17. This is a result of Forster; see [103, p. 175]. In Example 8.2.7 we obtain a similiar result when X is a compact Riemann surface.

Exercise 3.1.7 (Bounded derivations) (i) Let $C^1(S^1,\mathbf{C})$ be the Banach algebra of continuously differentiable functions on the circle with pointwise multiplication and the norm

$$\|f\|_{C^1} = \|f\|_\infty + \|f'\|_\infty. \tag{3.1.9}$$

Show that $C(S^1,\mathbf{C})$ is a $C^1(S^1,\mathbf{C})$ module for the pointwise multiplication and $\delta\colon C^1(S^1,\mathbf{C}) \to C(S^1,\mathbf{C})\colon\ f \mapsto f'$ is a module map.

(ii) *(Disc algebra)* Let A be the disc algebra of continuous complex functions f on the closed disc $\bar{\mathbf{D}} = \{z \in \mathbf{C} \colon |z| \leq 1\}$ such that f is holomorphic on the open disc $\mathbf{D} = \{z \in \mathbf{C} \colon |z| < 1\}$ with pointwise multiplication and $\|f\| = \sup_z\{|f(z)| \colon |z| \leq 1\}$. For $0 < r < 1$, let $\delta f(z) = f'(rz)$. Show that $f(z) \mapsto f(rz)$ is a homomorphism $A \to A$ and $\delta \colon A \to A$ is a continuous derivation with respect to this homomorphism.

(iii) Let A be a commutative semisimple Banach algebra. Then there are no non-zero continuous derivations $\delta \colon A \to A$; see [2, 5.8]. Discuss how this does not contradict (i) or (ii).

Exercise 3.1.8 (Traces) Let A be a Banach algebra.

(i) Show that every continuous multiplicative functional $\phi \colon A \to \mathbf{C}$ gives a continuous trace $\tau \colon A \to \mathbf{C}$ such that $\tau|[A, A] = 0$.
(ii) Show that $M_2(\mathbf{C})$ has a non-trivial trace, but the only continuous multiplicative linear functional is zero.
(iii) For A commutative, show that the continuous traces are the continuous linear functionals $\phi \colon A \to \mathbf{C}$.
(iv) For X a compact metric space, and $A = C(X, \mathbf{C})$, show that the traces on A are given by bounded measures μ on X via $\tau(f) = \int_X f(x)\mu(dx)$.
(v) Let $A = C_0(\mathbf{R}; \mathbf{C})$ be the space of continuous functions $f \colon \mathbf{R} \to \mathbf{C}$ such that $f(x) \to 0$ as $x \to \pm\infty$. By considering the one-point compactification $\mathbf{R} \cup \{\infty\}$ as in [101], show that $A \oplus \mathbf{C}$ is isomorphic to $C(\mathbf{R} \cup \{\infty\}; \mathbf{C})$. Hence find the traces on A.
(vi) In Exercise 3.5.4, we introduce the Toeplitz algebra $\mathcal{T}$ such that $\mathcal{T}/[\mathcal{T}, \mathcal{T}]$ is isomorphic to $C(S^1, \mathbf{C})$. Find the traces on $\mathcal{T}$.

Exercise 3.1.9 (Banach limits) Let $\ell^\infty(\mathbf{Z})$ be the space of bounded complex sequences $(a_n)_{n=-\infty}^{\infty}$ with coordinate-wise addition and the norm $\|(a_n)\| = \sup_{n \in \mathbf{Z}} |a_n|$.

(i) Let $(p_n)_{n=-\infty}^{\infty}$ be a complex sequence such that $\sum_{n=-\infty}^{\infty} |p_n|$ converges. Show this gives a trace on $\ell^\infty(\mathbf{Z})$ via $(a_n) \mapsto \sum_{n=-\infty}^{\infty} p_n a_n$.

(ii) Let LIM be the Banach limit as in [66, p. 31]. Show that LIM defines a trace on $\ell^\infty(\mathbf{Z})$ which is not given by any sequence in (i).

3.2 Ideals of Compact Operators on Hilbert Space

We express the essential facts as succinctly as possible in this section and refer the reader to texts such as [31] and [32] for explanation.

Definition 3.2.1 (1) *(Hilbert space)* Let H be a complex separable Hilbert space with inner product $\langle \cdot \mid \cdot \rangle \colon H \times H \to \mathbf{C}$, so that for all $f, g, h \in H$ and $\mu, \lambda \in \mathbf{C}$,

(i) $\langle f \mid g \rangle = \overline{\langle g \mid f \rangle}$;
(ii) $\langle f \mid \mu g + \lambda h \rangle = \mu \langle f \mid g \rangle + \lambda \langle f \mid h \rangle$, so the inner product is linear in the second variable;
(iii) $\langle h \mid h \rangle \geq 0$, and $\langle h \mid h \rangle = 0 \Rightarrow h = 0$;
(iv) H is complete for the metric associated with the norm $\|f - g\| = \langle f - g \mid f - g \rangle^{1/2}$;
(v) H is infinite dimensional, but separable for the norm of (iv).

One can prove that there exists a complete orthonormal basis $(z_j)_{j=0}^{\infty}$, such that any $f \in H$ may be expressed as a convergent series $f = \sum_{j=0}^{\infty} a_j z_j$ in the norm of H, where $\|f\|^2 = \sum_{j=0}^{\infty} |a_j|^2$, and the a_j are uniquely determined by $a_j = \langle z_j \mid f \rangle$. Another basic result is the Cauchy–Schwarz inequality, which gives $|\langle f \mid g \rangle| \leq \|f\| \|g\|$ for all $f, g \in H$.

Example 3.2.2 (Fourier coefficients) Consider the circle $S^1 = \{z \in \mathbf{C} \colon |z| = 1\}$ with normalized Lebesgue measure $d\theta/(2\pi)$. With $z = e^{i\theta}$, let $H = L^2(S^1; d\theta/(2\pi))$ be the Hilbert space of square integrable functions with the inner product

$$\langle g \mid f \rangle = \int_0^{2\pi} f(\theta) \bar{g}(\theta) \frac{d\theta}{2\pi}. \tag{3.2.1}$$

The Fourier coefficients of $f \in H$ are $\hat{f}(n) = \int_0^{2\pi} f(\theta) e^{-in\theta} d\theta/(2\pi)$ for $n \in \mathbf{Z}$. By the Riesz–Fischer theorem, H has a complete orthonormal basis $(z^n)_{n=-\infty}^{\infty}$.

Definition 3.2.3 (Bounded linear operators) Let $T \colon H \to H$ be a $\mathbf{C}$-linear map. Then T is said to be bounded with operator norm $\|T\|_{\mathcal{L}(H)}$ where

$$\|T\|_{\mathcal{L}(H)} = \sup_f \{\|Tf\| \colon f \in H; \|f\| \leq 1\} \tag{3.2.2}$$

is finite; the set of all such f is $\mathcal{L}(H)$. Then $\mathcal{L}(H)$ forms an algebra under multiplication given by composition, and $\|ST\|_{\mathcal{L}(H)} \leq \|S\|_{\mathcal{L}(H)} \|T\|_{\mathcal{L}(H)}$. We often write $\|T\|$ for $\|T\|_{\mathcal{L}(H)}$.

Definition 3.2.4 (Finite-rank operators) The space $\mathbf{F}$ of finite-rank operators on H is $\{T = \sum_{j=1}^{N} \xi_j \otimes \eta_j; \xi_j, \eta_j \in H\}$ so $T\zeta = \sum_{j=1}^{N} \langle \eta_j \mid \zeta \rangle_H \xi_j$. Each $T \in \mathbf{F}$ determines a bounded linear operator on H. Also, $\mathbf{F}$ determines an ideal in $\mathcal{L}(H)$ since $ST \in \mathbf{F}$ and $TS \in \mathbf{F}$ for all $T \in \mathbf{F}$ and $S \in \mathcal{L}(H)$.

(4) Again let $T: H \to H$ be a **C**-linear map. Then T is said to be compact if $\{Tf: f \in H; \|f\| \leq 1\}$ has compact closure in H for the norm topology; in particular, such a T is bounded.

Lemma 3.2.5 (Compact operators) *Let $\mathcal{K}(H)$ be the space of compact operators. Then $\mathcal{K}(H)$ is an ideal in $\mathcal{L}(H)$ for the usual multiplication operation, so that*

(i) $ST \in \mathcal{K}(H)$ and $TS \in \mathcal{K}(H)$ for all $T \in \mathcal{K}(H)$ and $S \in \mathcal{L}(H)$;
(ii) $\mathbf{F} \subset \mathcal{K}(H)$;
(iii) $T \in \mathcal{K}(H)$ if and only if $T^* \in \mathcal{K}(H)$;
(iv) $T \in \mathcal{K}(H)$ if and only if $T^*T \in \mathcal{K}(H)$;
(v) *if $T_n \in \mathcal{K}(H)$ and $T \in \mathcal{L}(H)$ has $\|T - T_n\| \to 0$ as $n \to \infty$ then* $T \in \mathcal{K}(H)$.

Also $\mathcal{K}(H)$ is uniquely determined as the smallest set that satisfies (i)–(v).

Proof See [29]. □

Definition 3.2.6 (Hilbert–Schmidt operators) A linear operator $T: H \to H$ is Hilbert–Schmidt if $\|T\|^2_{\mathcal{L}^2} = \sum_{j=0}^{\infty} \|Tz_j\|^2$ is finite. Then T is compact, and the set $\mathcal{L}^2(H)$ of all Hilbert–Schmidt operators forms an ideal in $\mathcal{L}(H)$ under the usual addition and multiplication; indeed,

$$\|T\| \leq \|T\|_{\mathcal{L}^2}; \quad \|UTV\|_{\mathcal{L}^2} \leq \|U\|\|T\|_{\mathcal{L}^2}\|V\|. \tag{3.2.3}$$

Example 3.2.7 (Hilbert–Schmidt integral operators) When $H = L^2(S^1, d\theta/(2\pi))$, we can express any $T \in \mathcal{L}^2(H)$ as an integral operator

$$Tf(x) = \int_{S^1} K(x,y) f(y) \frac{dy}{2\pi} \tag{3.2.4}$$

where $K(x,y)$ is confusingly called the kernel and $K \in L^2(S^1 \times S^1; dxdy/(2\pi)^2)$; indeed

$$\|T\|^2_{\mathcal{L}^2} = \iint_{S^1 \times S^1} |K(x,y)|^2 \frac{dx}{2\pi}\frac{dy}{2\pi} \tag{3.2.5}$$

by the Hilbert–Schmidt theorem. This criterion is easy to check in applications.

Definition 3.2.8 (Trace class operators) Let $I = \mathcal{L}^2(H)$, then $\mathcal{L}^1(H) = I^2 = \{ST: S, T \in \mathcal{L}^2(H)\}$ is the ideal of trace class operators. We can take the norm to be $\|T\|_{\mathcal{L}^1} = \inf\{\|U\|_{\mathcal{L}^2}\|V\|_{\mathcal{L}^2}: T = UV; U, V \in \mathcal{L}^2\}$. Then $\mathcal{L}^1(H)$ forms an ideal in $\mathcal{L}(H)$ for the usual multiplication

$$\|T\| \leq \|T\|_{\mathcal{L}^2} \leq \|T\|_{\mathcal{L}^1}; \quad \|UTV\|_{\mathcal{L}^1} \leq \|U\|\|T\|_{\mathcal{L}^1}\|V\|. \tag{3.2.6}$$

Let (z_j) be any complete orthonormal basis of H. Then on $\mathcal{L}^1(H)$, there is a continuous trace: $\mathcal{L}^1(H) \to \mathbf{C}$ defined by

$$\mathrm{trace}(T) = \sum_{j=0}^{\infty} \langle z_j \mid T z_j \rangle, \tag{3.2.7}$$

such that $|\mathrm{trace}(T)| \leq \|T\|_{\mathcal{L}^1}$; also

$$\mathrm{trace}(ST) = \mathrm{trace}(TS) \qquad (S, T \in \mathcal{L}^2(H));$$

$$\mathrm{trace}(UV) = \mathrm{trace}(VU) \qquad (U \in \mathcal{L}^1(H), V \in \mathcal{L}(H)). \tag{3.2.8}$$

Given a linear operator $T = \sum_{j=1}^{\infty} \xi_j \otimes \eta_j$ with $\sum_{j=1}^{\infty} \|\xi_j\| \|\eta_j\| < \infty$, then $T \in \mathcal{L}^1(H)$ and we can analogously define

$$\tau(T) = \sum_{j=1}^{\infty} \langle \xi_j \mid \eta_j \rangle_H .$$

Proposition 3.2.9 (Traces) *The trace class operators I with $\tau : I \to \mathbf{C}$ given by $\tau(T) = \mathrm{trace}(T)$ satisfy the following:*

(i) *I is a self-adjoint ideal in $\mathcal{L}(H)$ that contains all the finite-rank operators*

$$\left\{ T = \sum_{j=1}^{N} \xi_j \otimes \eta_j ; \xi_j, \eta_j \in H \right\};$$

(ii) *I is a Banach space for some suitable norm so that*

$$\|UTV\|_I \leq \|U\| \|T\|_I \|V\| \tag{3.2.9}$$

for all $T \in I$ and $U, V \in \mathcal{L}(H)$, so multiplication is norm continuous;

(iii) *$\tau : I \to \mathbf{C}$ is a continuous linear functional such that $\tau | [I, \mathcal{L}(H)] = 0$, so $\tau | [I, I] = 0$, and τ is a trace on I.*

Further, up to scalar multiples, τ is the unique trace with the properties (i), (ii) and (iii).

Proof We can use (3.2.6) and the choice $\mathcal{L}^1(H) = \mathcal{L}^2(H)\mathcal{L}^2(H)$ to prove existence. To see uniqueness, we consider τ satisfy on (i), (ii) and (iii), and observe that by continuity, we need only determine the value of τ on finite-rank T. Then for any finite-rank operator $T \in I$, we can write $T = (1/2)(T + T^*) + (1/2i)(iT - iT^*)$, where $T + T^*$ and $iT - iT^*$ are self-adjoint and finite-rank operators in I. Also, for any unitary $U \in \mathcal{L}(H)$, we have

$$\tau(UTU^* - T) = \tau(U(TU^*) - (TU^*)U) = 0, \tag{3.2.10}$$

so τ is invariant under unitary conjugation. Given a finite-rank and self-adjoint $S \in I$, we introduce the eigenvalues $\lambda_1, \dots, \lambda_m$ of S and the orthogonal projection e of rank m onto the range of S, so we can present S as an $m \times m$ complex matrix on eH with respect to some orthonormal basis, and the trace on eH is uniquely determined as in Proposition 1.3.3. (In some contexts, discontinuous traces are used on $\mathcal{L}^1(H)$, but we do not use them here.) □

When $H = L^2(S^1, d\theta/(2\pi))$, and $T \in \mathcal{L}^1(H)$, then T also belongs to $\mathcal{L}^2(H)$ and hence may be expressed as an integral operator with kernel $K(x,y)$. Suppose that K is continuous. Then by [66, p. 344].

$$\operatorname{trace}(T) = \int_{S^1} K(x,x)\frac{dx}{2\pi}. \tag{3.2.11}$$

(6) We can regard the operator ideals as contained

$$\mathcal{L}(H) \supset \mathcal{K}(H) \supset \mathcal{L}^2(H) \supset \mathcal{L}^1(H) \supset \mathbf{F} \tag{3.2.12}$$

with the trace defined on the smallest two of them. The ideals I here are all self-adjoint in the sense that $A \in I \Rightarrow A^* \in I$. Note that $\mathbf{F}$ and $\mathcal{L}^1(H)$ are dense linear subspaces of $\mathcal{K}$; the trace τ is densely defined but unbounded on $\mathcal{K}$.

Exercise 3.2.10 (Homomorphisms and traces) Let B, C be algebras over a field $\mathbf{k}$, and let $\operatorname{Hom}(B,C)$ be the space of $\mathbf{k}$-algebra homomorphisms $B \to C$.

(i) Show that there is a natural map $\operatorname{Hom}(C,\mathbf{k}) \times \operatorname{Hom}(B,C) \to \operatorname{Hom}(B,\mathbf{k})$ $(\phi, T) \mapsto \phi \circ T$.

(ii) Let $\operatorname{Tr}(C) = \{\tau\colon C \to \mathbf{k}\colon \tau(ab) = \tau(ba); a,b \in C\}$ be the space of $\mathbf{k}$-valued traces on C. Show that $\operatorname{Hom}(C,\mathbf{k}) \subseteq \operatorname{Tr}(C)$ and that there is a natural map $\operatorname{Tr}(C) \times \operatorname{Hom}(B,C) \to \operatorname{Tr}(B)$ $(\phi, T) \mapsto \phi \circ T$.

(iii) One can show that the only norm continuous trace on the algebra $B = \mathcal{K}(H)$ is the zero trace. Deduce that the spaces of continuous homomorphisms $\operatorname{Hom}(B, M_n)$ and $\operatorname{Hom}(B, C(X,\mathbf{C}))$ are also trivial.

Exercise 3.2.11 (Logarithmic kernel) (i) Show that

$$-\log 4\sin^2\frac{\theta-\phi}{2} = \sum_{n\neq 0;\, n=-\infty}^{\infty} \frac{e^{in(\theta-\phi)}}{|n|}.$$

(ii) Show that the integral operator T on $L^2(S^1, d\theta/2\pi)$ with this kernel function is Hilbert–Schmidt, but not trace class. Indeed, $T \in \mathcal{L}^p(L^2(S^1, d\theta/2\pi))$ for all $p > 1$.

(iii) Describe the operator $F = (-id/d\theta)T$ in terms of Fourier series; prove that $F \in \mathcal{L}(L^2(S^1, d\theta/2\pi))$.

3.3 Algebras of Operators on Hilbert Space

Definition 3.3.1 (C^*-algebra) Let A be a complex Banach $*$-algebra so that A is equipped with an adjoint operation $*$ such that $a^* \in A$ for all $a \in A$,

$$(a^*)^* = a; \qquad (ab)^* = b^*a^* \qquad (\lambda a)^* = \bar{\lambda}a^* \qquad (a,b \in A, \lambda \in \mathbf{C}); \tag{3.3.1}$$

and a norm $\|\cdot\|: A \to [0,\infty)$ that satisfies the usual Banach norm axioms

$$\|a+b\| \leq \|a\| + \|b\|; \|\lambda a\| = |\lambda|\|a\|; \tag{3.3.2}$$

$$\|a\| = 0 \Rightarrow a = 0; \tag{3.3.3}$$

A is complete for the norm; and A satisfies the special assumptions

$$\|a^*\| = \|a\|; \quad \|a^*a\| = \|a\|^2; \quad \|ab\| \leq \|a\|\|b\|. \tag{3.3.4}$$

Example 3.3.2 (C^*-algebras) (i) Let $\mathcal{A}$ be a closed $*$-subalgebra of $\mathcal{L}(H)$ where H is a Hilbert space. Then $\mathcal{A}$ is a C^* algebra, as one can easily check. Remarkably, the converse is also true, as described in [81] and Theorem 4.2.3. The first step towards the converse is Theorem 3.3.3.

(ii) There is at most one norm on a $*$-algebra that makes it a C^*-algebra; see [80].

(iii) Let (X,d) be a compact metric space, μ a probability measure on X such that $\mu(U) > 0$ for all non-empty open subsets of X. Let $H = L^2(X,\mu;\mathbf{C})$ be the Hilbert space of square integrable functions $f: X \to \mathbf{C}$ with inner product $\langle g \mid f\rangle_H = \int f(x)\bar{g}(x)\,\mu(dx)$. Let $A = C(X;\mathbf{C})$ the space of continuous functions on X with the usual pointwise multiplication, conjugation $f^*(x) = \bar{f}(x)$ and $\|f\| = \sup_x\{|f(x)|: x \in \mathbf{X}\}$. Then each $f \in A$ is associated with a multiplication operator $M_f: g \mapsto fg$, so $\|M_f\|_{\mathcal{L}(H)} = \|f\|$ and $M_f^* = M_{\bar{f}}$. Thus A is realized as a norm closed * subalgebra of $\mathcal{L}(H)$, so A is a commutative unital C^*-algebra.

The continuous traces $\tau: A \to \mathbf{C}$ on A are precisely the bounded linear functionals on A, and by Riesz's theorem have the form $\tau(f) = \int_X f(x)\nu(dx)$ where ν is a measure on X of bounded total variation. Each point $x \in X$ determines a linear functional $\phi_x: f \mapsto f(x)$ such that $\phi_x(fg) = \phi_x(f)\phi_x(g)$ for all $f,g \in A$; one can express $\phi_x(f) = \int_X f(y)\delta_x(dy)$ where δ_x is the unit point mass at $x \in X$. One can establish an equivalence between multiplicative linear functionals on A, maximal $*$-ideals in A and points of X, as follows.

Theorem 3.3.3 (Gelfand–Naimark) *Let $\mathcal{A}$ be a commutative and unital C^* algebra with maximal ideal space X. Then the Gelfand transform $T \mapsto \hat{T}$*

is a bijective unital $*$ homomorphism from $\mathcal{A}$ onto $C(X,\mathbf{C})$ such that $\|T\|_{\mathcal{A}} = \|\hat{T}\|_\infty$.

Proof We already know that the Gelfand transform is a homomorphism. One checks that the map is a $*$-homomorphism, that $\|T\|_{\mathcal{A}} = \|\hat{T}\|_\infty$ for all $T \in \mathcal{A}$ and that the image separates the points of X. By the Stone–Weierstrass theorem, a self-adjoint closed subalgebra of $C(X,\mathbf{C})$ that separates the points of X and contains the constant functions must be all of $C(X,\mathbf{C})$; see [66]. □

Exercise 3.3.4 (Normal operators) Let N be a bounded normal operator on Hilbert space such that $NN^* = N^*N$. Let $\mathcal{A}$ be the closed algebra generated by I, N and N^*. Show that $\mathcal{A}$ is isomorphic as a C^*-algebra to $C(X,\mathbf{C})$, where X is the spectrum of N. Define a linear functional $\tau\colon \mathcal{A} \to \mathbf{C}$ such that $\tau(I) = 1$ and $|\tau(A)| \leq \|A\|$ for all $A \in \mathcal{A}$.

Definition 3.3.5 (i) *(Compactification)* First we observe that the one point compactification of the line $\mathbf{R}$ is homeomorphic to the circle S^1, while the one-point compactification of the plane $\mathbf{R}^2$ is homeomorphic to the sphere S^2. Let $C_0(\mathbf{R};\mathbf{C})$ be the space of continuous $fs\mathbf{R} \to \mathbf{C}$ such that $f(x) \to 0$ as $x \to \pm\infty$; such f are bounded so we use the $\|\cdot\|_\infty$ norm. There is an exact sequence

$$0 \longrightarrow C_0(\mathbf{R};\mathbf{C}) \longrightarrow C(S^1;\mathbf{C}) \longrightarrow \mathbf{C} \longrightarrow 0, \tag{3.3.5}$$

and we deduce that we can adjoin a unit to $C_0(\mathbf{R}\colon \mathbf{C})$ and obtain a unital C^* algebra $C_0(\mathbf{R}\colon \mathbf{C})^+ \cong C(S^1;\mathbf{C})$.

(ii) *(Suspension)* Let A be a C^* algebra. Then for any topological space X, the space $C_b(X;A)$ of bounded and continuous functions $f\colon X \to A$ gives a C^* algebra for pointwise multiplication and addition, $f^*(x) = (f(x))^*$ and $\|f\| = \sup\{\|f(x)\|_A \colon x \in X\}$. This may be identified with $C_b(X) \otimes A$, where the tensor norm is completed for the injective tensor product norm, and hence an exact sequence

$$0 \longrightarrow C_0(\mathbf{R};\mathbf{C}) \otimes A \longrightarrow C(S^1;\mathbf{C}) \otimes A \longrightarrow A \longrightarrow 0,$$

where $C(S^1;A)$ is the space of loops in A, and $\{f \in C(S^1;A)\colon f(1) = 0\}$ is the subalgebra of loops that vanish at 1. The latter algebra, or equivalently $C_0(\mathbf{R};\mathbf{C})\otimes A$, is called the suspension of A. Hence $C_0(\mathbf{R}^2;\mathbf{C})\otimes A$ is the double suspension of A, denoted S^2A.

Definition 3.3.6 Let A be a commutative C^*-algebra. Then we define $\tilde{K}_0(A)$ as in Section 2.9, and then $K_1(A) = \tilde{K}_0(SA)$.

If A is a commutative and unital C^*-algebra, then by Exercises 2.9.9 and 2.9.13, $K_0(A)$ is a commutative and unital ring. This does not extend easily to noncommutative C^*-algebras. One of the main achievements of Connes's noncommutative geometry is the notion of the Chen character, which is defined for noncommutative C^*-algebras and extends the ring structure of topological K-homology.

There are more natural and illuminating ways of defining K_1, as discussed in Exercise 4.4.4, [90] and [103]. One shows that $K_1(\mathbf{C}) = 0$. Continuing the process of suspension, we can progress to define $\tilde{K}_2(A) = K_0(S^2A)$, and so on. This process does not produce new groups indefinitely, due to the following fundamental result.

Theorem 3.3.7 (Bott's periodicity theorem) *There is a natural isomorphism between $\tilde{K}_0(A)$ and $\tilde{K}_0(S^2A)$.*

The Gelfand–Naimark theory does not readily extend to noncommutative C^*-algebras, for reasons evidenced in the following examples.

(i) *A unital C^*-algebra has no nilpotent ideals other than 0. For if J is an ideal such that $J^n = 0$, and $T \in J$, then $(T^*T)^{2n} = 0$, so $\|(T^*T)^{2^n}\| = 0$. We deduce that $\|T^*T\| = 0$, so $T = 0$. This is important when we come to consider extensions of C^*-algebras later in this chapter, since the discussion in Section 1.8 is not directly applicable.*

(ii) *Let $M_m(\mathbf{C})$ be the space of linear transformations on $\mathbf{C}^m$ with the usual matrix norm, matrix multiplication and adjoint $[a_{jk}]^* = [\bar{a}_{kj}]$. Then $M_n(\mathbf{C})$ is a C^*-algebra. The only traces on $M_m(\mathbf{C})$ have the form trace$([a_{jk}]) = c\sum_{j=1}^m a_{jj}$; see Proposition 1.3.3. We temporarily write τ_m for the usual matrix trace with $\tau_m([a_{jk}]) = \sum_{j=1}^m a_{jj}$.*

(iii) *Any C^*-algebra A that is finite dimensional over $\mathbf{C}$ has the form*

$$A = \oplus_{j=1}^N M_{n_j}(\mathbf{C}) \tag{3.3.6}$$

for some $N, n_j \in \mathbf{N}$, by Wedderburn's theorem as in Complement 1.3.5. The traces are given by Proposition 1.3.3, and have the form

$$\tau = \sum_{j=1}^N c_j \tau_{n_j}. \tag{3.3.7}$$

(iv) *The space $\mathbf{F}$ of finite-rank operators on H is $\{T = \sum_{j=1}^N \xi_j \otimes \eta_j; \xi_j, \eta_j \in H\}$ so $T\zeta = \sum_{j=1}^N \langle \zeta, \eta_j\rangle_H \xi_j$. Then T has a trace $\tau(T) = \sum_{j=1}^n \langle \eta_j \mid \xi_j\rangle_H$ which matches with the notion of trace used for matrices in linear algebra. The space $\mathbf{F}$ is a *-subalgebra of $\mathcal{L}(H)$ and an ideal in $\mathcal{L}(H)$, but $\mathbf{F}$ is not closed for the operator norm, hence is not a C^*-algebra.*

(v) *The space $\mathcal{K}(H)$ of compact operators on H is a C^*-algebra. Indeed, $\mathcal{K}(H)$ is an ideal in $\mathcal{K}(H)$, and is the smallest norm-closed ideal that contains* **F**. *This C^*-algebra does not have a unit, so we consider also*

$$\mathcal{K}(H)^+ = \mathbf{C}I + \mathcal{K}(H) = \{\lambda I + K : K \in \mathcal{K}(H), \lambda \in \mathbf{C}\}, \tag{3.3.8}$$

which is now a unital C^*-algebra. However, $\mathcal{K}(H)$ does not have a non-trivial trace that is continuous for the norm topology.

(vi) *The compact operators form a norm closed subspace of $\mathcal{L}(H)$, so the quotient $\mathcal{C}(H) = \mathcal{L}(H)/\mathcal{K}(H)$ is a Banach algebra called the Calkin algebra, which is also a C^*-algebra, and there is an exact sequence of C^*-algebras and $*$-homomorphisms*

$$0 \hookrightarrow \mathcal{K}(H) \longrightarrow \mathcal{L}(H) \overset{\pi}{\longrightarrow} \mathcal{C}(H) \longrightarrow 0. \tag{3.3.9}$$

Exercise 3.3.8 (Reducing subspaces) (i) Let E be a closed linear subspace of a Hilbert space H, then let $E^\perp$ be the orthogonal complement $E^\perp = \{\xi \in H : \langle \eta \mid \xi \rangle = 0, \ \forall \eta \in E\}$. The orthogonal projection $e \in \mathcal{L}(H)$ with range E and nullspace $E^\perp$ is represented as

$$e = \begin{bmatrix} I & 0 \\ 0 & 0 \end{bmatrix} \quad \begin{matrix} E \\ E^\perp . \end{matrix} \tag{3.3.10}$$

Show that there is an exact sequence of Hilbert spaces

$$0 \longrightarrow E^\perp \longrightarrow H \longrightarrow E \longrightarrow 0. \tag{3.3.11}$$

(ii) Let $A \in \mathcal{L}(H)$. Say that E is invariant for A if $Ae = eAe$. Show that this is equivalent to A having the block form

$$A = \begin{bmatrix} A_1 & A_2 \\ 0 & A_4 \end{bmatrix} \quad \begin{matrix} E \\ E^\perp . \end{matrix} \tag{3.3.12}$$

Let $\mathcal{A}(E) = \{A \in \mathcal{L}(H) : Ae = eAe\}$. Show that $\mathcal{A}(E)$ forms an algebra under operator multiplication.

(iii) Show that E and $E^\perp$ are invariant for A, if and only if $[A, e] = 0$, or equivalently A has the block form

$$A = \begin{bmatrix} A_1 & 0 \\ 0 & A_4 \end{bmatrix} \quad \begin{matrix} E \\ E^\perp . \end{matrix} \tag{3.3.13}$$

In this case, we say that E is reducing for A.

(iv) Let $\mathcal{E} = \{A \in \mathcal{L}(H) : [A, e] = 0\}$. Show that $\mathcal{E}$ is a C^*-algebra.

Exercise 3.3.9 (Dirichlet space) Let A be the space of $f \in C(S^1; \mathbf{C})$ that have Fourier series $f(\theta) = \sum_{n=-\infty}^{\infty} a_n e^{in\theta}$, where $a_n = \int_0^{2\pi} f(e^{i\theta}) e^{-in\theta} d\theta /$

(2π), and $\sum_{n=-\infty}^{\infty} |n||a_n|^2$ converges. This is sometimes known as the Dirichlet space, especially when written in the style of Exercise 3.7.7. See also [85, p. 82].

(i) Show that

$$K(\theta,\phi) = \frac{f(\theta) - f(\phi)}{e^{i\theta} - e^{i\phi}} \tag{3.3.14}$$

defines the kernel of a Hilbert–Schmidt integral operator on $L^2(S^1;\mathbf{C})$ with

$$\|K\|^2_{\mathcal{L}^2} = \sum_{n=-\infty}^{\infty} |n||a_n|^2.$$

(ii) Let $F\colon L^2(S^1;\mathbf{C}) \to L^2(S^1;\mathbf{C})$ be the operator

$$Fg(\theta) = \mathrm{PV}\int_0^{2\pi} g(\theta - \phi)\cot\frac{\phi}{2}\frac{d\phi}{2\pi}, \tag{3.3.15}$$

and M_f the multiplication operator $M_f g = fg$. Find the kernel of $[M_f, F]$ and show that $[M_f, F]$ is Hilbert–Schmidt.

(iii) By considering $\delta(f) = [M_f, F]$, show that A is an algebra.

(iv) For $f_0, f_1, f_2 \in A$, show that

$$\varphi(f_0, f_1, f_2) = \iint_{S^1\times S^1} f_0(\theta_1)\frac{f_1(\theta_1) - f_1(\theta_2)}{\tan(\theta_1 - \theta_2)/2}\frac{f_2(\theta_2) - f_2(\theta_1)}{\tan(\theta_2 - \theta_2)/2}\frac{d\theta_1}{2\pi}\frac{d\theta_2}{2\pi} \tag{3.3.16}$$

converges and

$$\varphi(f_0, f_1, f_2) = \mathrm{trace}\big(M_{f_0}[M_{f_1}, F][M_{f_2}, F]\big). \tag{3.3.17}$$

Exercise 3.3.10 (Restricted general linear group) Let $\mathcal{A}$ be a subalgebra of $\mathcal{L}(H)$ and $\delta\colon \mathcal{A} \to \mathcal{L}(H)$ a derivation of the form $\delta(A) = [A, e]$ for some orthogonal projection e. Show that the following give subalgebras of $\mathcal{A}$:

(i) $\mathcal{A}_0 = \{A \in \mathcal{D}\colon \delta(A) = 0\}$;

(ii) $\mathcal{A}_\infty = \{A \in \mathcal{D}\colon \delta(A) \in \mathcal{K}(H)\}$.

(iii) Deduce that there is an algebra

$$\begin{bmatrix}\mathcal{L}(H) & \mathcal{K}(H)\\ \mathcal{K}(H) & \mathcal{L}(H)\end{bmatrix} = \left\{\begin{bmatrix}R & S\\ T & U\end{bmatrix} : S, T \in \mathcal{K}(H); R, U \in \mathcal{L}(H)\right\}. \tag{3.3.18}$$

(iv) Likewise, show that

$$\mathcal{A}_2 = \begin{bmatrix}\mathcal{L}(H) & \mathcal{L}^2(H)\\ \mathcal{L}^2(H) & \mathcal{L}(H)\end{bmatrix} = \left\{\begin{bmatrix}R & S\\ T & U\end{bmatrix} : S, T \in \mathcal{L}^2(H); R, U \in \mathcal{L}(H)\right\}$$

gives a unital algebra $\mathcal{A}_2$, and the set $G(\mathcal{A}_2)$ of invertible elements in $\mathcal{A}_2$ gives a group, so that $\delta(U^{-1}) = -U^{-1}\delta(U)U^{-1}$ for all $U \in G(\mathcal{A}_2)$. This is the restricted general linear group, as in [85, p. 80].

Exercise 3.3.11 (Idempotents) Let $E \in \mathcal{L}(H)$ satisfy $E^2 = E$. Show that $E \in \mathcal{K}$ if and only if E has finite rank, and $\mathrm{trace}(E) = \mathrm{rank}(E)$.

3.4 Fredholm Operators

Theorem 3.4.1 (Atkinson) *For $T \in \mathcal{L}(H)$, the following are equivalent:*

(i) *there exists $S \in \mathcal{L}(H)$ such that $1 - ST$ and $1 - TS$ are compact;*
(ii) *the image of T is closed in H, $\dim \mathit{Ker}(T)$ is finite and $\dim \mathit{Ker}(T^*)$ is finite;*
(iii) *$\pi(T)$ is invertible in the Calkin algebra $\mathcal{C}(H) = \mathcal{L}(H)/\mathcal{K}(H)$.*

See [90] for discussion. An operator satisfying these equivalent conditions is called Fredholm. By condition (iii), the space of Fredholm operators forms a multiplicative semigroup under composition. Also, the set of invertible elements in $\mathcal{C}(H)$ is open in the norm topology on $\mathcal{C}(H)$, and the inverse image of any open subset of $\mathcal{C}(H)$ is open in $\mathcal{L}(H)$. All separable infinite-dimensional complex Hilbert spaces are unitarily equivalent. However, it is sometimes necessary to deal with Hilbert spaces that have a fixed interpretation as function spaces, so we formulate the following definition for pairs of Hilbert spaces.

Definition 3.4.2 (Fredholm operator) A bounded linear operator $P\colon H_1 \to H_2$ between separable and infinite-dimensional complex Hilbert spaces is said to be Fredholm if there exists a bounded linear operator $Q\colon H_2 \to H_1$ such that $QP - 1$ and $PQ - 1$ are compact; then Q is called a parametrix. The index of a Fredholm operator is

$$\mathrm{index}(P) = \dim \mathrm{Ker}(P) - \dim \mathrm{Coker}(P). \tag{3.4.1}$$

Here $\mathrm{Coker}(P) = (PH_1)^{\perp} = \ker(P^*)$. The index does not depend upon the particular choice of Q, so we can select any convenient Q. This notion of index is consistent with the terminology of Section 2.9. One can introduce a functor K_0 on C^*-algebras such that $K_0(\mathcal{K}) = \mathbf{Z}$. Theorem 3.4.1 shows that the set $\mathcal{F}(H)$ of Fredholm operators on H is open, the index is well-defined on $\mathcal{F}(H)$ and constant on connected components of $\mathcal{F}(H)$. In the next section, we show how to compute the index. For a more analytical presentation, see appendix 1 of [18].

Exercise 3.4.3 (Cokernels) (i) Let A_j be finite-dimensional complex vector spaces with dimension a_j such that

$$0 \longrightarrow A_0 \longrightarrow A_1 \longrightarrow A_2 \longrightarrow A_3 \longrightarrow 0$$

is an exact sequence. Show that $a_0 + a_2 = a_1 + a_3$, hence that $A_0 \oplus A_2 \cong A_1 \oplus A_3$.

(ii) Let $T\colon H_0 \to H_1$ be a Fredholm operator and F a finite-dimensional subspace of H_1 such that $\mathrm{im}(T) + F = H_1$, where the sum is not necessarily direct. Show that there is an exact sequence

$$0 \longrightarrow \mathrm{Ker}(T) \longrightarrow T^{-1}(F) \longrightarrow F \longrightarrow \mathrm{Coker}(T) \longrightarrow 0, \tag{3.4.2}$$

and deduce that

$$\mathrm{Ker}(T) \oplus F \cong T^{-1}(F) \oplus \mathrm{Coker}(T).$$

(iii) Deduce that $F \cong T^{-1}(F)$ if and only if the index of T is zero.

As in Section 2.1, a complex is a sequence of **k**-vector spaces K_j, linked by **k**-linear maps $d_j\colon K_j \to K_{j+1}$ such that $d_{j+1}d_j = 0$, which we write in the following array:

$$\cdots \overset{d_{j-2}}{\longrightarrow} K_{j-1} \overset{d_{j-1}}{\longrightarrow} K_j \overset{d_j}{\longrightarrow} K_{j+1} \cdots \tag{3.4.3}$$

Let $Z_j = \mathrm{Ker}(d_j)$ and $B_j = \mathrm{Im}(d_{j-1})$, so that $B_j \subseteq Z_j$, and we define the homology group $H^j = Z_j/B_j$ to be the quotient vector space. By an endomorphism T of a complex, we mean a sequence $T = (T_j)$ of **k**-linear maps $T_j\colon K_j \to K_j$ such that $T_{j+1}d_j = d_{j+1}T_j$, and so the following diagram commutes:

$$\begin{array}{ccc} K_j & \overset{d_j}{\longrightarrow} & K_{j+1} \\ T_j \downarrow & & T_{j+1} \downarrow \\ K_j & \overset{d_j}{\longrightarrow} & K_{j+1} \end{array} \tag{3.4.4}$$

Such an endomorphism induces an endomorphism $\hat{H}^i(T)$ of the homology groups H^i. See [5]. When the K_j are complex separable Hilbert spaces, we suppose that the d_j are Fredholm operators. Then the H^j are finite dimensional, so we can express the endomorphism as a finite matrix, and then take the trace.

Definition 3.4.4 (Lefschetz number) Suppose that $\dim K_j = 0$ for all $j < 0$ and all $j > n$, and that $\dim H^j$ is finite for all $j = 0, \dots, n$ so the complex reduces to

$$0 \longrightarrow K_0 \longrightarrow K_1 \longrightarrow \cdots \longrightarrow K_{n-1} \longrightarrow K_n \longrightarrow 0.$$

Then the Lefschetz number of T is

$$\Lambda(T) = \sum_{j=0}^{n} (-1)^j \operatorname{trace} \hat{H}^j(T). \tag{3.4.5}$$

The following result gives a means for computing the index in terms of parametrices. We use this in specific examples in 3.5.2, 3.6.2, 3.8.6 and 3.8.8.

Proposition 3.4.5 *Suppose that P is Fredholm with parametrix Q, so $QP - 1$ and $PQ - 1$ are both trace class. Then*

$$\operatorname{index}(P) = -\operatorname{trace}(1 - PQ) + \operatorname{trace}(1 - QP). \tag{3.4.6}$$

Proof The proof that follows is not the most direct, but is intended to link with other theories. Given that we have a parametrix Q, it is natural to use the trace formula (3.2.18) to compute the index. There is a diagram

$$\begin{array}{ccccccc} 0 & \longrightarrow & H_1 & \overset{P}{\longrightarrow} & H_2 & \longrightarrow & 0 \\ & 1-QP & \downarrow & Q \swarrow & \downarrow & 1-PQ & \\ 0 & \longrightarrow & H_1 & \underset{P}{\longrightarrow} & H_2 & \longrightarrow & 0 \end{array} \tag{3.4.7}$$

with induced maps on the homology groups

$$\begin{array}{ccc} \operatorname{Ker}(P) & \longrightarrow & \operatorname{Coker}(P) \\ \downarrow & & \downarrow \\ \operatorname{Ker}(P) & \longrightarrow & \operatorname{Coker}(P) \end{array} \tag{3.4.8}$$

The Lefschetz number is

$$\Lambda = \operatorname{trace}(1 - QP) - \operatorname{trace}(1 - PQ), \tag{3.4.9}$$

and when $1 - QP$ and $1 - PQ$ are trace class, this is

$$\Lambda = \dim \operatorname{Ker}(P) - \dim \operatorname{Coker}(P). \tag{3.4.10}$$

The analytical details are described in [56, theorem 19.1.15]. □

3.5 Index Theory on the Circle via Toeplitz Operators

In the context of Example 3.2.2 with $H = L^2(S^1; d\theta/(2\pi))$, let the Hardy space H_+ be

$$H_+ = \left\{ \sum_{n=0}^{\infty} a_n z^n : \sum_{n=0}^{\infty} |a_n|^2 < \infty \right\} \tag{3.5.1}$$

and

$$H_- = \left\{ \sum_{n=-\infty}^{-1} a_n z^n : \sum_{n=-\infty}^{-1} |a_n|^2 < \infty \right\} \tag{3.5.2}$$

so that $H = H_+ \oplus H_-$ where the complementary subspaces H_+ and H_- are orthogonal. Now H_+ is the image of the orthogonal projection $e\colon H \to H_+$, known as the Hardy projection

$$e(z^n) = \begin{cases} 0, & \text{for } n < 0; \\ z^n, & \text{for } n = 0, 1, 2, \dots \end{cases} \tag{3.5.3}$$

The image of $1 - e$ is H_-, the orthogonal complement of H_+ in L^2. There are equivalent descriptions of H_+ as the space of square summable power series as in (3.5.2), the space of $f \in L^2(S^1; d\theta/(2\pi))$ such that $\hat{f}(n) = 0$ for $n < 0$; and the space of holomorphic functions $g\colon D \to \mathbf{C}$ such that $\sup_{0<r<1} \int_0^{2\pi} |g(re^{i\theta})|^2 d\theta/(2\pi)$ is finite; in the literature, H_+ is often written as H^2.

Let $C = C(S^1; \mathbf{C})$ be the space of continuous functions and $H = L^2(S^1; \mathbf{C})$. Each $f \in C$ is associated with a multiplication operator $\theta(f) \in \mathcal{L}(H)$ given by $\theta(f)\colon g \mapsto fg$. Observe that H is a module over C for the pointwise multiplication $(f, g) \mapsto fg$. When f is not constant, $\theta(f)$ is not compact, and will not commute with the orthogonal projection $e\colon H \to H_+$. We abbreviate by writing $f = \theta(f) \in \mathcal{L}(H)$.

Definition 3.5.1 (i) *(Unitary)* An operator $U \in \mathcal{L}(H)$ is called unitary if $U^*U = UU^* = I$. Let $\mathcal{U}(H)$ be the space of unitary operators, which forms a multiplicative group.

(ii) *(Isometry)* Say that $V \in \mathcal{L}(H)$ is an isometry if $V^*V = I$, or equivalently $\|Vf\| = \|f\|$ for all $f \in H$. Also, $W \in \mathcal{L}(H)$ is called a coisometry if $W = V^*$ where V is an isometry.

(iii) *(Partial Isometry)* A bounded linear operator U on H is a partial isometry if U is an isometry on the orthogonal complement of its kernel. Equivalently, U is a partial isometry if U^*U is a projection.

Exercise 3.5.2 (Unilateral shift operator) (i) The unilateral shift operator $S\colon H_+ \to H_+$ is given on power series $f(z) = \sum_{n=0}^{\infty} a_n z^n$, by $S\colon f(z) \mapsto zf(z)$. Show that the adjoint S^* satisfies

$$S^*\colon f(z) \mapsto \frac{f(z) - f(0)}{z} \qquad (z \in D) \tag{3.5.4}$$

so that $S^*S = I$. Deduce that S is an isometry, and that $[S^*, S]$ is the rank-one projection $f(z) \mapsto f(0)$, or equivalently $\sum_{n=0}^{\infty} a_n z^n \mapsto a_0$, so S is a Fredholm operator.

(ii) Show that $\cap_{n=1}^{\infty} S^n(H_+) = 0$.

(iii) Let $u \in H_+$ have $|u(e^{i\theta})| = 1$ for almost all $e^{i\theta} \in S^1$, and let $E = uH_+ = \{uf : f \in H_+\}$. Show that E is invariant for S. By a theorem of Beurling, all non-zero invariant subspaces E of H_+ for S have this form; see [66, p. 515].

(iv) Let E be a reducing subspace of H_+ for S. Then $E = 0$ or $E = H_+$. Otherwise, E and $E^\perp$ would both be non-zero invariant subspaces of H_+, so Beurling's theorem would give u, v as in (ii) such that $E = uH_+$ and $E^\perp = vH_+$. Show that $\langle S^n u, v\rangle = 0$ for all $n = 0, 1, \ldots$, and likewise $\langle S^n v, u\rangle = 0$ for all $n = 0, 1, \ldots$, so $u\bar{v}$ would have all Fourier coefficients zero, hence $u\bar{v} = 0$.

Definition 3.5.3 (Toeplitz operator) The Toeplitz operator $T_f \in \mathcal{L}(H_+)$ associated with $f \in C$ is

$$T_f(g) = e(fg) \qquad (g \in H_+). \tag{3.5.5}$$

We can write $\theta(f)$ as a block operator matrix

$$\theta(f) = \begin{bmatrix} e\theta(f)e & e\theta(f)(1-e) \\ (1-e)\theta(f)e & (1-e)\theta(f)(1-e) \end{bmatrix} \quad \begin{matrix} H_+ \\ H_- \end{matrix}, \tag{3.5.6}$$

so we recognize $T_f = e\theta(f)e$ as the top left corner. One can write this as

$$\theta(f) = \begin{bmatrix} T_f & H_f \\ K_f & L_f \end{bmatrix} \tag{3.5.7}$$

where $\theta(fg) = \theta(f)\theta(g)$, so

$$T_{fg} = T_f T_g + H_f K_g; \tag{3.5.8}$$

and $T_{fg} - T_f T_g = H_f K_g$, whence $f \mapsto T_f$ is not a homomorphism $C \to \mathcal{L}(H_+)$. However, this map is very close to being a homomorphism, in a sense which we will elaborate over the next few pages.

Exercise 3.5.4 (Compact Hankel operators) Let $H^\infty = \{f \in L^\infty(S^1; \mathbf{C}) : \hat{f}(n) = 0; n < 0\}$, and let $C = C(S^1; \mathbf{C})$. Then H^∞ can be identified with the space of bounded analytic functions $f : D \to \mathbf{C}$, so H^∞ is a norm closed subalgebra of L^∞ for $\|f\|_\infty = \sup_{z \in D}\{|f(z)| : z \in D\}$.

(i) Show that there is a homomorphism $H^\infty \to \mathcal{L}(H_+)$: $f \mapsto e\theta(f)e$. The operator $T_f = e\theta(f)e$ is often called an analytic Toeplitz operator.

(ii) Show also that $(1-e)\theta(f)e = 0$ for all $f \in H^\infty$. Deduce that $\theta(f)$ is an upper triangular block matrix for all $f \in H^\infty$.

(iii) Let $\overline{H^\infty} = \{\bar{f} : f \in H^\infty\}$. Show that $e\theta(f)(1-e) = 0$ for all $f \in \overline{H^\infty}$.

(iv) It has been shown that $C + \overline{H^\infty}$ is a norm closed subalgebra of L^∞. Now let $A = C \cap H^\infty$ be the disc algebra, and $\overline{A} = C \cap \overline{H^\infty}$. Show that there is an isomorphism of vector spaces

$$(C + \overline{H^\infty})/\overline{H^\infty} \cong C/\overline{A}.$$

(v) Deduce that for $f \in A$, the matrix $\theta(f)$ is block upper triangular, with the top-right corner H_f compact.

Let $A = C^\infty(S^1; \mathbf{C})$ be the space of infinitely differentiable functions $f : S^1 \to \mathbf{C}$.

Lemma 3.5.5 *Let $f, g \in C^\infty(S^1; \mathbf{C})$. Then*

(i) *$[e, g]$ is a trace class operator on L^2 with a smooth Schwartz kernel.*

(ii) *The trace of $f[e, g]$ on $H = L^2$ is*

$$\mathrm{trace}_H\big(f[e,g]\big) = \frac{1}{2\pi i}\int_0^{2\pi} f(\theta)g'(\theta)\,d\theta. \tag{3.5.9}$$

Proof Let $F = 2e - 1$, so that $F = F^*$, $F^2 = 1$ and

$$F(z^n) = \begin{cases} -z^n, & \text{for } n < 0; \\ z^n, & \text{for } n = 0, 1, 2, \ldots. \end{cases} \tag{3.5.10}$$

This F is a variant of the classical conjugate function operator (Hilbert transform). For $0 < r < 1$, we consider

$$\begin{aligned}\sum_{n=-\infty}^{\infty} r^{|n|}F(z^n) &= \frac{1 + r^2 - 2r/z}{1 + r^2 - r(z + 1/z)} \\ &\to 1 + i\cot\frac{\theta}{2} \qquad (r \to 1-).\end{aligned} \tag{3.5.11}$$

Hence F has kernel $F(\theta, \phi) = 1 + i\cot(1/2)(\theta - \phi)$ and $[e, g] = (1/2)[F, g]$ has kernel

$$[e, g] \leftrightarrow -g(\theta) + g(\phi) - \frac{i}{2}\big(g(\theta) - g(\phi)\big)\cot\frac{\theta - \phi}{2}, \tag{3.5.12}$$

where the singularity of $\cot(1/2)(\theta - \phi)$ on $\theta = \phi$ is cancelled by the factor $(g(\theta) - g(\phi))$ in parentheses; so $f[e, g]$ has a smooth function as its integral kernel

$$f[e,g] \leftrightarrow -f(\theta)\big(g(\theta) - g(\phi)\big) - \frac{i}{2} f(\theta)\big(g(\theta) - g(\phi)\big) \cot \frac{\theta - \phi}{2}; \tag{3.5.13}$$

so the diagonal is $-if(\theta)g'(\theta)$, and by the Schwartz kernel theorem, as in (3.2.18),

$$\mathrm{trace}_H\big(f[e,g]\big) = \frac{-i}{2\pi} \int_0^{2\pi} f(\theta) g'(\theta)\, d\theta. \tag{3.5.14}$$

□

Exercise 3.5.6 (Toeplitz algebra) (i) Let $f(z) = z$ for $z \in S^1$. Show that T_f is the shift operator $S\colon H_+ \to H_+$.

(ii) Let $\mathcal{T}$ be the Toeplitz algebra, namely the C^*-subalgebra of $\mathcal{L}(H_+)$ generated by $\{T_f : f \in C\}$. Show that $\mathcal{T}$ contains the rank one projection $P = [S^*, S]$, so $P \in \mathcal{T} \cap \mathcal{K}(H_+)$. One can deduce that $\mathcal{K}(H) \subseteq \mathcal{T} \cap \mathcal{K}(H_+)$, so $\mathcal{K}(H_+) \subset \mathcal{T}$.

(iii) Show that the space $\{T_f + \mathcal{K}(H_+) : f \in C(S^1)\}$ forms a commutative subalgebra of $\mathcal{L}(H_+)/\mathcal{K}(H_+)$.

(iv) Let $\mathcal{I}$ be the ideal in $\mathcal{T}$ that is generated by the commutator subspace $[\mathcal{T}, \mathcal{T}]$. Show that $[\mathcal{T}, \mathcal{T}] \subseteq \mathcal{K}(H_+)$, and hence that $\mathcal{I} = \mathcal{K}(H_+)$. Deduce that $\mathcal{T}/\mathcal{K}(H_+)$ is a commutative C^*-algebra.

(v) Show that there is a map $\rho\colon C(S^1) \to \mathcal{T}$ which is a homomorphism modulo $\mathcal{K}$, as in Definition 1.8.2, and deduce that there is an $*$-isomorphism $C(S^1) \to \mathcal{T}/\mathcal{K}(H_+)$.

(vi) Obtain the exact sequence of C^*-algebras and $*$-homomorphisms

$$0 \longrightarrow \mathcal{K}(H_+) \longrightarrow \mathcal{T} \longrightarrow C(S^1) \longrightarrow 0.$$

Exercise 3.5.7 (Bott periodicity) (i) Show that there is an exact sequence of C^*-algebras

$$0 \longrightarrow C_0(\mathbf{R}; \mathbf{C}) \longrightarrow C(S^1; \mathbf{C}) \longrightarrow \mathbf{C} \longrightarrow 0.$$

(ii) By considering the map $T_f \mapsto f(1)$, obtain an exact sequence of C^*-algebras

$$0 \longrightarrow \mathcal{T}_0 \longrightarrow \mathcal{T} \longrightarrow \mathbf{C} \longrightarrow 0.$$

(iii) Show that $\mathcal{T}_0$ is generated by $I - S$, and by considering $[I - S, I - S^*]$ show that $\mathcal{K}(H) \subset \mathcal{T}_0$.

(iv) Deduce that there is an exact sequence of C^*-algebras

$$0 \longrightarrow \mathcal{K}(H) \longrightarrow \mathcal{T}_0 \longrightarrow C_0(\mathbf{R}) \longrightarrow 0.$$

Cuntz proved that $K_j(\mathcal{T}_0) = 0$ $(j = 0, 1)$, and deduced that

$$K_j(C_0(\mathbf{R}^2; \mathbf{C}) \otimes \mathcal{A}) \cong K_j(\mathcal{A}) \qquad (j = 0, 1)$$

for all C^*-algebras $\mathcal{A}$. This is the proof of Bott periodicity via Toeplitz operators; see [106].

Exercise 3.5.8 (Index of a partial isometry) Let U be a partial isometry that is a Fredholm operator. Show that

$$\mathrm{trace}(UU^* - U^*U) = \mathrm{dimker}(U) - \mathrm{dimcoker}(U).$$

3.6 The Index Formula for Toeplitz Operators

Now with $R = \mathcal{L}(H)$ and $I = \mathcal{L}^1(H)$, there is a linear map $\rho\colon A \to R\colon f \mapsto T_f$ such that ρ is a homomorphism modulo I, indeed

$$\begin{aligned}\rho(f)\rho(g) - \rho(fg) &= efeege - efge \\ &= ef[e, g]e \end{aligned} \qquad (3.6.1)$$

is trace class. Then as in Proposition 2.8.2, we define $\varphi = \mathrm{tr}_\tau(\delta\rho + \rho^2)$ to be the 1-cocycle

$$\begin{aligned}\varphi(a_0, a_1) &= \mathrm{tr}_\tau(\delta\rho + \rho^2)(a_0, a_1) \\ &= \mathrm{trace}_H(\delta\rho + \rho^2)(a_0, a_1) - \mathrm{trace}_H(\delta\rho + \rho^2)(a_1, a_0) \\ &= \mathrm{trace}_H\big(\rho(a_0)\rho(a_1) - \rho(a_0 a_1)\big) - \mathrm{trace}_H\big(\rho(a_1)\rho(a_0) - \rho(a_1 a_0)\big).\end{aligned} \qquad (3.6.2)$$

Proposition 3.6.1 *The cyclic cocycle satisfies*

$$\varphi(f, g) = \mathrm{trace}_H\big(f[e, g]\big) = \frac{1}{2\pi i}\int_0^{2\pi} f(\theta)g'(\theta)d\theta. \qquad (3.6.3)$$

Proof The index formula can be recast in several ways. Let $D\colon \mathcal{L}(H) \to \mathcal{L}(H)$ be any derivation. Then from $e^2 = e$, we deduce that $(De)e + eDe = De$, so $eDe = (De)(1 - e)$, and $(1 - e)De = (De)e$. In particular, with $Df = [f, g]$, the inner derivation associated with g, we have

$$e[e, g] = [e, g](1 - e). \qquad (3.6.4)$$

The trace may be expressed in terms of the operators on Hardy space eH as

$$\mathrm{trace}_{eH}(\rho(g)\rho(f) - \rho(gf)) = \mathrm{trace}_{eH}(ef[e,g]e). \tag{3.6.5}$$

To see this, write

$$\begin{aligned}\mathrm{trace}_{eH}(\rho(g)\rho(f) - \rho(gf)) &= \mathrm{trace}_{eH}(egefe - egfe)\\ &= \mathrm{trace}_{eH}(eg(e-1)fe)\\ &= \mathrm{trace}_{H}(eg(e-1)f)\\ &= \mathrm{trace}_{H}([e,g](e-1)f)\\ &= \mathrm{trace}_{H}((e-1)f[e,g]),\end{aligned} \tag{3.6.6}$$

and likewise

$$\begin{aligned}\mathrm{trace}_{eH}(\rho(f)\rho(g) - \rho(gf)) &= \mathrm{trace}_{eH}(ef[e,g]e)\\ &= \mathrm{trace}_{H}((ef[e,g])\end{aligned} \tag{3.6.7}$$

hence

$$\begin{aligned}\mathrm{trace}_{H}(f[e,g]) = &-\mathrm{trace}_{eH}(\rho(g)\rho(f) - \rho(gf))\\ &+ \mathrm{trace}_{eH}(\rho(f)\rho(g) - \rho(fg)),\end{aligned} \tag{3.6.8}$$

so

$$\varphi(f,g) = \mathrm{trace}_{H}(f[e,g]), \tag{3.6.9}$$

where the right-hand side is given by the Lemma. □

Corollary 3.6.2 (Gohberg) *Suppose that $f \in C^\infty(S^1;\mathbf{C})$ has $1/f \in C^\infty(S^1;\mathbf{C})$. Then T_f is a Fredholm operator with index*

$$\mathrm{Index}(T_f) = -\text{winding number of } f. \tag{3.6.10}$$

Proof We can use $Q = \rho(1/f)$ as a parametrix for $P = \rho(f)$ since $\rho(1/f)\rho(f) - \rho(1)$ is trace class on eH. Here $f: S^1 \to \mathbf{C}$ has an image which is a smooth curve in $\mathbf{C} \setminus \{0\}$ that goes from $f(e^{i0})$ back to $f(e^{2\pi i})$, so we can count the number of times this curve winds around the origin. Hence

$$\begin{aligned}\mathrm{Index}(T_f) &= \mathrm{trace}_{eH}(1 - \rho(1/f)\rho(f)) - \mathrm{trace}_{eH}(1 - \rho(f)\rho(1/f))\\ &= \varphi(1/f, f)\\ &= -\frac{1}{2\pi i}\int_0^{2\pi} \frac{f'(\theta)}{f(\theta)} d\theta,\end{aligned} \tag{3.6.11}$$

which we recognize as the negative of the winding number of f. □

3.7 Wallach's Formula

In this section we express the results of Section 3.6 in a more symmetrical way using cocycles based upon the Poisson bracket. We begin with some results on commutators.

Proposition 3.7.1 *Let J be an ideal in a complex algebra $\mathcal{E}$ and suppose that $X_j \in \mathcal{E}$ $(j = 0,\dots,k)$ with $X_0 = I$ satisfy $[X_j, X_\ell] \in J$ for all $j,\ell = 0,\dots,k$. Let A be the subalgebra of $\mathcal{E}$ generated by X_j for $j = 0,\dots,k$. Then there is an exact sequence*

$$0 \longrightarrow J \longrightarrow A + J \longrightarrow C_0 \longrightarrow 0 \tag{3.7.1}$$

where C_0 is a commutative and finitely generated algebra.

Proof Recall Exercise 1.6.9. We observe that A is spanned by monomials of the form $Y_1 \dots Y_m$ where $Y_j \in \{X_0,\dots,X_k\}$. Also

$$[S, TU] = [S,T]U + T[S,U] \in J \qquad (S,T,U \in A);$$

so given two such monomials, we have

$$\begin{aligned}[Y_1 \dots Y_m, Z_1 \dots Z_r] &= \sum_{j=1}^{m} Y_1 \dots Y_{j-1}[Y_j, Z_1 \dots Z_r]Y_{j+1} \dots Y_m \\ &= \sum_{j=1}^{m}\sum_{\ell=1}^{r} Y_1 \dots Y_{j-1}Z_1 \dots Z_{\ell-1}[Y_j, Z_k]Z_{k+1} \dots Z_r Y_{j+1} \dots Y_m \end{aligned} \tag{3.7.2}$$

where $[y_j, z_k] \in J$. Now $A \cap J$ is an ideal in A, and by the preceding calculation $[a,b] \in A \cap J$ for all $a,b \in A$, so $A/(A \cap J)$ is a commutative algebra. Also $A + J$ is a subalgebra of $\mathcal{E}$ such that $(A + J)/J$ is canonically isomorphic to $A/(A \cap J)$ by [47, theorem 2.10], and we define $C_0 = A/(A \cap J)$ to make the exact sequence. There is a surjective homomorphism on the polynomial algebra $\mathbf{C}[x_1,\dots,x_k] \to (A + J)/J$ given by $1 \mapsto X_0 + J$ and $x_j \mapsto X_j + J$ for $j = 1,\dots,k$; so C_0 is a finitely generated algebra. □

The algebra A has subalgebras spanned by product

$$a_0[b_1,c_1]a_1[b_2,c_2]\dots[b_n,c_n]a_n \in (J \cap A)^n \qquad (a_j,b_j,c_j \in A) \tag{3.7.3}$$

involving n commutators. So by Exercise 1.1.4, there is a filtration

$$g_{A\cap J}(A) = (A/(A \cap J)) \oplus (A \cap J)/(A \cap J)^2 \oplus \cdots \tag{3.7.4}$$

with respect to powers of the ideal $A \cap J$ of A. In the case $\mathcal{L} = \mathcal{L}(H)$ and $J = \mathcal{L}^n(H)$, there is a trace defined on $(A \cap J)^n \subseteq \mathcal{L}^1(H)$, and in Corollary 9.6.2 we give such an example. See [60, Proposition 1.5.1] for further discussion. In Section 6.1, we describe algebras such as A_1 in terms of Kähler differentials, and the case $k = 2$ is already interesting.

Corollary 3.7.2 *Suppose that $X_j \in \mathcal{L}(H)$ are self-adjoint operators for $j = 0, \dots, k$ with $X_0 = I$, such that $[X_j, X_\ell] \in \mathcal{K}(H)$ for $j, \ell = 0, \dots, k$; let $\mathcal{E}$ be the C^*-algebra generated by $\mathcal{K}$ and the X_j.*

(i) *Then there exists a commutative C^*-algebra C_0 such that the sequence*

$$0 \longrightarrow \mathcal{K}(H) \longrightarrow \mathcal{E} \longrightarrow C_0 \longrightarrow 0$$

is exact, and C_0 is isomorphic to $C(\Omega; \mathbf{C})$ for some compact metric space Ω.

(ii) *The map $\rho\colon \mathbf{C}[x_1, \dots, x_k] \to \mathcal{E}$*

$$\sum_{j_1,\dots,j_k}^{n} a_{j_1 j_2 \dots j_k} x_1^{j_1} x_2^{j_2} \dots x_k^{j_k} \mapsto \sum_{j_1,\dots,j_k}^{n} a_{j_1 j_2 \dots j_k} X_1^{j_1} X_2^{j_2} \dots X_k^{j_k}$$

is a homomorphism modulo $\mathcal{K}(H)$, with dual ρ' giving a natural map from linear functionals on $\mathcal{E}/\mathcal{K}(H)$ to linear functionals on $\mathbf{C}[x_1, \dots, x_n]$.

Proof (i) By Proposition 3.7.1, we can take $J = \mathcal{K}$, so that $[\mathcal{E}, \mathcal{E}] \subseteq \mathcal{K}(H) \cap \mathcal{E}$; hence $A_0 = \mathcal{E}/\mathcal{K}(H)$ is a commutative C^*-algebra. The result follows from the Gelfand–Naimark theorem.

(ii) This follows from Proposition 3.7.1 and Lemma 1.8.3. □

Definition 3.7.3 (i) *(Joint essential spectrum)* The set Ω is called the joint essential spectrum of $(X_1, \dots, X_k)$. (This set implicitly depends upon the choice of ideal $J = \mathcal{K}(H)$ as well as the X_j; see Exercise 1.6.9.)

(ii) *(Symbol)* For $X \in \mathcal{E}$ the image of $X + \mathcal{K}$ in $C(\Omega; \mathbf{C})$ under the Gelfand map is called the symbol $\sigma(X)$ of X. (To resolve a possible notational confusion, one can think of the spectrum of X in $\mathcal{E}/\mathcal{K}$ as the range of the function $\sigma(X)$, as in Exercise 3.1.1.)

Proposition 3.7.4 (Wallach) *Let $X, Y \in \mathcal{L}(H)$ be self-adjoint operators such that $[X, Y] \in \mathcal{L}^1(H)$, and let $\rho\colon \mathbf{C}[x, y] \to \mathcal{L}(H)\colon\ f(x, y) \mapsto f(X, Y)$ be as in Corollary 3.7.2.*

(i) *Then there is bilinear form $\beta\colon \mathbf{C}[x, y] \times \mathbf{C}[x, y] \to \mathbf{C}$:*

$$\beta(f, g) = \operatorname{trace}([f(X, Y), g(X, Y)).$$

(ii) *Let $\varphi\colon \mathbf{C}[x, y] \times \mathbf{C}[x, y] \to \mathbf{C}$ be a bilinear form such that for all singly generated subalgebras R of $\mathbf{C}[x, y]$ the restriction is $\varphi | R \times R = 0$. Then there exists a linear functional $\phi\colon \mathbf{C}[x, y] \to \mathbf{C}$ such that*

$$\varphi(f,g) = \phi(\{f,g\}) = \phi\left(\frac{\partial f}{\partial x}\frac{\partial g}{\partial y} - \frac{\partial f}{\partial y}\frac{\partial g}{\partial x}\right) \qquad (f,g \in \mathbf{C}[x,y]). \quad (3.7.5)$$

Proof (i) This follows easily from Proposition 3.7.1, where we take $J = \mathcal{L}^1(H)$.

(ii) See [51, 52]. Such a φ is called a Pincus form. □

Example 3.7.5 (Unilateral shift as Toeplitz operator) For instance, we can choose $H_+ = eH$ as in Exercise 3.5.2 and write $T_z = S$, so the shift is the Toeplitz operator generated by the identity function $z \mapsto z$ on the circle. Let

$$X = \frac{S + S^*}{2}, \qquad Y = \frac{S - S^*}{2i},$$

so that

$$i[T,T^*] = 2[X,Y] = S^*S - SS^*, \tag{3.7.6}$$

which is finite rank, hence compact. In the case of the Toeplitz operators over the circle, the formula for the index involves the derivative of the symbol, so the index involves an additional smooth structure of the circle which involves unbounded operators on $L^2(S^1)$. This is an essential ingredient of the theory, in that we need to exploit or introduce additional smooth structure to obtain usable index formulas for families of Fredholm operators.

As an instance of these results, we can write the cocycle formula (3.6.3) in a more symmetrical form. Given $f,g \in C^1(S^1, \mathbf{C})$, we can extend f to a harmonic function on the disc $\mathbf{D} = \{z \in \mathbf{C} : |z| < 1\}$ via the Poisson integral

$$f(re^{i\phi}) = \int_{-\pi}^{\pi} \frac{(1-r^2)f(e^{i(\phi-\theta)})}{1+r^2-2r\cos\theta}\frac{d\theta}{2\pi} \tag{3.7.7}$$

so that the original f gives the boundary values of the extended function, and likewise we extend g with the Poisson integral.

Corollary 3.7.6 (Wallach) *The cocycle φ of Proposition 3.6.1 is given by the Poisson bracket of f and g as in*

$$\varphi(f,g) = \frac{1}{2\pi i}\iint_{\mathbf{D}} \left(\frac{\partial f}{\partial x}\frac{\partial g}{\partial y} - \frac{\partial f}{\partial y}\frac{\partial g}{\partial x}\right) dxdy. \tag{3.7.8}$$

Proof From the calculations of Section 3.6, we have

$$\begin{aligned}\varphi(f,g) &= \mathrm{trace}_{eH}\big(\rho(f)\rho(g) - \rho(g)\rho(f)\big)\\ &= \frac{1}{2\pi i}\int_{S^1} f(\theta)g'(\theta)\,d\theta\\ &= \frac{1}{2\pi}\iint_{D}\left(\frac{\partial f}{\partial x}\frac{\partial g}{\partial y} - \frac{\partial f}{\partial y}\frac{\partial g}{\partial x}\right) dxdy, \end{aligned} \tag{3.7.9}$$

where we have used Green's theorem in the plane to introduce the Poisson bracket of f and g. □

The following exercises are taken from results in [51].

Exercise 3.7.7 (Hilbert–Schmidt commutators) Let $f \in L^2(S^1;\mathbf{R})$ have Fourier series $f(\psi) = \sum_{n=-\infty}^{\infty} a_n e^{in\psi}$, where $\sum_{n=\infty}^{\infty} |n||a_n|^2$ converges.

(i) Show that the harmonic extension of f to D is given in polar coordinates $z = re^{in\psi}$ by

$$f(z) = \sum_{n=-\infty}^{\infty} a_n r^{|n|} e^{in\theta}, \tag{3.7.10}$$

and compute

$$\frac{1}{2\pi i}\iint_D \left(\left(\frac{\partial f}{\partial x}\right)^2 + \left(\frac{\partial f}{\partial y}\right)^2\right) dxdy \tag{3.7.11}$$

by changing to polar coordinates.

(ii) Express the integral of (3.7.11) in terms of the gradient of f and a suitable 2×2 matrix.

(iii) Show that the integral operator with kernel

$$\frac{f(\psi) - f(\phi)}{\tan((\psi - \phi)/2)}$$

is self-adjoint and Hilbert–Schmidt on $L^2(S^1, d\psi/(2\pi))$, and deduce that $[F, \theta(f)]$ is Hilbert–Schmidt.

Exercise 3.7.8 (Functionals on commutator products) Let $\mathcal{E}$ be a $*$-subalgebra of $\mathcal{L}(H)$ such that $\mathcal{L}^n(H) \subset \mathcal{E}$; let C^∞ be a unital and commutative $*$-algebra, and $\rho\colon C^\infty \to \mathcal{E}$ a homomorphism modulo $\mathcal{L}^n(H)$ such that $\rho(f^*) = \rho(f)^*$.

(i) By considering Lemma 1.8.3, show that $[\rho(C^\infty), \rho(C^\infty)] \subset \mathcal{L}^n(H)$.

(ii) Show that the following functionals are well-defined on $(C^\infty)^{\otimes(2n+1)}$:

$$\begin{aligned}&\mathrm{trace}\big(\rho(f_0)\omega(f_1, f_2)\omega(f_3, f_4)\dots\omega(f_{2n-1}, f_{2n})\big);\\ &\mathrm{trace}\big(\rho(f_0)[\rho(f_1),\rho(f_2)][\rho(f_3),\rho(f_4)]\dots[\rho(f_{2n-1}),\rho(f_{2n})]\big).\end{aligned} \tag{3.7.12}$$

Exercise 3.7.9 (Wick ordered calculus) Let $X, Y \in \mathcal{L}(H)$ be self-adjoint such that $[X, Y] \in \mathcal{L}^1(H)$. For $f, g\colon \mathbf{R}^2 \to \mathbf{C}$ Schwartz class functions, define the Fourier transform by

$$\hat{f}(\xi, \eta) = \iint_{\mathbf{R}^2} \hat{f}(x, y) e^{-i\xi x} e^{-i\eta y}\, dxdy,$$

and likewise with $\hat{g}$. Then let

$$f(X,Y) = \frac{1}{(2\pi)^2} \iint_{\mathbf{R}^2} \hat{f}(\xi,\eta) e^{i\xi X} e^{i\eta Y}\, d\xi d\eta. \tag{3.7.13}$$

(i) Using this formula for $f(X,Y)$ and likewise $g(X,Y)$, prove that

$$\text{trace}\big[f(X,Y), g(X,Y)]\big) = \frac{1}{(2\pi)^4} \int_{\mathbf{R}^4} \hat{f}(\xi,\eta)\hat{g}(\lambda,\mu),$$

$$\text{trace}\big([e^{i\xi X}e^{i\eta Y}, e^{i\lambda X}e^{i\mu Y}]\big) d\xi d\eta d\lambda d\mu. \tag{3.7.14}$$

(ii) Let $P = I + f(X,Y)$ and $Q = I + g(X,Y)$ have $PQ - I$ and $QP - I$ trace class. Show that

$$\text{trace}\,(I - QP) - \text{trace}\,(I - PQ) = \text{trace}\,\big([f(X,Y), g(X,Y)]\big). \tag{3.7.15}$$

(iii) Deduce a formula for the index of P.

In Proposition 9.4.4, we obtain a more subtle version of this result which applies to pseudo-differential operators.

3.8 Extensions of Commutative C^*-Algebras

Definition 3.8.1 (Extension) Let A and B be C^*-algebras. An extension of A by B is an exact sequence of C^*-algebras and $*$-homomorphisms

$$0 \longrightarrow B \overset{\iota}{\longrightarrow} E \overset{q}{\longrightarrow} A \longrightarrow 0. \tag{3.8.1}$$

The extension is said to be trivial if there exists a $*$-homomorphism $j: A \to E$ such that $q \circ j = id$. Then j is injective, and we can split the sequence by a $*$-homomorphism.

For instance by Exercise 3.5.4, the Toeplitz algebra arises as

$$0 \longrightarrow \mathcal{K}(H) \overset{\iota}{\longrightarrow} \mathcal{T} \overset{\varphi}{\longrightarrow} C(S^1, \mathbf{C}) \longrightarrow 0, \tag{3.8.2}$$

which is exact but does not split.

Definition 3.8.2 (Spectrum) (i) For $T \in \mathcal{L}(H)$, the spectrum $\sigma(T)$ consists of $\lambda \in \mathbf{C}$ such that $\lambda I - T$ does not have an inverse in $\mathcal{L}(H)$.

(ii) The essential spectrum $\sigma_e(T)$ consists of $\lambda \in \mathbf{C}$ such that $\lambda I - \pi(T)$ does not have an inverse in the Calkin algebra $\mathcal{C} = \mathcal{L}(H)/\mathcal{K}(H)$.

From Gelfand's Theorem 3.1.2, it follows that σ_e and σ are non-empty compact subsets of $\mathbf{C}$ such that $\sigma_e \subseteq \sigma$. By Tietze's extension theorem, the restriction map $C(\sigma; \mathbf{C}) \to C(\sigma_e; \mathbf{C}) : f \mapsto f|\sigma_e$ is a metric surjection.

Example 3.8.3 (Spectra of operators) (i) For $T \in \mathcal{L}(H)$ of finite rank, linear algebra shows that the spectrum of T is finite and consists of eigenvalues, and the non-zero eigenvalues have finite geometric multiplicity.

(ii) For $T \in \mathcal{K}(H)$ of infinite rank, the essential spectrum is $\{0\}$ and the spectrum consists of $\{0\}$ together with a finite or countably infinite sequence of eigenvalues, which has no limit point other than possibly 0.

(iii) The spectrum of a self-adjoint operator in $\mathcal{L}(H)$ is a compact subset of $\mathbf{R}$; one can show this by proving $\|((x + iy)I + T)\xi\| \geq |y|\|\xi\|$ for $x, y \in \mathbf{R}$ and $\xi \in H$. Weyl showed that the $\sigma(T) \setminus \sigma_e(T)$ consists of isolated eigenvalues of T of finite multiplicity. That is, for any self-adjoint T and compact K the spectrum of $T + K$ is equal to the spectrum of T, apart from isolated eigenvalues of finite multiplicity. So the points in the essential spectrum are either eigenvalues of infinite multiplicity, or limit points of a sequence of distinct points in the spectrum. (Be aware that the term 'essential self-adjointness' is used in the theory of differential equations for a concept which is rather different from this.)

Example 3.8.4 (Spectra of operators and algebras) (i) Let $T \in \mathcal{L}(H)$ be self-adjoint. Then the C^*-algebra generated by I and T is canonically isometrically isomorphic to $C(\sigma(T); \mathbf{C})$. This C^*-algebra tells us about $\sigma(T)$ as a topological space, whereas it tells us nothing about the spectral multiplicity of T.

(ii) Let $T \in \mathcal{L}(H)$ be self-adjoint, and let $\mathcal{E}$ be the C^*-algebra generated by $\mathcal{K}(H)$, I and T. Then $\mathcal{E}/\mathcal{K}(H)$ is a commutative unital C^*-algebra, and there is an exact sequence

$$0 \longrightarrow \mathcal{K}(H) \overset{\iota}{\longrightarrow} \mathcal{E} \overset{q}{\longrightarrow} C(\sigma_e(T); \mathbf{C}) \longrightarrow 0. \tag{3.8.3}$$

(iii) Recall that $N \in \mathcal{L}(H)$ is normal if $N^*N - NN^* = 0$. We say that $T \in \mathcal{L}(H)$ is essentially normal if $TT^* - T^*T \in \mathcal{K}(H)$. For such a T, there is an almost homomorphism $\rho \colon \mathbf{C}[z, \bar{z}] \to \mathcal{L}(H)$ modulo $\mathcal{K}(H)$

$$\sum_{j,k} a_{jk} z^j \bar{z}^k \mapsto \sum_{jk} a_{jk} T^j T^{*k}. \tag{3.8.4}$$

The crucial idea is that ρ is an almost homomorphism modulo $\mathcal{K}(H)$, and we can then determine whether ρ gives an exact sequence which can be split using a $*$-homomorphism. An obvious example of an essentially normal operator is $T = N + K$, where N is normal and K is compact. A classical problem is to determine which essentially normal T can be written as $T = N + K$ with N normal and K compact. The following result was achieved for self-adjoint operators by Weyl, von Neumann and Berg, then extended to normal operators, as in [13].

Theorem 3.8.5 (Weyl, von Neumann, Berg) *Let N_1 and N_2 be normal operators. Then there exist a unitary operator U and a compact operator K such that $N_2 = U^* N_1 U + K$, if and only if $\sigma_e(N_1) = \sigma_e(N_2)$.*

Theorem 3.8.6 *Suppose that T is essentially normal.*

(i) *Then I, T, T^* and $\mathcal{K}(H)$ generate a C^*-subalgebra $\mathcal{E}$ of $\mathcal{L}(H)$ and an extension*

$$0 \longrightarrow \mathcal{K}(H) \overset{\iota}{\longrightarrow} \mathcal{E} \overset{\pi}{\longrightarrow} C(\sigma_e(T); \mathbf{C}) \longrightarrow 0. \tag{3.8.5}$$

(ii) *The extension has a splitting $*$-homomorphism, if and only if $T = N + K$ for some normal N and compact K.*

Proof (i) We apply Gelfand's theorem to the commutative C^*-algebra $\mathcal{E}/\mathcal{K}(H)$ and use the map $\pi(T) = z$ with $\pi(T^*) = \bar{z}$ and $\pi L^\infty(H)$ to determine a $*$-homomorphism. By the Stone–Weierstrass theorem, $C(\sigma_e(T); \mathbf{C})$ is the norm closure over $\sigma_e(T)$ of the complex polynomials in z and $\bar{z}$, hence $\mathcal{E}/\mathcal{K}(H)$ is isomorphic to $C(\sigma_e(T); \mathbf{C})$.

(ii) Suppose that the extension is trivial, and let $j: C(\sigma_e(T); \mathbf{C}) \to \mathcal{E}$ be the splitting $*$-homomorphism. Then $N = j(z)$ satisfies $NN^* - N^*N = j(z)j(\bar{z}) - j(\bar{z})j(z) = 0$, so N is normal, and $\pi(T) - \pi(N) = z - \pi \circ j(z) = 0$, so $T - N = K$ for some $K \in \mathcal{K}(H)$.

Conversely, suppose that $T = N + K$. By adjusting K, we can suppose that $\sigma(N) = \sigma_e(N) = \sigma_e(T)$. Then the map $j: \sum a_{jk} z^j \bar{z}^k \mapsto \sum_{jk} a_{jk} N^j (N^*)^k$, initially defined on complex polynomials $p(z, \bar{z})$ by $j: p(z, \bar{z}) \to p(N, N^*)$, extends to define a $*$-homomorphism on $C(\sigma_e(T); \mathbf{C})$. Furthermore, $p(N, N^*) + \mathcal{K}(H) = p(T, T^*) + \mathcal{K}(H)$ holds in the Calkin algebra, so

$$\pi \circ jp(z, \bar{z}) = \pi p(N, N^*) = \pi p(T, T^*) = p(z, \bar{z}); \tag{3.8.6}$$

hence j extends to a $*$-homomorphism such that $\pi \circ j = Id$, and we have a trivial extension. □

The following result refines Proposition 3.7.4 and explains the significance of the set $\sigma(T) \setminus \sigma_e(T)$.

Theorem 3.8.7 *Let $X, Y \in \mathcal{L}(H)$ be self-adjoint such that $[X, Y] \in \mathcal{L}^1(H)$, and let $T = X + iY$, so T is essentially normal. Then there exists a measure μ on $\sigma(T)$ such that*

$$\operatorname{trace}\big[f(X,Y),g(X,Y)]\big) = \iint_{\sigma(T)} \frac{\partial(f,g)}{\partial(q,p)}\mu(dqdp) \qquad (f,g \in \mathbf{C}[x,y]) \tag{3.8.7}$$

such that μ is a constant multiple of Lebesgue area measure on each connected component of $\sigma(T) \setminus \sigma_e(T)$, namely

$$d\mu = \frac{-1}{2\pi i}\operatorname{ind}(T - \lambda I)dqdp \qquad (\lambda \in \sigma(T) \setminus \sigma_e(T)). \tag{3.8.8}$$

Proof This was obtained in certain cases by Pincus, and in generality by Helton and Howe; see [51, 52]. □

Example 3.8.8 (Toeplitz algebra) Section 3.6 leads to a significant instance in which the K-theory is non-trivial. There is an exact sequence of C^*-algebras and $*$-homomorphisms

$$0 \longrightarrow \mathcal{K}(H_+) \longrightarrow \mathcal{T} \longrightarrow C(S^1;\mathbf{C}) \longrightarrow 0,$$

where $\mathcal{T}$ is the unital C^*-subalgebra generated by the Toeplitz operators T_f for $f \in C(S^1, \mathbf{C})$ and $\mathcal{K}(H_+)$. The map $f \mapsto T_f$ is a homomorphism modulo $\mathcal{K}(H_+)$. For $f \in C(S^1, \mathbf{C})$ such that $1/f \in C(S^1, \mathbf{C})$, the Toeplitz operator T_f is Fredholm with index equal to minus the winding number of f. The extension is trivial if and only if the K_0-class is zero. In particular,

$$K_0(C(S^1;\mathbf{C})) \cong H^1(S^1;\mathbf{Z}) \cong \mathbf{Z}. \tag{3.8.9}$$

Example 3.8.9 (K_0 is not exact) [90, Ex. 4.3.6] There is an exact sequence of commutative algebras and algebra homomorphisms

$$0 \longrightarrow C_0(0,1) \longrightarrow C[0,1] \longrightarrow \mathbf{C} \oplus \mathbf{C} \longrightarrow 0; \tag{3.8.10}$$

whereas $K_0(C[0,1]) = \mathbf{Z}$ and $K_0(\mathbf{C} \oplus \mathbf{C}) = \mathbf{Z} \oplus \mathbf{Z}$; hence $K_0(\cdot)$ is not an exact functor. The algebra $\mathbf{C} \oplus \mathbf{C}$ has a non-trivial idempotent, and features in Example 7.4.3.This issue is resolved by the six-term cyclic exact sequence which incorporates Bott periodicity. Let

$$0 \longrightarrow J \longrightarrow R \longrightarrow A \longrightarrow 0 \tag{3.8.11}$$

be an exact sequence of C^*-algebras. Then there is a cyclic exact sequence

$$\begin{array}{ccccc} K_0(J) & \longrightarrow & K_0(R) & \longrightarrow & K_0(A) \\ \uparrow & & & & \downarrow \\ K_1(A) & \longleftarrow & K_1(R) & \longleftarrow & K_0(J) \end{array} \tag{3.8.12}$$

3.9 Idempotents and Generalized Toeplitz Operators

Let H be a complex separable Hilbert space and $e \in \mathcal{L}(H)$ an idempotent. Then for $a \in \mathcal{L}(H)$, the associated Toeplitz operator is $eae\colon eH \to eH$. Suppose further that eH and $(1-e)H$ are both infinite dimensional, and let $J \in \mathcal{L}(H)$ satisfy $J^2 = 1$ and $J\colon eH \to (1-e)H$ is a unitary equivalence. Then the associated Hankel operator is $eaJe\colon eH \to eH$. We have

$$a = \begin{bmatrix} eae & ea(1-e) \\ (1-e)ae & (1-e)a(1-e) \end{bmatrix} \quad \text{acting on} \quad \begin{matrix} eH \\ (1-e)H \end{matrix}. \tag{3.9.1}$$

Example 3.9.1 (Hankel and Wiener–Hopf integral operators) Let $H = L^2(\mathbf{R})$ and $eH = L^2(0,\infty)$ for $e\colon f(x) \mapsto \mathbf{I}_{(0,\infty)}(x)f(x)$. Let $J\colon H \to H$ be the flip map $Jf(x) = f(-x)$. Then given $k \in C_c^\infty(\mathbf{R};\mathbf{C})$ introduce the following operators on $L^2(0,\infty)$

(i) the convolution operator

$$a\colon f(x) \mapsto \int_{-\infty}^{\infty} k(x-y)f(y)\,dy;$$

(ii) the Wiener–Hopf (or Toeplitz) integral operator

$$eae\colon f(x) \mapsto \int_0^\infty k(x-y)f(y)\,dy; \tag{3.9.2}$$

(iii) the Hankel operator

$$eaJe\colon f(x) \mapsto \int_0^\infty k(x+y)f(y)\,dy. \tag{3.9.3}$$

Historically, the theories of Wiener–Hopf and Toeplitz operators developed in parallel before they were unified.

4

GNS Algebra

4.1 Idempotents and Dilations

The results of Chapter 3 can be generalized in two different ways:

(i) The GNS construction, which develops $\rho\colon A \to R$ of the form $\rho(a) = eae$, for a wider class of algebras than the extensions in Section 3.8;

(ii) the index can be introduced using K_0.

This chapter is concerned with algebraic versions of the Gelfand–Naimark–Segal construction. We review the original GNS result and Stinespring's extension in the theory of C^*-algebras, and prove the algebraic version. Let A and B be unital algebras over $\mathbf{k}$, and $\rho\colon A \to B$ a linear map such that $\rho(1_A) = 1_B$; such a ρ is said to be unital or based. As in previous examples, such a ρ can be produced by the following construction. Let R be an algebra with idempotent e, so $e^2 = e$, and let $u\colon A \to R$ be a unital algebra homomorphism. Then $B = eRe$ defines an algebra with unit $1_B = e$, but B is generally not a unital subalgebra of R. We define $\rho\colon A \to B$ by $\rho(a) = eu(a)e$ to obtain a linear $\rho\colon A \to B$ such that $\rho(1_A) = 1_B$. Can any linear $\rho : A \to B$ such that $\rho(1_A) = 1_B$ be produced in this way? This process is known as dilation. The following diagram is a counterpart to the diagram for projective modules in Section 2.10:

$$\begin{array}{ccc} u & & R \\ & \nearrow & \downarrow \\ A & \longrightarrow & B \\ & \rho & \downarrow \\ & & 0 \end{array} \qquad (4.1.1)$$

Here we can choose R, but we are required to make u a homomorphism. In analysis, one speaks of dilating the linear map ρ to a homomorphism or representation of A. A $*$ representation on Hilbert space is a homomorphism $\pi : A \to \mathcal{L}(H)$ such that $\pi(a^*) = \pi(a)^*$. This journey is not straightforward, since there is a subtle difference between liftings and dilations, and [30] contains several unanswered questions about modules over function algebras which follow from the precise distinction.

The chapter begins with discussion of the Gelfand–Naimark–Segal theorem which characterizes states on a C^*-algebra. Then there are proofs of the Stinespring dilation theorem for completely positive maps, and its extension by Kasparov to Hilbert modules over C^*-algebras. Quillen proved an abstract GNS theorem, which provides a framework for all of these results, following his reading of the C^* version in [10]. The crucial idea is to construct for any algebra A and unital linear map $\rho\colon A \to B$, a GNS(ρ) algebra which is characterized by universal mapping properties.

The notion of a Fredholm module is important in cyclic theory, especially when one constructs significant examples of cyclic cocycles. Fredholm modules are introduced at the end of this chapter, alongside the definition of the Cuntz algebra QA.

4.2 GNS Theorem for States on a C^*-Algebra

As motivation, we consider the case $\mathbf{k} = \mathbf{C}$ and review the original Gelfand–Naimark–Segal construction.

Definition 4.2.1 (State) Let A be a C^*-algebra. A state is a linear functional $\phi\colon A \to \mathbf{C}$ such that

$$\phi(a^*a) \geq 0 \qquad (a \in A) \tag{4.2.1}$$

and is normalized so that $\sup\{|\phi(a)|\colon \|a\| \leq 1\} = 1$. One can show that any C^*-algebra A can be isometrically embedded in a unital C^*-algebra A^+ that is unital, and that any state on A extends to a state on A^+ such that $\phi(1) = 1$.

Proposition 4.2.2 *Let A be a unital C^*-algebra and let $\phi\colon A \to \mathbf{C}$ be a state. Then there exist a Hilbert space E, a homomorphism $\pi : A \to \mathcal{L}(E)$ such that $\pi(1_A) = I$ and $\pi(a^*) = \pi(a)^*$, and a unit vector $v \in E$ such that*

$$\phi(a) = \langle v \mid \pi(a)v\rangle_E \qquad (a \in A). \tag{4.2.2}$$

Proof Note that $\phi(1_A) = 1$. We let $M = A \otimes A$ and introduce a product $M \times M \to M$ by defining

$$(a_1, b_1) \cdot (a_2, b_2) = (a_1, \phi(b_1 a_2) b_2) \tag{4.2.3}$$

and extending this linearly to obtain an associative complex algebra M which is an A-bimodule, and $M \otimes_A M \to M$ is an A-bimodule map. This is clear by inspection. We write $\langle u \mid v \rangle = \phi(u^* v)$ which is linear in v and conjugate linear in u, and $\langle u \mid u \rangle = \phi(u^* u) \geq 0$ since ϕ is a state.

Observe that $E_0 = \{u \in A : \phi(u^* u) = 0\}$ is a left A module since for all $a \in A$, where exists $w \in A$ such that $w^* w = \|a\|^2 1_A - a^* a$ and $0 \leq \phi(w^* w)$, so for $u \in E_0$,

$$0 \leq \phi(u^* a^* a u) \leq \|a\|^2 \phi(u^* u) \leq 0 \tag{4.2.4}$$

hence $au \in E_0$. Let E be the Hilbert space that is formed by completing A/E_0 so there are natural inclusion maps

$$0 \to E_0 \to A \to E. \tag{4.2.5}$$

Let v be the image of 1_A, and define $\pi : A \to \mathcal{L}(E)$ by $\pi(a)w = aw$, so $\pi(1_A) = 1_{\mathcal{L}(E)}$ and $\pi(a^*) = \pi(a)^*$ for the usual adjoint on $\mathcal{L}(E)$ since

$$\langle w \mid \pi(a^*) v \rangle_E = \phi(w^* a^* u) = \overline{\phi(u^* a w)} = \overline{\langle u \mid \pi(a) w \rangle_E} = \langle \pi(a) w \mid u \rangle_E. \tag{4.2.6}$$

Let $\iota : \mathbf{C} \to E$ be $t \mapsto tv$ and $\iota^* : E \to \mathbf{C}$ be $w \mapsto \langle v \mid w \rangle_E$; then $\iota^* \iota = Id$.

Thus the state ϕ is given by a $*$-representation on Hilbert space and a unit vector state. □

Theorem 4.2.3 (Gelfand–Naimark) *Let A be a unital C^*-algebra. Then there exists a Hilbert space E, and a unital isometric $*$-homomorphism $\pi : A \to \mathcal{L}(E)$.*

Proof Let $a \in A$ have $\|a\| = 1$, and consider the commutative and unital C^*-subalgebra $B = C^*\{a^a, 1\}$ generated by 1 and $a^* a$ in A. Then $B \cong C(\sigma(a^* a); \mathbf{C})$ where $\sigma(a^* a) \subseteq [0,1]$ contains 1 since $1 = \|a\|$. Then we introduce a normalized state $\tilde{\phi} : B \to \mathbf{C}$ such that $1 = \tilde{\phi}(a^* a) = \tilde{\phi}(1)$. By the Hahn–Banach theorem, $\tilde{\phi}$ extends to a state $\phi : A \to \mathbf{C}$ such that $1 = \phi(1)$.

By Proposition 4.2.2, there exists a Hilbert space E_ϕ a unit vector $v_\phi \in E_\phi$ and a $*$-representation $\pi_\phi : A \to \mathcal{L}(E_\phi)$ such that $\|\pi_\phi(a)\| = \|a\|$ and $\phi(b) = \langle v_\phi \mid \pi_\phi(b) v_\phi \rangle_{E_\phi}$ for all $b \in A$. Now let $E = \oplus_{\phi \in S(A)} E_\phi$ be the Hilbert space that arises from taking the orthogonal sum over all normalized states on A, and let $\pi = \oplus_{\phi \in S(A)} \pi_\phi$ be the diagonal sum of the $*$-representations over all normalized states on A. Now π is a unital $*$-representation, hence

$\|\pi(a)\| \leq \|a\|$ for all $a \in A$, and $\|a\| = \|\pi_\phi(a)\| \leq \|\pi(a)\|$ by the choice of ϕ; hence π is isometric. □

Exercise 4.2.4 (Separable case of GNS) Suppose that A is a unital C^*-algebra that is separable for the norm topology. Show that E in Theorem 4.2.3 can be replaced by a separable Hilbert space.

4.3 GNS Algebra

In the previous section, we introduced a new algebra M from A and represented A as multiplication operators on this space. In this section we follow through the process more generally. Let $\mathbf{k}$ have characteristic zero, and let $\rho: A \to B$ be a unital $\mathbf{k}$-linear map between unital $\mathbf{k}$-algebras. We introduce the free A-bimodule $M = A \otimes_{\mathbf{k}} B \otimes_{\mathbf{k}} A$, and define a product $M \times M \to M$ by

$$(a_1, b_1, c_1) \cdot (a_2, b_2, c_2) = (a_1, b_1\rho(c_1a_2)b_2, c_2) \tag{4.3.1}$$

and extending linearly.

Lemma 4.3.1 *Then M is an associative (non-unital) algebra over* $\mathbf{k}$*, and an A-bimodule such that the product $M \otimes_A M \to M$ is an A-bimodule map.*

Proof This boils down to checking that

$$\big(a_1'(a_1, b_1, c_1)a_1''\big) \cdot \big(a_2(a_2, b_2, c_2)a_2''\big) = \big(a_1'a_1, b_1\rho(c_1a_1''a_2'a_2)b_2, c_2a_2''\big) \tag{4.3.2}$$

makes sense however one interprets the expressions. Then we define $R = \mathrm{GNS}(\rho)$ to be the semi-direct product $R = A \oplus M$ with $\hat{u}: A \to R$: $\hat{u}(a) = (a, 0)$ and $\hat{v}: B \to R$: $\hat{v}(b) = (0, 1 \otimes b \otimes 1)$. The purpose of $\hat{v}$ is to provide a splitting map then extended to normal operators for $R \to B$. Let $e = (0, \hat{e}) \in R$ where $\hat{e} = (1, 1, 1) \in M$. Then $e^2 = e$, $e\hat{u}(a)e = \hat{v}(\rho(b))$ and $eRe \cong B$. Indeed, we have

$$\hat{e}^2 = (1, 1, 1) \cdot (1, 1, 1) = (1, 1\rho(1 \cdot 1)1, 1) = (1, 1, 1) = \hat{e}; \tag{4.3.3}$$

$$\hat{e}(a\hat{e}) = (1, 1, 1) \cdot (a, 1, 1) = (1, \rho(a), 1) \in 1 \otimes B \otimes 1; \tag{4.3.4}$$

and

$$\begin{aligned}\hat{e}(a, b, c)\hat{e} &= (1, 1, 1) \cdot (a, b, c) \cdot (1, 1, 1)\\ &= (1, \rho(a)b, c) \cdot (1, 1, 1)\\ &= (1, \rho(a)b\rho(c), 1) \in 1 \otimes B \otimes 1.\end{aligned} \tag{4.3.5}$$

Hence

$$eRe = (0, \hat{e}M\hat{e}) \cong (0, 1 \otimes B \otimes 1) \cong B, \tag{4.3.6}$$

and

$$\begin{aligned} a \mapsto & e\hat{u}(a)e \\ &= \big((0,\hat{e})(a,0)\big)(0,\hat{e}) \\ &= (0, \hat{e}a\hat{e}) \\ &= \big(0, (1, \rho(a), 1)\big), \end{aligned} \tag{4.3.7}$$

so $a \mapsto \rho(a)$ under this isomorphism. □

Proposition 4.3.2 (Universal property of GNS(ρ)) *Let A, B and S be unital* **k***-algebras, let $u\colon A \to S$ be an algebra homomorphism and let $v\colon B \to S$ be a linear map satisfying*

$$v(b_1)u(a)v(b_2) = v(b_1\rho(a)b_2) \qquad (a \in A; b_1, b_2 \in B) \tag{4.3.8}$$

if and only if

$$v(b_1b_2) = v(b_1)v(b_2) \quad \text{and} \quad v(1)u(a)v(1) = v(\rho(a)) \quad (a \in A; b_1, b_2 \in B). \tag{4.3.9}$$

Then there exists a unique algebra homomorphism $\omega : GNS(\rho) \to S$ carrying $(\hat{u}, \hat{v})$ to (u, v) and $\hat{v}(b) = (1, b, 1)$.

Proof Let $\omega(a, 0) = u(a)$ and $\omega(0, (a, b, c)) = u(a)v(b)u(c)$, then check that this works. □

4.4 Stinespring's Theorem

Let H be a Hilbert space and $\mathcal{L}(H)$ the space of bounded linear operators on H with the operator norm $\|T\| = \sup\{|\langle \eta \mid T\xi\rangle_H| : \xi, \eta \in H; \|\xi\|, \|\eta\| \leq 1\}$. Then $\mathcal{L}(H)$ is a C^*-algebra. Let $M_n(\mathcal{L}(H)) = M_n \otimes \mathcal{L}(H)$ be the space of $n \times n$ matrices with entries in $\mathcal{L}(H)$ so that a typical element is $T = [T_{jk}]$. We form the Hilbert space $\mathbf{C}^n \otimes H$ of $n \times 1$ column vectors with entries in H, and let $M_n(\mathcal{L}(H))$ operate on $\mathbf{C}^n \otimes H$ by having $[T_{jk}]$ multiply columns on the left. The operator norm and adjoint $*$ are defined so $[T_{jk}]^* = [T_{kj}^*]$. This makes $M_n(\mathcal{L}(H)) = \mathcal{L}(\mathbf{C}^n \otimes H)$ into a C^*-algebra; indeed, it is the only way to do so.

Let A be a C^*-algebra. By Theorem 4.2.3, there exists a Hilbert space H_0 and a unital $*$-homomorphism $\pi : A \to \mathcal{L}(H_0)$ such that π is an isometry; thus

A may be regarded as a unital subalgebra of $\mathcal{L}(H_0)$. Let $M_n(A)$ be the space of $n \times n$ matrices with entries in A; then $M_n(A)$ is also a C^*-algebra, with the norm and adjoint inherited from $M_n(\mathcal{L}(H_0))$.

Let $\Phi\colon A \to \mathcal{L}(H)$ be a linear map; we say that Φ is unital if $\Phi(1) = I$. Then we dilate to $\Phi_n\colon M_n(A) \to M_n(\mathcal{L}(H))$ by

$$\Phi_n\colon [a_{jm}] \to [\Phi(a_{jm})]. \tag{4.4.1}$$

Definition 4.4.1 (Completely positive) Let A be a unital C^*-algebra and $\Phi\colon A \to \mathcal{L}(H)$ be a linear map. Say that Φ is positive if $\langle \xi \mid \Phi(a^*a)\xi\rangle_H \geq 0$ for all $a \in A$ and $\xi \in H$. Moreover, we say that Φ is n-positive if $\Phi_n\colon M_n(A) \to M_n(\mathcal{L}(H))$ is positive for some $n = 1, 2, \dots$. If Φ is n-positive for all $n = 1, 2, \dots,$ then we say that Φ is completely positive. See [82] for more discussion.

Proposition 4.4.2 (Stinespring's dilation theorem) *Let A be a unital C^*-algebra and $\Phi\colon A \to \mathcal{L}(H)$ be unital and completely positive. Then there exist a Hilbert space E, a $*$-homomorphism $\pi\colon A \to \mathcal{L}(H)$, and a bounded linear map $V\colon H \to E$ such that*

$$\rho(a) = V^*\pi(a)V \qquad (a \in A). \tag{4.4.2}$$

Proof Let $E_1 = A \otimes H$ and consider the sesquilinear form on E_1 defined by

$$\begin{aligned}\left\langle \sum_{j=1}^n a_j \otimes \xi_j \,\middle|\, \sum_{\ell=1}^n b_\ell \otimes \eta_\ell \right\rangle &= \sum_{j,\ell=1}^n \langle \xi_j \mid \rho(a_j^* b_\ell)\eta_\ell\rangle_H \\ &= \left\langle \begin{bmatrix}\xi_1\\ \vdots \\ \xi_n\end{bmatrix} \,\middle|\, \begin{bmatrix}\rho(a_j^* b_\ell)\end{bmatrix} \begin{bmatrix}\eta_1\\ \vdots \\ \eta_n\end{bmatrix}\right\rangle_{\mathbf{C}^n \otimes H}.\end{aligned} \tag{4.4.3}$$

Note that $\langle \sum_{j=1}^n a_j \otimes \xi_j \mid \sum_{j=1}^n a_j \otimes \xi_j\rangle \geq 0$ since $[a_j^* a_k] \in M_n(A)$ has $[a_j^* a_k] \geq 0$ and hence $[\rho(a_j^* a_k)] = \rho_n([a_j^* a_k]) \geq 0$.

Also, E_1 is a left A-module for the operation $A \otimes E_1 \to E_1\colon (a, \sum_j a_j \otimes \xi_j) \mapsto \sum_j a a_j \otimes \xi_j$. Let E_0 be the left A-submodule of E_1 given by

$$E_0 = \left\{ \sum_{j=1}^n a_j \otimes \xi_j : \left\langle \sum_{j=1}^n a_j \otimes \xi_j \,\middle|\, \sum_{j=1}^n a_j \otimes \xi_j \right\rangle = 0 \right\}; \tag{4.4.4}$$

then let E be the completion of E_1/E_0 for the sesquilinear form, so that E is a Hilbert space and

$$0 \to E_0 \to E_1 \to E \to 0$$

is an exact sequence of left A-modules, up to the completion. We introduce $\pi: A \to \mathcal{L}(E)$ by $\pi(a)\colon \sum_{j=1}^n a_j \otimes \xi_j = \sum_{j=1}^n aa_j \otimes \xi_j$, which gives a unital $*$-homomorphism. Then with $V\colon H \to E\colon \xi \mapsto 1 \otimes \xi$ we obtain a linear map, which is an isometry since

$$\langle V\xi \mid V\xi \rangle_E = \langle \xi \mid \rho(1)\xi \rangle_H = \langle \xi \mid \xi \rangle_H,$$

and $V^*\pi(a)V = \rho(a)$ since

$$\langle V\eta \mid \pi(a)V\xi \rangle_E = \langle \eta \mid \rho(a)\xi \rangle_H. \tag{4.4.5}$$

Note that $V^*V = 1_{\mathcal{L}(H)}$ so $e = VV^*$ is an idempotent in $\mathcal{L}(E)$, indeed an orthogonal projection onto a subspace of E. □

Theorem 4.4.3 *Let A be a commutative, unital and norm separable C^*-algebra, and let $\mathcal{E}$ be an extension of A by $\mathcal{K}$, so*

$$0 \longrightarrow \mathcal{K}(H) \overset{\iota}{\longrightarrow} \mathcal{E} \overset{q}{\longrightarrow} A \longrightarrow 0. \tag{4.4.6}$$

Then there exists a linear lifting map $\rho\colon A \to \mathcal{L}(H)$ such that $q \circ \rho = id$ and ρ is a homomorphism modulo $\mathcal{K}(H)$.

Proof By [3, p. 50], there exists a compact metric space X such that A is isometrically $*$ isomorphic to $C(X;\mathbf{C})$. This is a variant of theorem 1.23 of [13]. Let X be a compact metric space and $\mathcal{E}$ an extension of $C(X;\mathbf{C})$ by $\mathcal{K}(H)$, as in the following diagram.

$$\begin{array}{ccccccccc} 0 & \longrightarrow & \mathcal{K}(H) & \overset{\iota}{\longrightarrow} & \mathcal{E} & \overset{q}{\longrightarrow} & C(X;\mathbf{C}) & \longrightarrow & 0 \\ & & =\downarrow & & & \swarrow_{\rho} & & & \\ 0 & \longrightarrow & \mathcal{K}(H) & \longrightarrow & \mathcal{L}(H) & \longrightarrow & \mathcal{C} & \longrightarrow & 0 \end{array} \tag{4.4.7}$$

There exists a unital positive $\rho\colon C(X\colon \mathbf{C}) \to \mathcal{L}(H)$, which hence is completely positive since the domain is a commutative algebra. So by Stinespring's dilation theorem, there exist a Hilbert space E, an isometric embedding $V\colon H \to E$ and a $*$-representation $\pi: C(X;\mathbf{C}) \to \mathcal{L}(E)$ such that $\rho(f) = V^*\pi(f)V$. In terms of matrices, we have

$$\pi(f) = \begin{bmatrix} \rho(f) & \pi_{12}(f) \\ \pi_{21}(f) & \pi_{22}(f) \end{bmatrix} \quad \text{acting on} \quad \begin{matrix} H \\ E \ominus H' \end{matrix} \tag{4.4.8}$$

where $\pi_{12}(\bar{f}) = \pi_{21}(f)^*$. Now $\pi(fg) = \pi(f)\pi(g)$, so

$$\rho(fg) = \rho(f)\rho(g) + \pi_{12}(f)\pi_{21}(g); \tag{4.4.9}$$

in particular, $\rho(f\bar{f}) - \rho(f)\rho(\bar{f}) = \pi_{12}(f)\pi_{12}(f)^*$ is compact, so $\pi_{12}(f)$ is compact. Hence ρ is a homomorphism modulo the compact operators, with curvature

$$\omega(f,g) = \rho(fg) - \rho(f)\rho(g) = \pi_{12}(f)\pi_{21}(g). \tag{4.4.10}$$

The crucial point is that the extension gives rise to a linear splitting map which is a homomorphism modulo an ideal. In later sections, we use this idea repeatedly. □

Exercise 4.4.4 (Bass–Whitehead group) Let A be a unital Banach subalgebra of $\mathcal{L}(H)$.

(i) Show that $M_n \otimes A$ is a unital Banach subalgebra of $M_n \otimes \mathcal{L}(H)$.

(ii) Let $GL_n(A)$ be the set of $u \in M_n(A)$ such that $uv = uv = I$ for some $v \in M_n(A)$. Show that $GL_n(A)$ is an open subset of $M_n(A)$, and a multiplicative group.

(iii) Let $[GL_n(A), GL_n(A)] = \langle\{uvu^{-1}v^{-1} : u, v \in GL_n(A)\}\rangle$ be the subgroup of $GL_n(A)$ generated by the commutators. Show that

$$GL_n(A)/[GL_n(A), GL_n(A)]$$

is an abelian group.

(iv) By considering the map $GL_n(A), \to GL_{n+1}(A)$ as in (2.9.7), show that $GL(A) = \cup_{n=1}^{\infty} GL_n(A)$ is a group, and hence define the Bass–Whitehead group

$$K_1(A) = GL(A)/[GL(A), GL(A)].$$

See [68, p. 343] for further discussion.

4.5 The Generalized Stinespring Theorem

A Hilbert module over a C^*-algebra B consists of a complex vector space E and right module operation $E \times B \to E$ given by $(\xi, b) \mapsto \xi b$; when B is unital we also require $\xi 1_B = \xi$ for all $\xi \in E$. Furthermore, we suppose that there is a product $\langle \cdot \mid \cdot \rangle : E \times E \to B$, so that for all $f, g, h \in E$ and $\mu, \lambda \in \mathbf{C}$

(i) $\langle f \mid g \rangle = \langle g \mid f \rangle^*$;
(ii) $\langle f \mid \mu g + \lambda h \rangle = \mu \langle f \mid g \rangle + \lambda \langle f \mid h \rangle$;
(iii) $\langle h \mid h \rangle \geq 0$, and $\langle h \mid h \rangle = 0 \Rightarrow h = 0$;
(iv) $\langle f \mid gb \rangle = \langle f \mid g \rangle b$ for all $b \in B$;
(v) E is complete for the metric associated with $\|f - g\| = \langle f - g \mid f - g \rangle^{1/2}$.

Evidently, we can define the notion of positive, n-positive and completely positive for maps between C^*-algebras. See [65] for details.

The following is a version of Stinespring's theorem suitable for Hilbert C^*-modules.

Proposition 4.5.1 *Let A and B be C^*-algebras, let E be a Hilbert B-module and $\rho\colon A \to \mathcal{L}(E)$ a unital completely positive map. Then there exists a Hilbert B-module E_ρ, a $*$-homomorphism $\pi : A \to \mathcal{L}(E_\rho)$, and a linear map $V\colon E \to E_\rho$ such that $VV^* \in \mathcal{L}(E_\rho)$ is a self-adjoint idempotent,*

$$\rho(a) = V^*\pi(a)V \tag{4.5.1}$$

and $A \otimes B \otimes A$ acts on E_ρ via $(a,b,c) \mapsto \pi(a)VbV^*\pi(c)$.

Proof Here $A\otimes E$ is an $A\otimes B$ module for the operation $a\otimes b\colon c\otimes\xi \to ac\otimes\xi b$, so that B operates on the right of E and has the B-valued inner product

$$\langle c \otimes \eta \mid a \otimes \xi\rangle = \langle \eta \mid \rho(c^*a)\xi\rangle_E, \tag{4.5.2}$$

with $\pi(a) \in \mathcal{L}(A \otimes E)$ given by $\pi(a)(\alpha \otimes \xi) = a\alpha \otimes \xi$. Let E_0 be the right B-module given by

$$E_0 = \left\{\sum_{j=1}^{n} a_j \otimes \xi_j : \left\langle \sum_j a_j \otimes \xi_j \,\middle|\, \sum_j a_j \otimes \xi_j \right\rangle = 0\right\}, \tag{4.5.3}$$

and let E_ρ be the completion of $(A \otimes E)/E_0$ for this sesquilinear form. Let $V\colon E \to E_\rho\, \xi \mapsto 1\otimes\xi$ with $V^*(a\otimes\xi) = \rho(a)\xi$ so that V is a right B-module map with $V^*V = 1$, so we regard E as a subspace of E_ρ. Also note that

$$\langle V\eta \mid \pi(a)V\xi\rangle = \langle 1 \otimes \eta \mid a \otimes \xi\rangle = \langle \eta \mid \rho(a)\xi\rangle, \tag{4.5.4}$$

so $V^*\pi(a)V = \rho(a)$.

Let $M = A \otimes B \otimes A$. Then the GNS$(\rho)$ algebra $R = A \oplus M$ has a left module action on $A \otimes E$ via M for

$$(a,b,c)\colon \alpha \otimes \xi \mapsto a \otimes \rho(c\alpha)\xi b \tag{4.5.5}$$

and we have

$$(a,b,c)(\alpha \otimes \xi) = \pi(a)VbV^*(c\alpha \otimes \xi); \tag{4.5.6}$$

thus $(a,b,c) \mapsto \pi(a)VbV^*\pi(c)$. Now VV^* is a self-adjoint idempotent in $\mathcal{L}(E_\rho)$, namely the orthogonal projection with range E, and ρ is identified with the compression of the $*$-representation π to the complemented submodule E of E_ρ. The triple (E_ρ, V, V^*) is associated with a factorization of $A \otimes B^{op}$ modules $\tilde{\rho}\colon A \otimes B \to \mathrm{Hom}(A,B)\colon\ \tilde{\rho}(a \otimes b)(\alpha) = \rho(a\alpha)b$. In the context of C^*-algebras, one usually wishes to specify the operation of the adjoint on the maps in question. □

4.6 Uniqueness of GNS(ρ)

Proposition 4.6.1 *Given (R, ρ, u, v), where R is an algebra, e is an idempotent in R, $u: A \to R$ is an algebra homomorphism and $v: B \to eRe$ is an algebra homomorphism such that $v(\rho(a)) = eu(a)e$. Then there exists a unique algebra homomorphism $\phi: GNS(\rho) \to R$ such that ϕ takes $(\hat{e}, \hat{u}, \hat{v})$ to (e, u, v).*

Proof Uniqueness follows from $\phi(a) = u(a)$ and $\phi(a, b, c) = u(a)v(b)u(c)$. □

4.7 Projective Hilbert Modules

In Lemma 2.9.4, we characterized the finitely generated projective $\mathbf{C}[X]$ modules. The infinite-dimensional continuous version is more complicated, as we discuss now. There are three main issues: ensuring that various maps are defined everywhere; ensuring subspaces are the images of projections; ensuring that certain maps commute with module operations. In a Hilbert space, every closed linear subspace is the range of a bounded linear projection operator, namely the orthogonal projection. No such result holds for general Banach spaces. So it is natural to consider the modules that are Hilbert spaces. For a given Banach algebra $\mathcal{A}$ and Hilbert space H, it can be a difficult issue to produce any non-zero projective module operation of $\mathcal{A}$ on H. In this section, we produce projective Hilbert modules over the space of continuous functions on the circle, and the disc algebra; these correspond to significant results in operator theory, but leave some questions open. We also provide examples of the Stinespring dilation theorem and GNS(ρ).

Definition 4.7.1 (i) *(Hilbert module)* Let $\mathcal{A}$ be a unital complex Banach algebra and H a separable Hilbert space. We say that H is a Hilbert module over $\mathcal{A}$ if there is a bilinear map $\mathcal{A} \times H \to H$ $(a, \xi) \mapsto \Phi(\xi)$ that obeys the axioms for a left module, and which is continuous in the norm topology on $\mathcal{A} \times H$. By the uniform boundedness theorem, this is equivalent to having a unital homomorphism $\Phi: \mathcal{A} \to \mathcal{L}(H)$ and a constant K such that

$$\|\Phi(a)\xi\|_H H \leq K\|a\|_{\mathcal{A}}\|\xi\|_H \qquad (a \in \mathcal{A}, \xi \in H). \tag{4.7.1}$$

When $K = 1$, we say that the module is contractive.

(ii) *(Projective Hilbert module)* A Hilbert module H_0 is projective if for every pair of Hilbert modules H_1 and H_2, and all continuous A-module maps

$\psi : H_0 \to H_1$ and $\phi : H_2 \to H_1$, where ϕ is surjective, there exists a continuous A-module map $\tilde{\psi} : H_0 \to H_2$ such that $\phi \circ \tilde{\psi} = \psi$.

Let S^1 be the unit circle and U be the bilateral shift, the unitary operator on $L^2(S^1, d\theta/2\pi)$ which is realized by multiplication as $Uf(z) = zf(z)$ for $f \in L^2(S^1, d\theta/2\pi)$ and $z \in S^1$. Note that S^1 is the dual group of $\mathbf{Z}$, and $\ell^2(\mathbf{Z})$ is naturally isomorphic to $L^2(S^1, d\theta/2\pi)$ via $(a_j) \mapsto \sum_{j=-\infty}^{\infty} a_j z^j$. Thus the C^*-algebra generated by U can be interpreted as $C(S^1; \mathbf{C})$, the continuous complex-valued functions on the circle, or as $C^*(\mathbf{Z})$, namely the C^*-algebra generated by the left regular representation of the group $\mathbf{Z}$ on $\ell^2(\mathbf{Z})$. A state $\phi : C(S^1; \mathbf{C}) \to \mathbf{C}$ is equivalent to a probability measure on S^1, as in

$$\phi(f) = \int_{S^1} f(z)\mu(dz). \tag{4.7.2}$$

Proposition 4.7.2 (Sz. Nagy) *Let $\mathcal{A} = C(S^1; \mathbf{C})$.*

(i) *For all $U \in \mathcal{U}(H)$ the map $f \mapsto f(U)$ determines a contractive Hilbert module.*

(ii) *Let $\Phi : \mathcal{A} \to \mathcal{L}(H)$ give a Hilbert module on H. Then there exists $U \in \mathcal{U}(H)$ and an invertible $V \in \mathcal{L}(H)$ such that $\Phi(f) = Vf(U)V^{-1}$.*

Proof (i) We define $\Phi(z) = U$ and extend to a map $\Phi(\sum_{j=-N}^{N} a_j z^j) = \sum_{j=-N}^{N} a_j U^j$. Then by the spectral theorem, for all $\xi, \eta \in H$ with $\|\xi\|, \|\eta\| = 1$, there exists a measure with norm one such that

$$\langle f(U)\xi, \eta\rangle = \int_{S^1} f(e^{i\theta})\mu_{\xi,\eta}(d\theta), \tag{4.7.3}$$

hence the result.

(ii) With $\Phi(z) = T$ we have $\Phi(z)\Phi(1/z) = \Phi(1/z)\Phi(z) = I$, so T is invertible, and $\|T^n\| = \|\Phi(z^n)\| \le K$, for all integers n. We introduce the norm on H by applying the Banach limit

$$\|\xi\|_{H_1} = \mathrm{LIM}_n \|T^n \xi\|_H, \tag{4.7.4}$$

so that

$$K^{-1}\|\xi\|_H \le \|\xi\|_{H_1} \le K\|\xi\|_H \tag{4.7.5}$$

and the parallelogram law

$$\|\eta + \xi\|_{H_1}^2 + \|\eta - \xi\|_{H_1}^2 = 2\|\eta\|_{H_1}^2 + 2\|\xi\|_{H_1}^2; \tag{4.7.6}$$

so H_1 gives a Hilbert space norm on H. Let $V : H \to H_1$ be the identity map $\xi \mapsto \xi$ which is invertible. Also

$$\|T\xi\|_{H_1} = \mathrm{LIM}_n \|T^{n+1}\xi\|_H = \mathrm{LIM}_n \|T^{n+1}\xi\|_H = \|\xi\|_{H_1}, \tag{4.7.7}$$

so T is an invertible isometry, hence a unitary U on H_1. We deduce that $T = V^{-1}UV$ and

$$\Phi\left(\sum_{j=-N}^{N} a_j z^j\right) = V^{-1}\left(\sum_{j=-N}^{N} a_j U^j\right) V^{-1}, \qquad (4.7.8)$$

as required. □

Corollary 4.7.3 *All Hilbert modules over $C(S^1;\mathbf{C})$ are projective.*

Proof By Proposition 4.7.2(ii), it suffices to consider contractive modules. Let H_0, H_1 and H_2 be contractive Hilbert modules over $C(S^1;\mathbf{C})$, as in the definition, so that we have

$$\begin{array}{ccc} & & H_2 \\ & \nearrow & \downarrow \\ H_0 & \longrightarrow & H_1 \\ & & \downarrow \\ & & 0. \end{array}$$

By the proposition, there exist $U_j \in \mathcal{U}(H_j)$ such that the module operation on H_1 is given via $f \mapsto f(U_j)$ as in (i) for $j = 1,2$. Then we introduce $H_1^{\perp} = H_2 \ominus H_1$ and write U_2 as the unitary block matrix

$$U_2 = \begin{bmatrix} U_1 & 0 \\ 0 & \tilde{U}_2 \end{bmatrix} : \begin{matrix} H_1 \\ H_2 \ominus H_1 \end{matrix} \rightarrow \begin{matrix} H_1 \\ H_2 \ominus H_1 \end{matrix} \qquad (4.7.9)$$

so that $\tilde{U}_2 \in \mathcal{U}(H_2 \ominus H_1)$; then H_2 is a Hilbert module with complementary submodules H_1 and $H_2 \ominus H_1$. The sequence $H_2 \to H_1 \to 0$ splits, and we have a lifting via the map $f \mapsto f(U_2) \in \mathcal{L}(H_2)$.

Whereas the disc algebra A is a closed linear subspace of $C(S^1;\mathbf{C})$ by Exercise 3.1.7, its linear topological properties are different. Combining results of Grothendieck and Paley, one can show that there does not exist a continuous, surjective and linear map $C(S^1;\mathbf{C}) \to A$. The following results circumvent this obstacle.

Lemma 4.7.4 *Let $f(e^{i\theta}) = \sum_{j=-N}^{N} a_j e^{ij\theta}$ be a trigonometric polynomial with $a_j \in \mathbf{C}$ such that $f(e^{i\theta}) \geq 0$ for all $f \in [0,2\pi]$. Then there exists a polynomial $r(z) = \sum_{j=0}^{N} b_j z^j$ such that*

$$f\left(e^{i\theta}\right) = |r\left(e^{i\theta}\right)|^2 \qquad (\theta \in [0,2\pi]). \qquad (4.7.10)$$

□

Proposition 4.7.5 (von Neumann) *Let H be a complex separable Hilbert space, and A the disc algebra from Exercise 3.1.7. Let $C \in \mathcal{L}(H)$ have $\|C\|_{\mathcal{L}(H)} \leq 1$. Then H is a contractive Hilbert module via the homomorphism $f \mapsto f(C)$.*

Proof We consider the space

$$\mathcal{S} = \{f(e^{i\theta}) = p(e^{i\theta}) + \overline{q(e^{i\theta})} : p(z), q(z) \in \mathbf{C}[z]\}, \tag{4.7.11}$$

which is a complex vector subspace of $C(S^1; \mathbf{C})$ which contains 1 and is closed under $f \mapsto f^*$. Given the contraction $C \in \mathcal{L}(H)$, we write

$$\Phi(p + \bar{q}) = p(C) + q(C)^*. \tag{4.7.12}$$

Then $\Phi : \mathcal{S} \to \mathcal{L}(H)$ is a positive linear map which extends to a positive map $\Phi : C(S^1; \mathbf{C}) \to \mathcal{L}(H)$, such that $\|\Phi\| = \|\Phi(1)\|_{\mathcal{L}(H)} = 1$. To deduce this from the Lemma, we write

$$\begin{aligned} f(e^{i\theta}) &= \sum_{j=0}^{N} b_j e^{ij\theta} \sum_{j=0}^{N} \bar{b}_j e^{-ij\theta} \\ &= \sum_{j,k=-N}^{N} b_j \bar{b}_k e^{i(j-k)\theta} \end{aligned}$$

so

$$\Phi(f) = \sum_{j,k=0}^{N} b_j \bar{b}_k C^{(j-k)} \tag{4.7.13}$$

where we have written $C^{(j)} = C^j$ for $j = 0, 1, 2, \ldots$ and $C^{(j)} = C^{*(-j)}$ for $j = -1, -2, \ldots$. The operator matrix $[C^{(j-k)}]_{j,k=0}^{N}$ is self-adjoint and non-negative, so $\Phi(f) \geq 0$ for all $f \geq 0$. Hence Φ is non-negative, and $\Phi(1) = I$.

By restricting Φ to the subspace of analytic trigonometric polynomials $p \in \mathcal{S}$, we deduce that for all $p(z) \in \mathbf{C}[z]$, $p(C) = \Phi(p)$ satisfies

$$\|p(C)\|_{\mathcal{L}(H)} \leq \|p\|_{C(S^1;\mathbf{C})}; \tag{4.7.14}$$

and the polynomials are dense in A, so there is a contractive homomorphism $A \to \mathcal{L}(H) : f \mapsto f(C)$. For a Hilbert space H, we regard H as column vectors and H^t as rows, so there is a multiplication $(H, H^t) \to \mathcal{L}(H) : (\xi, \eta^t) \mapsto \xi \otimes \eta^t$, the operator on H of rank ≤ 1. The Hilbert module H over the disc algebra is equivalent to the bounded homomorphism

$$\begin{bmatrix} A & H \\ H^t & \mathbf{C} \end{bmatrix} \rightarrow \begin{bmatrix} \mathcal{L}(H) & H \\ H^t & \mathbf{C} \end{bmatrix} : \quad \begin{bmatrix} f & \xi \\ \eta^t & c \end{bmatrix} \mapsto \begin{bmatrix} f(C) & \xi \\ \eta^t & c \end{bmatrix}$$

where $C \in \mathcal{L}(H)$ determines the module operation via $(z, \xi) \mapsto C\xi$. □

Example 4.7.6 (i) *(The GNS algebra for $C^1(S^1; \mathbf{C})$)* With $\mathcal{A} = C(S^1; \mathbf{C})$, we have

$$\mathrm{GNS}(\Phi) = \mathcal{A} \otimes \mathcal{L}(H) \otimes \mathcal{A} \tag{4.7.15}$$

with $\mathcal{A} \rightarrow \mathrm{GNS}(\Phi)$ given by

$$a \mapsto (1,1,1)(a,1,1)(1,1,1) = (1, \Phi(a), 1). \tag{4.7.16}$$

Note that the image of Φ is contained in the $\mathcal{A}$-bimodule $\mathcal{A} \otimes \mathcal{S} \otimes \mathcal{A}$. In terms of the Stinespring theorem, we have $\mathcal{A} \otimes H$ with the sesquilinear form defined by

$$\left\langle \sum_{j=1}^{n} a_j \otimes \xi_j \,\Big|\, \sum_{\ell=1}^{n} b_\ell \otimes \eta_\ell \right\rangle = \sum_{j,\ell=1}^{n} \langle \xi_j \mid \rho(a_j^* b_\ell)\eta_\ell \rangle_H \tag{4.7.17}$$

and $U : \sum_{j=1}^{n} a_j \otimes \xi_j \mapsto \sum_{j=1}^{n} z a_j \otimes \xi_j$ so that E is the Hilbert space that arises from the completion of $\mathcal{A} \otimes H$, and U is a unitary on E. Then $V : H \rightarrow E : \xi \mapsto 1 \otimes \xi$ gives

$$\Phi(f) = V^* f(U) V. \tag{4.7.18}$$

(ii) *(Projective modules over the disc algebra)* Let A be the disc algebra and

$$0 \longrightarrow H_0 \longrightarrow J \longrightarrow H_1 \longrightarrow 0 \tag{4.7.19}$$

an exact sequence of Hilbert A-modules. We consider $J = H_0 \oplus H_1$, as a direct sum of Hilbert spaces, and consider whether this is also equivalent to a direct sum of Hilbert modules. We have a bounded linear map $\Phi : A \rightarrow \mathcal{L}(H_0 \oplus H_1)$, as in

$$\Phi(f) = \begin{bmatrix} \phi_0(f) & \delta(f) \\ 0 & \phi_1(f) \end{bmatrix} \begin{matrix} H_0 \\ H_1 \end{matrix} \qquad (f \in A), \tag{4.7.20}$$

where $\phi_0(f) \in \mathcal{L}(H_0)$, $\phi_1(f) \in \mathcal{L}(H_1)$ and $\delta(f) \in \mathcal{L}(H_1, H_0)$ for all $f \in A$. Evidently Φ is a bounded homomorphism, if and only if

$$\begin{aligned} \phi_0(fg) = \phi_0(f)\phi_0(g), \quad \phi_1(fg) = \phi_1(f)\phi_1(g), \\ \delta(fg) = \delta(f)\phi_1(g) + \phi_0(f)\delta(g), \end{aligned} \tag{4.7.21}$$

so $\phi_0 : A \rightarrow \mathcal{L}(H_0)$ and $\phi_1 : A \rightarrow \mathcal{L}(H_1)$ are bounded homomorphisms, while δ is a bounded derivation. In [14], Carlson et al. show that the exact sequence

(4.7.19) splits, if and only if δ is an inner derivation, in the sense that $\phi_0(z) = T_0$, $\phi_1(z) = T_1$ and $\delta(z) = LT_1 - T_0L$ for some $L \in \mathcal{L}(H_1, H_0)$. This data determines $\delta(z^n)$ and hence $\delta(f)$ for all $f \in A$ via the Taylor series. For given $\Delta \in \mathcal{L}(H_1, H_0)$, $T_0 \in \mathcal{L}(H_0)$ and $T_1 \in \mathcal{L}(H_1)$, the equation

$$\Delta = LT_1 - T_0L$$

for unknown L is called Sylvester's equation. Given a solution, the crucial observation is that

$$\begin{bmatrix} \phi_0(f) & \delta(f) \\ 0 & \phi_1(f) \end{bmatrix} \begin{bmatrix} I & L \\ 0 & I \end{bmatrix} = \begin{bmatrix} I & L \\ 0 & I \end{bmatrix} \begin{bmatrix} \phi_0(f) & 0 \\ 0 & \phi_1(f) \end{bmatrix} \qquad (f \in A), \tag{4.7.22}$$

so we can reduce to a diagonal sum of block operator matrices. The following result is from [14].

Proposition 4.7.7 *Every unitary Hilbert module over the disc algebra is projective.*

Proof By assumption, we have a unitary operator $T_1 \in \mathcal{L}(H_1)$ and bounded derivation $\delta\colon A \to \mathcal{L}(H_1, H_0)$, so for all $\xi \in H_1$ and $\eta \in H_0$, the sequence $(\langle \eta \mid \delta(z^n)T_1^{*n}\xi\rangle)_{n=0}^{\infty}$ is bounded. We define $\langle \eta \mid L\xi\rangle = \mathrm{LIM}_n\langle \eta \mid \delta(z^n)T_1^{*n}\xi\rangle$, which determines $L \in \mathcal{L}(H_1, H_0)$. Also, by translation invariance of the limit, we have

$$\begin{aligned}\langle \eta \mid L\xi\rangle &= \mathrm{LIM}_n\langle \eta \mid \delta(z^{n+1})T_1^{*(n+1)}\xi\rangle \\ &= \mathrm{LIM}_n\langle \eta \mid T_0\delta(z^n)T_1^{*(n+1)}\xi\rangle + \mathrm{LIM}_n\langle \eta \mid \delta(z)T_1^nT_1^{*(n+1)}\xi\rangle \\ &= \langle \eta \mid T_0LT_1^*\xi\rangle + \langle \eta \mid \delta(z)T_1^*\xi\rangle,\end{aligned}$$

so $L = T_0LT_1^* + \delta(z)T_1^*$, hence $\delta(z) = LT_1 - T_0L$, and δ is an inner derivation. Hence the exact sequence splits and H_1 is a projective A-module. □

It remains an open problem to determine the full category of projective Hilbert modules over A. For given homomorphisms ϕ_0 and ϕ_1 the set

$$\begin{aligned}\mathrm{Der}(A, \mathcal{L}(H_1, H_0)) = \big\{\delta\colon A &\to \mathcal{L}(H_1, H_0)\colon \\ &\delta(fg) = \delta(f)\phi_1(g) + \phi_0(f)\delta(g), \forall f, g \in A\big\}\end{aligned}$$

is an additive group under addition of functions, with a subgroup that is given by the set of inner derivations $\delta(f) = L\rho_1(f) - \rho_0(f)L$ for $L \in \mathcal{L}(H_1, H_0)$, so we can form an abelian quotient group $H^1(A, \mathcal{L}(H_1, H_0))$ as in (2.6.6). Due to the possible lack of projective Hilbert modules, we cannot define a useful functor Ext for A via the general approach set out in [49, p. 233]. In Section 6.4, we will return to the notion of extensions of modules.

Exercise 4.7.8 (Projective modules over the continuous functions) Extend the statements and proofs of Proposition 4.7.2 and Corollary 4.7.3 to the case of $\mathcal{A} = C(X; \mathbf{C})$, where X is a compact metric space. The key observation is that the multiplicative group $C(X; S^1)$ has a functional with similar properties to LIM.

4.8 Algebras Associated with the Continuous Functions on the Circle

Given unital $\mathbf{k}$-algebras A and B, the free product $A \star B$ is generated by A and B with no additional relations, so elements of $A \star B$ are $\mathbf{k}$-linear sums of $a_1 b_1 a_2 \dots b_n$ with $a_j \in A$ and $b_\ell \in B$. We identify $1_A \star 1_B$ with $1_B \star 1_A$ as the unit of $A \star B$, and $\lambda \star b = a \star \lambda b = \lambda(a \star b)$. Now $A \star B$ is isomorphic as an algebra to $B \star A$.

To clarify the various algebras that can arise, we consider a basic example. We have seen that $C(S^1, \mathbf{C}) = C^*(\mathbf{Z})$, the C^*-algebra generated by the unitary representations of $\mathbf{Z}$. Now we distinguish between the following:

(i) $C(S^1; \mathbf{C}) \bar{\otimes} C(S^1; \mathbf{C}) = C(S^1 \times S^1; \mathbf{C})$, the space of continuous functions $S^1 \times S^1 \to \mathbf{C}$.
(ii) $C(S^1; \mathbf{C}) \oplus C(S^1; \mathbf{C}) = C(S^1; \mathbf{C}^2)$, the space of continuous functions $S^1 \to \mathbf{C} \oplus \mathbf{C}$.
(iii) $C(S^1; \mathbf{C}) \star C(S^1; \mathbf{C}) = C_r^*(\mathbf{Z} \star \mathbf{Z})$, the C^*-algebra associated with the left regular representation of the free group on two generators $\mathbf{Z} \star \mathbf{Z}$.
(iv) We have also encountered the Toeplitz algebra

$$0 \to \mathcal{K} \to \mathcal{T} \to C(S^1; \mathbf{C}) \to 0, \tag{4.8.1}$$

the extension of $C(S^1; \mathbf{C})$ via the compact operators. For more detail on these constructions, see [11].

4.9 Algebras Described by Universal Mapping Properties

Let A be a unital $\mathbf{k}$-algebra, where $\mathbf{k}$ is a field of characteristic zero. In $M_2(\mathbf{k})$ we introduce

$$I = \begin{bmatrix} 1 & 0 \\ 0 & 1 \end{bmatrix}, \quad E = (1/2)\begin{bmatrix} 1 & 1 \\ 1 & 1 \end{bmatrix}, \quad F = \begin{bmatrix} 0 & 1 \\ 1 & 0 \end{bmatrix}$$

so $E^2 = E$, $F^2 = I$ and $E = (I + F)/2$. Hence the linear space $\mathbf{k}I \oplus \mathbf{k}E$ is an algebra, namely the universal algebra over $\mathbf{k}$ generated by an idempotent. Given $e \in \mathrm{Idem}(A)$, there exists an algebra homomorphism $\mathbf{k}I \oplus \mathbf{k}E \to \tilde{M}_\infty(A)$ given by $sI + tE \mapsto s1 + te$.

(i) Let $\mathbf{k} \oplus \mathbf{k}\hat{e}$ be the algebra that is generated by a single idempotent $\hat{e} = \hat{e}^2$. Given any unital $\mathbf{k}$-algebra R,

$$\mathrm{Hom}_{alg}(\mathbf{k} \oplus \mathbf{k}\hat{e}, R) = \{e \in R : e^2 = e\}, \tag{4.9.1}$$

since $a \oplus b\hat{e} \mapsto a1_R + be$ gives a homomorphism.

(ii) The free algebra $A \star B$ is so large that

$$\mathrm{Hom}_{alg}(A \star B, R) = \mathrm{Hom}_{alg}(A, R) \times \mathrm{Hom}_{alg}(B, R). \tag{4.9.2}$$

Given $\varphi \in \mathrm{Hom}(A, R)$ and $\psi \in \mathrm{Hom}(B, R)$, we can define a homomorphism $A \star B \to R$ by

$$a_1 \dots a_n b_n \mapsto \varphi(a_1)\psi(b_1) \dots \varphi(a_n)\psi(b_n). \tag{4.9.3}$$

(iii) The tensor algebra $T(V)$ for a $\mathbf{k}$-vector space was introduced in Section 1.4. We can identify

$$\mathrm{Hom}_{alg}(T(V), R) = \mathrm{Hom}_{\mathbf{k}}(V, R) \tag{4.9.4}$$

since a $\mathbf{k}$-linear map $\rho\colon V \to R$ extends to $\rho\colon V^{\otimes n} \to R$

$$\rho(a_1, \dots, a_n) = \rho(a_1)\rho(a_2) \dots \rho(a_n). \tag{4.9.5}$$

In particular, let A be a unital algebra and identify $1_{T(A)}$ with $\oplus_{n=0}^{\infty} 1_A^{\otimes n}$ and form the ideal $(1_{T(A)} - 1_A)$ in $T(A)$. Then $B = T(A)/(1_{T(A)} - 1_A)$ arises as the image of $\hat{\rho}\colon A \hookrightarrow T(A) \to B$, namely the linear map given by inclusion $A \mapsto A^{\otimes 1}$ followed by the quotient map. The universal mapping property is

$$\mathrm{Hom}_{alg}(B, R) = \{\rho \in \mathrm{Hom}_{\mathbf{k}}(A, R)\colon \rho(1_A) = 1_R\} \tag{4.9.6}$$

via $u \mapsto u\hat{\rho} = \rho$.

4.10 The Universal GNS Algebra of the Tensor Algebra

Using the results of the previous section, we can identify the GNS algebra for the tensor algebra.

Proposition 4.10.1 *Let $B = T(A)/(1_{T(A)} - 1_A)$, and let $\hat{\rho}\colon A \to B$ be the universal map as in Section 4.9(iii). Then GNS($\hat{\rho}$) is canonically isomorphic to $A \star (\mathbf{k} \oplus \mathbf{k}\hat{e})$.*

Proof An algebra homomorphism $\mathrm{GNS}(\hat{\rho}) \to R$ is specified by the datum (e, u, v) in the previous section, so given $\rho \in \mathrm{Hom}_{\mathbf{k}}(A, R)$ such that $\rho(1_A) = 1_R$, let $v(\hat{\rho}(a)) = eu(a)e$. An algebra homomorphism on $A \star (\mathbf{k} \oplus \mathbf{k}\hat{e})$ is equivalent to (e, u), where $u\colon A \to R$ is an algebra homomorphism and $e = e^2 \in R$. Then by the universal property of B we have $v\colon B \to eRe$ arising directly from u, but momentarily forgetting this, we have a canonical algebra homomorphism $A \star (\mathbf{k} \oplus \mathbf{k}\hat{e}) \to \mathrm{GNS}(\hat{\rho})$ via $(\hat{e}, \hat{u})$. To check that this is an isomorphism, we must show how to construct a unique v given (e, u). Let $\rho(a) = eu(a)e$, so $\rho\colon A \to eRe$ has $\rho(1_A) = e$. By the universal property of B and $\hat{\rho}$, there exists a unique algebra homomorphism $v\colon B \to eRe$ such that $v(\hat{\rho}(a)) = \rho(a) = eu(a)e$. □

4.11 The Cuntz Algebra

In C^*-algebra theory there are various algebras associated with Joachim Cuntz; see [90, p. 71] for $\mathcal{O}_n$. In cyclic theory, the Cuntz algebra particularly refers to the free product $QA = A \star A$, where A is an arbitrary unital C^*-algebra. There are two canonical homomorphisms $A \to QA$ given by $a \mapsto a \star 1$ and $a \mapsto \tilde{a} = 1 \star a$. There is a unique involution $\tilde{}$ on QA that interchanges these homomorphisms.

Definition 4.11.1 (Superalgebra) A superalgebra is an algebra R with an action of $\mathbf{Z}/(2) = \{\pm 1\}$, denoted $r \mapsto \varepsilon r$. There is a linear decomposition $R = R^+ \oplus R^-$ into the ± 1 eigenspaces. Then $\mathrm{Hom}_{alg}(A, R) = \mathrm{Hom}_{superalg}(QA, R)$. The algebra $\mathcal{E}(A) = Q(A) \star \mathbf{k}[\mathbf{Z}/(2)]$ arises from the free product of QA with the group algebra $\mathbf{k}[\mathbf{Z}/(2)]$. Let F be an involution such that $F^2 = 1$ so $\mathbf{k} \oplus \mathbf{k}F = \mathbf{k}[\mathbf{Z}/(2)]$ and $\mathcal{E}(A) = Q(A) \oplus Q(A)F$. The multiplication on $\mathcal{E}(A)$ has $F\omega F = \tilde{\omega}$.

The following result is a variant of Proposition 4.10.1.

Proposition 4.11.2 *Let $\hat{\rho}\colon A \to T(A)/(1_{T(A)} - 1_A)$ as in Proposition 4.10.1. Then GNS($\hat{\rho}$) is canonically isomorphic to $\mathcal{E}(A)$.*

Proof Observe that $A \star \mathbf{k}[\mathbf{Z}/(2)] = Q(A) \otimes k[\mathbf{Z}/(2)]$ since they have the same algebra homomorphisms to any algebra R; so

$$\mathrm{Hom}(A \star \mathbf{k}[F], R) = \big\{(u, F)\colon u\colon A \to R \text{ algebra homomorphism}, \ F^2 = 1\big\}. \tag{4.11.1}$$

Conversely, given such a pair (u, F), we obtain another algebra homomorphism $FuF\colon A \to R$, which we regard as $\tilde{u}$. Then there is a unique superalgebra homomorphism $A \star A \to R$ extending u. This further extends to $(A \star A) \otimes k[\mathbf{Z}/(2)]$ because the action of $\mathbf{Z}/(2)$ on R is inner.

Idempotents are in one-to-one correspondence with involutions $e \leftrightarrow F$ where $F = 2e - 1$. Hence $\mathbf{k} \oplus \mathbf{k}\hat{e} \cong k[F] = k[\mathbf{Z}/(2)]$. Hence GNS$(\hat{\rho})$ is isomorphic to $\mathcal{E}(A)$. □

4.12 Fredholm Modules

Definition 4.12.1 (Fredholm module) Let A be a unital $*$-algebra. A Fredholm module over A consists of a separable Hilbert space H and a $*$-homomorphism $\pi : A \to \mathcal{L}(H)$ and a self-adjoint involution $F \in \mathcal{L}(H)$ such that $[F, \pi(a)] \in \mathcal{K}(H)$ for all $a \in A$.

Note that $e = (F + 1)/2$ satisfies $e = e^*$ and $e^2 = e$, so we can introduce $H_1 = \mathrm{Im}(e)$ and $H_2 = \mathrm{Ker}(e)$ such that $H = H_1 \oplus H_2$ is an orthogonal decomposition. The case of greatest interest is when H_1 and H_2 are both infinite dimensional and hence unitarily equivalent.

A graded Fredholm module over A is a Fredholm module (H, π, F) with grading $\varepsilon \in \mathcal{L}(H)$

$$\varepsilon = \begin{bmatrix} 1 & 0 \\ 0 & -1 \end{bmatrix} : \begin{matrix} H_+ \\ H_- \end{matrix} \to \begin{matrix} H_+ \\ H_- \end{matrix} \tag{4.12.1}$$

such that $\varepsilon a = a \varepsilon$ for all $a \in A$ and $\varepsilon F = -F\varepsilon$.

Proposition 4.12.2 (i) *A Fredholm module over A is the same as a $*$-representation of the algebra $A \star \mathbf{C}[F]$ on some Hilbert space, such that the ideal Ker$\{A \star \mathbf{C}[F] \to A \otimes \mathbf{C}[F]\}$ is represented by compact operators. This ideal is generated by $\{[F, a] : a \in A\}$.*

(ii) *A graded Fredholm module is equivalent to a $*$-representation on Hilbert space of the algebra $Q(A) = A \star A$ such that the ideal Ker$\{A \star A \to A\}$ is represented by compact operators.*

Proof (i) With the grading operator ε such that $\varepsilon a = a\varepsilon$ and $\varepsilon F = -F\varepsilon$, we use the identity $F = F^*$ to write

$$F = \begin{bmatrix} 0 & T \\ T^* & 0 \end{bmatrix} \quad \text{for} \quad \varepsilon = \begin{bmatrix} 1 & 0 \\ 0 & -1 \end{bmatrix} : \begin{matrix} H_+ \\ H_- \end{matrix} \to \begin{matrix} H_+ \\ H_- \end{matrix}, \tag{4.12.2}$$

and $T : H_- \to H_+$ is unitary since $F^2 = 1$. The condition $\varepsilon a = a\varepsilon$ gives

$$a \leftrightarrow \begin{bmatrix} u(a) & 0 \\ 0 & v(a) \end{bmatrix} \tag{4.12.3}$$

hence

$$[a,F] \leftrightarrow \begin{bmatrix} 0 & u(a)T - Tv(a) \\ v(a)T^* - T^*u(a) & 0 \end{bmatrix}, \tag{4.12.4}$$

hence $[a, F]$ is compact if and only if $T^*u(a)T - v(a)$ is compact for all $a \in A$.

(ii) Now $A \star A \to \mathcal{L}(H)$ restricts to $\mathcal{T} = \mathrm{Ker}(A \star A \to A)$; a product $a_1\tilde{b}_1 \dots a_n\tilde{b}_n$ is associated with $u(a_1)v(\tilde{b}_1)u(a_2)\dots v(\tilde{b}_n)$.

$$\begin{array}{ccccccccc} 0 & \longrightarrow & \mathcal{T} & \longrightarrow & A \star A & \longrightarrow & A & \longrightarrow & 0 \\ & & \downarrow & & \downarrow & & & & \\ 0 & \longrightarrow & \mathcal{K}(H) & \longrightarrow & \mathcal{L}(H) & \longrightarrow & \mathcal{C}(H) & \longrightarrow & 0 \end{array} \tag{4.12.5}$$

□

Remarks 4.12.3 (i) Suppose that $A = C(X; \mathbf{C})$ where X is a compact metric space, and let $u, v\colon A \to \mathcal{L}(H)$ be $*$-homomorphisms. Then a Fredholm operator $T \in \mathcal{L}(H)$ is called elliptic relative to (u, v) if $Tu(a) - v(a)T \in \mathcal{K}(H)$ for all $a \in A$. We say that T intertwines the two actions of A on H, modulo the compact operators. This is the condition involved in the proof of Proposition 4.12.2(ii). See [13, 7.6].

(ii) In order to obtain usable index formulas, it is necessary to introduce extra structure on the Fredholm modules, particularly to replace $\mathcal{K}(H)$ by $\mathcal{L}^p(H)$, so that we can compute the trace. In Section 5.6, we introduce theta summable Fredholm modules, which host a wealth of cyclic cocycles.

Exercise 4.12.4 (Holomorphic calculus) Hille [53] gives a general discussion of resolvents and holomorphic calculus. For self-adjoint $F \in \mathcal{L}(H)$ and $1 \leq p < \infty$, let $\delta\colon \mathcal{L}(H) \to \mathcal{L}(H)\colon \delta(X) = [F, X]$, and let $\mathcal{A} = \{X \in \mathcal{L}(H)\colon \delta(X) \in \mathcal{L}^p(H)\}$.

(i) Show that $\mathcal{A}$ is a $*$-subalgebra of $\mathcal{L}(H)$, containing $\mathcal{L}^p$ and $\{\lambda I : \lambda \in \mathbf{C}\}$.

(ii) Choose $R > 0$ so that $\sigma(X) \subset \{z \in \mathbf{C}\colon |z| < R\}$. Verify the identity

$$\delta((\lambda I - X)^{-1}) = (\lambda I - X)^{-1}\delta(X)(\lambda I - X)^{-1} \qquad (|\lambda| \geq R) \tag{4.12.6}$$

and deduce that $(\lambda I - X)^{-1} \in \mathcal{A}$ for all $X \in \mathcal{A}$ and all $\lambda \in \mathbf{C}$ such that $|\lambda| \geq R$.

(iii) For $X \in \mathcal{A}$, let f be a holomorphic function inside and on $\{z \in \mathbf{C}\colon |z| = R\}$. Using the definition

$$f(X) = \frac{1}{2\pi i}\int_{|\lambda|=R} (\lambda I - X)^{-1} f(\lambda)\, d\lambda \tag{4.12.7}$$

and (ii), show that $f(X) \in \mathcal{A}$.

5

Geometrical Examples

One of the main ingredients of cyclic theory is index theory for elliptic operators over a compact Riemannian manifold M. In Section 5.2, we sketch how the heat kernel $e^{-t\Delta/2}$ can be defined on $L^2(M)$, and recall its basic properties. In Section 5.4, we introduce Maxwell's equations via vector calculus over $\mathbf{R}^3$, and extend to the Dirac operator over $\mathbf{R}^4$, and in Section 5.5 we define the Dirac operator over spin manifolds. The main concept of this chapter is that of a Fredholm module with spectral triple, which Connes introduced to extend previous index theorems. In approaches to index theory that are based upon the heat equation, the quantum harmonic oscillator is a useful tool. We give a brief introduction in Section 5.8, based upon creation and annihilation operators from [69].

5.1 Fredholm Modules over the Circle

(i) Let $H = L^2(S^1, d\theta/(2\pi))$ and let $A = C^\infty(S^1, \mathbf{C})$, with e the Hardy projection $e(z^n) = z^n$ for $n \geq 0$ and $e(z^n) = 0$ for $n < 0$; then let $F = 2e - 1$, so

$$F(z^n) = \begin{cases} z^n & \text{for } n = 0, 1, \dots; \\ -z^n & \text{for } n = -1, -2, \dots. \end{cases} \tag{5.1.1}$$

This F is a variant of the conjugate function operator (Hilbert transform). Then $[F, a]$ is trace class, hence compact, for all $a \in A$. With $D = -id/d\theta$ we have $De^{in\theta} = ne^{in\theta}$.

(ii) Now the heat kernel is represented by the classical Jacobi theta function

$$e^{-tD^2}(\phi,\psi) = \sum_{n=-\infty}^{\infty} e^{-n^2t+in(\phi-\psi)}$$
$$= \vartheta_3\left(\frac{\phi-\psi}{2\pi}\,\Big|\,\frac{it}{2\pi}\right). \tag{5.1.2}$$

For $f \in L^2(S^1; d\theta/(2\pi))$, the function

$$u(\phi,t) = \int_0^{2\pi} \vartheta_3\left(\frac{\phi-\psi}{2\pi}\,\Big|\,\frac{it}{2\pi}\right) f(\psi)\frac{d\psi}{2\pi} \tag{5.1.3}$$

satisfies the heat equation

$$\frac{\partial}{\partial t}u(\phi,t) = -D^2u(\phi,t) \tag{5.1.4}$$

with initial condition

$$u(\phi,t) \to f(\phi) \quad \text{in} \quad L^2 \quad \text{as} \quad t \to 0+. \tag{5.1.5}$$

(iii) Let $P\colon H \to H\colon Pf = \int_0^{2\pi} f(\psi)d\psi/(2\pi)$ be the projection onto the constant functions, or in terms of Fourier series $f(\psi) = \sum_{n=-\infty}^{\infty} a_n e^{in\psi}$, let $Pf = a_0$. Then Let $P^\perp = I - P$ be the complementary projection, so $P^\perp f = \sum_{n:n\neq 0} a_n e^{in\psi}$. Let

$$G(x,y) = \sum_{n=-\infty;n\neq 0}^{\infty} \frac{e^{in(x-y)}}{n^2} \tag{5.1.6}$$

and $Gf(\psi) = \int_0^{2\pi} G(\psi,\phi)f(\phi)d\pi/(2\pi)$; then

$$D^2G = GD^2 = P^\perp.$$

This G is Green's function for the operator D^2 over the circle, and $G\colon H \to H$ the corresponding Green's operator. Evidently we have $G = \int_0^\infty e^{-tD^2}P^\perp dt$.

5.2 Heat Kernels on Riemannian Manifolds

Let U be an open neighbourhood in $\mathbf{R}^n$. A second-order differential operator with smooth coefficients can be written as

$$Lf = \sum_{j,k=1}^{n} a^{jk}(x)\frac{\partial^2 f}{\partial x_j \partial x_k} + \sum_{k=1}^{n} b_k(x)\frac{\partial f}{\partial x_k} + c(x)f \qquad (x \in U) \tag{5.2.1}$$

or equivalently

$$Lf = \sum_{j,k=1}^{n} \frac{\partial}{\partial x_j}\left(a^{jk}(x)\frac{\partial f}{\partial x_k}\right) + \sum_{k=1}^{n} \tilde{b}_k(x)\frac{\partial f}{\partial x_k} + c(x)f \qquad (x \in U) \tag{5.2.2}$$

where $\tilde{b}_k(x) = b_k(x) - \sum_{j=1}^{n} \frac{\partial a^{jk}(x)}{\partial x_j}$. If the matrix $K = [a^{jk}(x)]$ is real symmetric for all $x \in U$, then we say that L is in divergence form. Also, if there exists $\kappa_U > 0$ such that

$$\sum_{j,k=1}^{n} a^{jk}(x)\xi_j\xi_k \geq \kappa_U \sum_{j=1}^{n} \xi_j^2 \qquad ((\xi_j)_{j=1}^{n} \in \mathbf{R}^n; x \in U) \tag{5.2.3}$$

then we say that K is uniformly positive definite and L is uniformly elliptic on U. For example, the Laplace operator for the standard metric on Euclidean space has $K = I_n$ and $b_j = c = 0$.

This is the situation which arises locally in Riemannian geometry. Generally, let $[a^{jk}]_{j,k=1}^{n}$ be uniformly positive definite on U, and write $[a^{jk}]_{j,k=1}^{n} = [g^{jk}]_{j,k=1}^{n}$, then let $g = [g_{jk}]_{j,k=1}^{n}$ be the inverse matrix of $[g^{jk}]_{j,k=1}^{n}$. The Riemannian metric on the coordinate patch U is denoted $\sum_{j,k=1}^{m} g_{jk}dx_jdx_k$, with the corresponding inner product on the tangent space $\langle X, Y\rangle = \sum_{j,k=1}^{n} g_{jk}X_jY_k$; the Christoffel symbols are

$$\Gamma_{i,j}^{k} = \frac{1}{2}\sum_{h=1}^{n} g^{kh}\left(\frac{\partial g_{hj}}{\partial x_i} + \frac{\partial g_{hi}}{\partial x_j} - \frac{\partial g_{ij}}{\partial x_h}\right). \tag{5.2.4}$$

Then the Laplace–Beltrami operator is Δ, where

$$-\Delta f = \sum_{i,j=1}^{n} g^{ij}\left(\frac{\partial^2 f}{\partial x_i \partial x_j} - \sum_{k=1}^{n} \Gamma_{i,j}^{k}\frac{\partial f}{\partial x_k}\right), \tag{5.2.5}$$

which may be written in divergence form as

$$-\Delta f = \sum_{i,j=1}^{n} \frac{1}{\sqrt{\det g}}\frac{\partial}{\partial x_i}\left(\sqrt{\det g}\, g^{ij}\frac{\partial f}{\partial x_j}\right) \tag{5.2.6}$$

for $f \in C_c^\infty(U, \mathbf{C})$, with $\operatorname{grad} f = (\sum_j g^{ij}\frac{\partial f}{\partial x_j})$.

Suppose that M is an n-dimensional smooth manifold, with coordinate patches such as U. The manifold M has a volume form $\mu = \sqrt{\det g}dx_1 \wedge \cdots \wedge dx_n$, hence there is a Hilbert space $H = L^2(M; \mu, \mathbf{C})$ which contains $A = C_c^\infty(M, \mathbf{C})$ as a dense linear subspace. There are various ways of introducing a heat kernel and Green's function for M, and proving that Δ is essentially self-adjoint on A. Since waves travel over the manifold at finite propagation speed, there are technical advantages in starting with the wave equation and constructing a distributional kernel $\cos(t\sqrt{\Delta})$. This enables us to consider only finitely many coordinate patches at any given $t > 0$, so we can consider issues of compactness later on.

Theorem 5.2.1 (Chernoff) *Let M be a complete Riemannian manifold without a boundary with Laplace operator Δ. Consider the initial value problem*

$$\frac{\partial}{\partial t}\begin{bmatrix} u \\ v \end{bmatrix} = \begin{bmatrix} 0 & I \\ -\Delta & 0 \end{bmatrix}\begin{bmatrix} u \\ v \end{bmatrix} \tag{5.2.7}$$

$$\begin{bmatrix} u \\ v \end{bmatrix}_{t=0} = \begin{bmatrix} f \\ 0 \end{bmatrix}, \tag{5.2.8}$$

where $f \in C_c^\infty(M; \mathbf{R})$. Then the initial value problem has a unique solution $u(x,t) = \cos(t\sqrt{\Delta})f(x)$ such that $u(x,t)$ is smooth for $(x,t) \in M \times (0,\infty)$, the function $x \mapsto u(x,t)$ has compact support in M and u satisfies the wave equation

$$\frac{\partial^2 u}{\partial x^2} + \Delta u = 0. \tag{5.2.9}$$

The operator Δ is densely defined, self-adjoint and non-negative in $L^2(M)$.

Proof See section 3 (A) of [17].

An alternative choice in (5.2.8) is

$$X = \begin{bmatrix} 0 & \sqrt{\Delta} \\ -\sqrt{\Delta} & 0 \end{bmatrix} \tag{5.2.10}$$

which is evidently skew-symmetric. □

For instance, in $\mathbf{R}^3$ we can convert to polar coordinates and write

$$-\Delta f = \frac{1}{r^2}\frac{\partial}{\partial r}\left(r^2\frac{\partial f}{\partial r}\right) + \frac{1}{r^2 \sin\theta}\frac{\partial}{\partial \theta}\left(\sin\theta\frac{\partial f}{\partial \theta}\right) + \frac{1}{r^2\sin^2\theta}\frac{\partial^2 f}{\partial \phi^2},$$

and the surface area measure on $S^2(t) = \{y \in \mathbf{R}^3 : \|y\| = t\}$ is $d\sigma_t = t^2 \sin\theta d\theta d\phi$. Then Poisson expressed the solution to the wave equation as

$$\cos(t\sqrt{\Delta})f(x) = \frac{\partial}{\partial t}\left(\frac{1}{4\pi t}\int_{S^2(t)} f(x+y)\sigma_t(dy)\right), \tag{5.2.11}$$

as one can verify by direct calculation using the divergence theorem; see Exercise 5.2.5. The solutions to the wave equation on $\mathbf{R}^n$ are related to one another via Hadamard's method of descent, which produces formulas which extend (5.2.11). The general case of the manifold M uses a parametrix for $\cos(t\sqrt{\Delta})(x,y)$ which likewise has support on $\{(x,y) \in M^2 : \rho(x,y) \leq t\}$, where ρ is the Riemannian distance. See section 17.5 of [56] for details and sharp results in this case. Given this solution of the wave equation, we can then define

$$\varphi(\sqrt{\Delta})f(x) = \frac{1}{2\pi}\int_{-\infty}^{\infty} \hat{\varphi}(s)\cos(s\sqrt{\Delta})f(x)\,ds \tag{5.2.12}$$

for any $\varphi \in C_c^\infty(\mathbf{R};\mathbf{R})$ such that $\varphi(u) = \varphi(-u)$ for all $u \in \mathbf{R}$ and $f \in L^2(M;\mathbf{C})$. This is a version of the Poisson summation formula. In particular we can define

$$T_t f(x) = \frac{1}{\sqrt{2\pi t}}\int_{-\infty}^{\infty} e^{-s^2/(2t)}\cos(s\sqrt{\Delta})\,f(x)\,ds, \tag{5.2.13}$$

which gives a solution of the heat equation.

Proposition 5.2.2 *The heat semigroup $T_t = e^{-t\Delta/2}$ on $H = L^2(M,\mu)$ is characterized by the following properties:*

(i) *$T_t \in \mathcal{L}(H)$ for all $t > 0$;*
(ii) *$T_t T_s = T_{t+s}$ for $t,s > 0$;*
(iii) *$T_t f \to f$ in H as $t \to 0$ for all $f \in H$;*
(iv) *$(T_t f - f)/t \to -\Delta f/2$ as $t \to 0+$ for all $f \in \mathcal{D}(\Delta)$, where $\mathcal{D}(\Delta)$ is a linear subspace containing $C_c^\infty(M)$;*
(v) *$T_t^* = T_t$ for all $t > 0$, so $T_t \geq 0$ as an operator;*
(vi) *T_t is positivity preserving, so that $f \geq 0 \Rightarrow T_t f \geq 0$, and T_t has a kernel $T_t(x,y) \in C^\infty(M \times M;[0,\infty))$ for all $t > 0$.*
(vii) *Suppose further that M is compact. Then also, $T_t \in \mathcal{L}^1(H)$ for $t > 0$ and*

$$\mathrm{trace}(T_t) = \int_M T_t(x,x)\mu(dx). \tag{5.2.14}$$

Proof Conditions (i)–(v) follow from Theorem 5.2.1 via the spectral theorem for self-adjoint operators on Hilbert space. Indeed, by the spectral theorem applied to Δ, there exists a family of $e_{(\lambda,\omega]} \in \mathcal{L}(H)$ for real $\lambda \leq \omega$ such that

(1) $e_{(\lambda,\omega]}e_{(\lambda,\omega]} = e_{(\lambda,\omega]} = e^*_{(\lambda,\omega]}$;
(2) $e_{(\lambda,\omega]} + e_{(\omega,\rho]} = e_{(\lambda,\rho]}$ for all $\lambda \leq \omega \leq \rho$;
(3) $e_{(\lambda,0]}$ constant for all $\lambda < 0$;
(4) $e_{(-1,\omega]}f \to f$ in H as $\omega \to \infty$ for all $f \in H$;
(5) $\Delta e_{(\lambda,\omega]} = e_{(\lambda,\omega]}\Delta$ and $e_{(\lambda,\omega]}e_{(\lambda',\omega']} = e_{(\lambda',\omega']}e_{(\lambda,\omega]}$ for all $\lambda \leq \omega$ and $\lambda' \leq \omega'$.

It follows from (1) and (2) that $e_{(\lambda,\omega]}e_{(\omega,\rho]} = 0$. The operators listed in (1)–(5) together generate a commutative C^*-algebra. We deduce that $(\lambda,\omega] \mapsto \langle e_{(\lambda,\omega]}f, f\rangle_H/\|f\|^2$ determines a Borel probability measure on $[0,\infty)$, and

$$\langle T_t f, f\rangle = \int_0^\infty e^{-t\lambda/2}\langle e(d\lambda)f, f\rangle_H \qquad (f \in H). \tag{5.2.15}$$

The domain of Δ consists of those $f \in H$ such that $\int_0^\infty \lambda^2\langle e(d\lambda)f, f\rangle$ converges. Also, $\langle\cos(t\sqrt{\Delta})f,\rangle_H$ is the Fourier cosine transform of $\langle e(d\lambda)f, f\rangle_H$.

(vi) Positivity requires a different mode of proof, which depends upon the positive maximum principle. We consider the one-point compactification $M \cup \{\infty\}$ and the Banach space $C(M \cup \{\infty\}; \mathbf{R})$ with the supremum norm, which contains $C_c^\infty(M;\mathbf{R})$ as a dense linear subspace. Suppose that f is twice continuously differentiable and attains its maximum at $x_0 \in M$ and $f(x_0) = \|f\|_\infty$. Then $\Delta f(x_0) \geq 0$. Indeed, let γ_t be a geodesic emanating from x_0, so that $\gamma_0 = x_0$, and pointing in the direction $\gamma_0' = v \in T_{x_0}M$; then the Hessian of f satisfies

$$f(\gamma_t) = f(x_0) + t\langle\nabla f(x_0), v\rangle + \frac{t^2}{2}\langle \mathrm{Hess}\, f(x_0)v, v\rangle + o(t^2) \tag{5.2.16}$$

as $t \to 0+$. Since $f(\gamma_t) \leq f(x_0)$, we deduce that $\nabla f(x_0) = 0$ and by averaging over the choices of v, we obtain $-\Delta f(x_0) \leq 0$.

We deduce that for all $\lambda > 0$,

$$\|(\lambda + \Delta)f\|_\infty \geq (\lambda I + \Delta)f(x_0) \geq \lambda f(x_0) = \lambda\|f\|_\infty. \tag{5.2.17}$$

It follows that $\lambda I + \Delta$ is densely defined and invertible in $C(M \cup \{\infty\}; \mathbf{R})$, with operator norm

$$\|\lambda(\lambda I + \Delta)^{-1}\| \leq 1 \qquad (\lambda > 0). \tag{5.2.18}$$

By taking $\lambda = m/t$, we can recover the heat semigroup via Yoshida's formula

$$e^{-t\Delta}f(x) = \lim_{m\to\infty}(I + t\Delta/m)^{-m}f(x) \tag{5.2.19}$$

so that $\|e^{-t\Delta}\| \leq 1$ and $e^{-t\Delta}$ is positivity preserving in the sense that $f \geq 0 \Rightarrow e^{-t\Delta}f \geq 0$.

To establish differentiability, we consider $\varphi_{2n}(x) = x^{2n}e^{-tx^2/2}$ and observe that $\hat{\varphi}_{2n}(s) = p_{2n}(s)e^{-s^2/(2t)}$, where p_{2n} is a real even polynomial of degree $2n$; indeed, in Section 5.8 we obtain $\hat{\varphi}_{2n}$ via the Hermite polynomials. This shows that $\Delta^n T_t = \varphi_{2n}(\sqrt{\Delta})$ is a bounded and self-adjoint operator on $L^2(M)$.

With $u(x,t) = T_t f(x)$ and $f \in L^2(M)$, we have $u, \Delta u \in L^2(M)$, and the divergence theorem gives

$$\begin{aligned}\int_M \|\nabla u\|^2 \mu(dx) &= \int_M (\Delta u)(x) u(x) \mu(dx)\\ &\leq \Big(\int_M (\Delta u)(x)^2 \mu(dx)\Big)^{1/2} \Big(\int_M u(x)^2 \mu(dx)\Big)^{1/2},\end{aligned} \tag{5.2.20}$$

by Cauchy–Schwarz, so $\nabla u \in L^2(TM)$. To proceed further, one can use versions of Garding's inequality.

For any $x_0 \in M$, we can introduce the exponential map $\exp_x : T_{x_0}M \to M$, which restricts to a diffeomorphism mapping some small neighbourhood of 0 to an open neighbourbood of x_0. On these neighbourhoods, we can carry out local analysis of the Laplace operator.

On $U = \{x \in \mathbf{R}^n : \|x\| < r\}$, the equation

$$(I + \Delta)^{2n} h = g$$

with $g \in L^2(U)$ has a solution

$$h(x) = \frac{1}{(2\pi)^n} \int_U \int_{\mathbf{R}^n} \frac{e^{i(x-y)\cdot\xi}\, d\xi}{(1 + \|\xi\|^2)^{2n}} g(y)\, dy \tag{5.2.21}$$

and we see that $h \in C^1(U)$.

Using this argument repeatedly, we see that $T_t(x,y)$ is C^∞ in x, and since $T_t(x,y) = T_t(y,x)$, also C^∞ in y. Hence T_t has a positive kernel $T_t(x,y) \in C^\infty(M \times M); [0,\infty))$. Details of this calculation with precise estimates are given in [16].

(vii) When M is compact, we can apply Mercer's trace formula [66, p. 343] to deduce that $T_t \in \mathcal{L}^1(H)$ for all $t > 0$, and the trace is given by integrating the continuous kernel $T_t(x,y)$ along the diagonal $x = y$. See [25] for more detailed statements. □

Exercise 5.2.3 (Christoffel symbols) Introduce canonical variables $(q_1, \dots, q_n; p_1, \dots, p_n)$, and let $[g_{jk}]_{j,k=1}^n$ be a positive definite matrix function of $(q_1, \dots, q_n)$. Derive the canonical equations of motion for the Hamiltonian

$$H(q_1, \dots, q_n; p_1, \dots, p_n) = \frac{1}{2} \sum_{j,k=1}^n g_{jk} p_j p_k,$$

compute $d^2 q_j/dt^2$ and hence derive the Christoffel symbols $\Gamma^k_{i,j}$ as above.

Exercise 5.2.4 (Jacobi's theta function) (i) Let $\Delta = -d^2/d\theta^2$ be the Laplacian on the circle S^1. Use (5.2.12) to find the heat kernel on S^1 as in (5.1.2).

(ii) Find an expression for the heat kernel $T_t(x, y)$ for the real torus $(S^1)^k$, and deduce an expression for trace(T_t).

Exercise 5.2.5 (Poisson's solution of wave equations) Show that

$$\int_0^t \int_{S^2(s)} \operatorname{div}\nabla f(x+y)\sigma_s(dy)\,ds$$
$$= t^2 \frac{\partial}{\partial t}\left(\frac{1}{t^2}\int_{S^2(t)} f(x+y)\sigma_t(dy)\right) \qquad (f \in C^\infty(\mathbf{R}^3; \mathbf{C})),$$

hence verify Poisson's formula (5.2.11) for the solution of wave equations.

5.3 Green's Function

In this section M is an n-dimensional, compact smooth and connected Riemannian manifold without a boundary. One can show that $\mu(M)$ is finite, and by scaling the metric, we assume for convenience that $\mu(M) = 1$. With this normalization, we let $H = L^2(\mu)$.

Lemma 5.3.1 (Hopf) *Suppose that $f \in C^2(M)$ satisfies $\Delta f = 0$. Then f is constant.*

Proof By Stokes's theorem,

$$-\int_M \Delta(f^2)d\mu = \int_M \operatorname{div}\nabla f^2\, d\mu = 0. \tag{5.3.1}$$

We deduce that

$$0 = -2\int_M f\Delta f\, d\mu + 2\int_M \sum_{j,k=1}^n g^{jk}\frac{\partial f}{\partial x_j}\frac{\partial f}{\partial x_k}d\mu \tag{5.3.2}$$

and since $\Delta f = 0$, and

$$\sum_{j,k=1}^n g^{jk}\frac{\partial f}{\partial x_j}\frac{\partial f}{\partial x_k} = 0 \tag{5.3.3}$$

where $[g^{jk}]$ is strictly positive definite, $\frac{\partial f}{\partial x_k} = 0$ for all k, so by integrating along differentiable paths in M, we see that f is constant. □

Let P be the projection onto the constant functions $P\colon H \to H$ given by $Pf = \int_M f d\mu$, and let $P^\perp = I = P$ be the complementary orthogonal projection.

Theorem 5.3.2 (Poincaré's inequality) (i) *There exists* $\lambda_1 > 0$ *such that*

$$\int_M \Big(f(x) - \int_M f d\mu\Big)^2 \mu(dx) \leq \frac{1}{\lambda_1}\int_M \|\nabla f(x)\|^2 \mu(dx). \tag{5.3.4}$$

(ii) *There exists a compact and self-adjoint operator* G *on* $L^2(M;\mu)$ *such that*

$$\Delta G = G\Delta = P^\perp. \tag{5.3.5}$$

Proof (i) For $\lambda > 0$, the bounded linear operator $(\lambda I + \Delta)^{-1}$ is defined via the integral

$$\begin{aligned}(\lambda I + \Delta)^{-1} f(x) &= \int_0^\infty e^{-\lambda t} e^{-t\Delta} f\, dt \\ &= \int_0^\infty e^{-\lambda t} \int_M T_t(x,y) f(y)\mu(dy)\, dt,\end{aligned} \tag{5.3.6}$$

so that the integral kernel of $(\lambda I + \Delta)^{-1}$ is a Hilbert–Schmidt operator with real positive eigenvalues $1/(\lambda + \lambda_0) \geq 1/(\lambda + \lambda_1) \geq \cdots$.

By Hopf's lemma, $\lambda_0 = 0$ has multiplicity one, so we have $0 = \lambda_0 < \lambda_1$, and as in the above proof

$$\int_M (\Delta P^\perp f)(P^\perp f)\, d\mu = \int_M \sum_{j,k=1}^n g^{jk} \frac{\partial f}{\partial x_j}\frac{\partial f}{\partial x_k} d\mu \geq \lambda_1 \int_M |P^\perp f(x)|^2 \mu(dx). \tag{5.3.7}$$

Evidently $\Delta P = 0$, so $P\Delta = 0$, hence the result.

(ii) We deduce that Δ is invertible on the subspace $P^\perp H$, so we can define

$$G = P^\perp \Delta^{-1} P^\perp. \tag{5.3.8}$$

□

In Section 7.1, we discuss a more sophisticated variant of this example.

Example 5.3.3 (Harmonic functions) (i) On non-compact manifolds, the space of harmonic functions can be rather large. The functions $f(x,y) = ax + by$ with a, b real constants are harmonic on $\mathbf{R}^2$. Suppose however that f is harmonic and $f(x,y) \geq 0$ for all $(x,y) \in \mathbf{R}^2$; then by Liouville's theorem, f is constant.

(ii) Let S^2 be the unit 2-sphere in $\mathbf{R}^3$. Then Δ has eigenvalues $\lambda_n = n(n+1)$ for $n = 0, 1, \dots$, where λ_n has geometric multiplicity $2n+1$. The eigenfunctions may be found by separating variables in polar coordinates and using spherical harmonics, as in [105]. Hopf's Lemma is therefore a special statement about the multiplicity of the lowest eigenvalue, which does not readily extend to larger eigenvalues.

(iii) Poincaré's inequality is not valid for disconnected compact manifolds such as the orthogonal group $O(3)$. For instance, $f(X) = \det X$ takes the value 1 on $SO(3)$ and -1 on the coset $-SO(3)$, so $\nabla f = 0$, although f is non-constant.

(iv) We can also interpret PH as $\{f \in H : T_t f = f \quad \forall t > 0\}$ and $Pf = \lim_{t\to\infty} T_t f$.

Exercise 5.3.4 (Electrostatics) Let $f : D \to \mathbf{C}$ be a meromorphic function with finitely many zeros z_j with multiplicities n_j and finitely many poles p_j with multiplicity m_j. Let δ_a represent the unit point charge at $a \in D$. By considering the case of $(z-a)^m$, show that

$$\frac{1}{2\pi}\left(\frac{\partial^2}{\partial x^2} + \frac{\partial^2}{\partial y^2}\right)\log|f(z)| = \sum_j n_j \delta_{z_j} - \sum_j m_j \delta_{p_j}, \tag{5.3.9}$$

in the sense of distributions, so that the total charge is given by

$$\frac{1}{2\pi}\int_D \left(\frac{\partial^2}{\partial x^2} + \frac{\partial^2}{\partial y^2}\right)\log|f(z)| dxdy = \sum_j n_j - \sum_j m_j. \tag{5.3.10}$$

We regard $\log|f(z)|$ as the electrostatic potential given by a field of positive point charges at z_j and negative point charges at p_j. The algebraic description in terms of divisors is given in Section 8.2.5.

5.4 Maxwell's Equation

Maxwell's equations are used in classical electrodynamics, where there is both an electric field and a magnetic field. In $\mathbf{R}^3$, we combine the time and space variables into a vector in $\mathbf{R}^4$, which is realized as a quaternion.

Exercise 5.4.1 (i) First we consider differential operators with quaternion coefficients, as in Example 1.3.2. We introduce the electric field $\mathbf{E} = (E_1, E_2, E_3)$, and the magnetic field $\mathbf{B} = (B_1, B_2, B_3)$, where the components are smooth functions of $(t, x, y, z) \in \mathbf{R}^4$. Then Maxwell's equations give

$$\frac{\partial \mathbf{B}}{\partial t} + \operatorname{curl} \mathbf{E} = \mathbf{0}, \tag{5.4.1}$$

$$\operatorname{div} \mathbf{B} = 0. \tag{5.4.2}$$

(ii) Compute D^2 where

$$D = \begin{bmatrix} 0 & \operatorname{div} \\ \operatorname{grad} & \operatorname{curl} \end{bmatrix}.$$

With $(t, x, y, z) = (x_0, x_1, x_2, x_3)$, we introduce the 2-form

$$\begin{aligned} F = & -E_1 dx_0 dx_1 - E_2 dx_0 dx_1 - E_3 dx_0 dx_1 \\ & + B_3 dx_1 dx_2 - B_2 dx_1 dx_3 + B_1 dx_2 dx_3. \end{aligned} \tag{5.4.3}$$

Proposition 5.4.2 (i) *The condition $dF = 0$ is equivalent to Maxwell's equation.*

(ii) *Maxwell's equations hold if and only if there exists a 1-form A called the potential field such that $dA = F$.*

Proof (i) By direct calculation of dF one checks that $dF = 0$ is equivalent to the pair of Maxwell equations.

(ii) In classical notation, A corresponds to $(A_0, \mathbf{A})$, where

$$\mathbf{E} = \nabla A_0 - \frac{\partial \mathbf{A}}{\partial t}, \qquad \mathbf{B} = \operatorname{curl} \mathbf{A}; \tag{5.4.4}$$

so we can find $(A_0, \mathbf{A})$ by vector calculus. Since $\operatorname{div} \mathbf{B} = 0$, there exists a vector potential $\mathbf{A}$ such that $\mathbf{B} = \operatorname{curl} \mathbf{A}$, and a possible choice is

$$\mathbf{A} = \left(\int_0^x B_3(dx', y, z) dx' - \int_0^z B_1(0, y, z') dz' \right) \mathbf{j} - \left(\int_0^x B_2(x', y, z)\, dx' \right) \mathbf{k}; \tag{5.4.5}$$

then

$$\begin{aligned} \operatorname{curl}\left(\frac{\partial \mathbf{A}}{\partial t} + \mathbf{E} \right) &= \frac{\partial}{\partial t} \operatorname{curl} \mathbf{A} + \operatorname{curl} \mathbf{E} \\ &= \frac{\partial \mathbf{B}}{\partial t} + \operatorname{curl} \mathbf{E} \\ &= \mathbf{0}; \end{aligned} \tag{5.4.6}$$

hence there exists a scalar field A_0 such that

$$\nabla A_0 = \frac{\partial \mathbf{A}}{\partial t} + \mathbf{E}; \tag{5.4.7}$$

a possible choice of A_0 is given by integration along the coordinate axes; details left to the reader. □

Exercise 5.4.3 To deal with fields in higher dimensions, the next step is to replace the scalar components of the fields by matrices, and increase the number of spatial dimensions. Let $\mathbf{K} = \mathbf{C}(x_1, \dots, x_n)$. For $j = 1, \dots, n$, let A_j be $m \times m$ matrices with entries in $\mathbf{K}$, and $A = \sum_{j=1}^n A_j dx_j$ be a matrix-valued 1-form. Then we define

$$F = dA + A \wedge A. \tag{5.4.8}$$

A direct calculation gives

$$F = \sum_{1 \le j < k \le n} \left(\frac{\partial A_k}{\partial x_j} - \frac{\partial A_j}{\partial x_k} + A_j A_k - A_k A_j \right) dx_j dx_k, \tag{5.4.9}$$

where

$$\frac{\partial A_k}{\partial x_j} - \frac{\partial A_j}{\partial x_k} + A_j A_k - A_k A_j = \left[\frac{\partial}{\partial x_j} + A_j, \frac{\partial}{\partial x_k} + A_k \right]. \tag{5.4.10}$$

Show that $F = 0$ if and only if the differential operators $\frac{\partial}{\partial x_j} + A_j$ commute.

In [97, 98] there is a discussion of how the vanishing of curvature is related to complete integrability of the systems of differential equations

$$\frac{\partial Y}{\partial x_j} + A_j Y = 0 \qquad (j = 1, \dots, n)$$

and how this is interpreted in terms of the Galois theory of linear differential equations.

Exercise 5.4.4 Let B_j, C_j be rational functions of x with values in $sl(2, \mathbf{C})$, and let

$$\begin{aligned} B(x, \lambda) &= B_0(x) + B_1(x)\lambda + B_2(x)\lambda^2, \\ C(x, \lambda) &= C_0(x) + C_1(x)\lambda + C_2(x)\lambda^2 \end{aligned} \tag{5.4.11}$$

be matrix functions, and then let

$$A = \left(\frac{\partial}{\partial \lambda} - B \right) d\lambda + \left(\frac{\partial}{\partial x} - C \right) dx. \tag{5.4.12}$$

Show that the curvature is

$$F = dA + A \wedge A = \left(\frac{\partial B}{\partial x} - \frac{\partial C}{\partial \lambda} + [B, C] \right) d\lambda dx. \tag{5.4.13}$$

Show also the pair of ordinary differential equations

$$\begin{aligned} \frac{d\Psi}{d\lambda} &= B\Psi, \\ \frac{d\Psi}{dx} &= C\Psi \end{aligned} \tag{5.4.14}$$

has a non-trivial solution, only if $F = 0$. The solutions in certain cases are discussed in [36, ch. 4].

Example 5.4.5 (Monodromy) Let $A_j \in M_m(\mathbf{C})$ be constant matrices such that $\sum_{j=1}^n A_j = 0$, and for distinct $a_j \in \mathbf{C}$ consider the differential equation

$$\frac{d\Psi}{dz} + \left(\sum_{j=1}^{n} \frac{A_j}{z - a_j}\right) \Psi(z) = 0 \qquad (z \in \mathbf{C} \cup \{\infty\}, \tag{5.4.15}$$

where Ψ is an $m \times m$ matrix. With $S = \{a_1, \dots, a_n\}$, we regard

$$A(z) = \sum_{j=1}^{n} \frac{A_j}{z - a_j} \tag{5.4.16}$$

as a rational matrix function with possible poles in S; the condition $\sum_{j=1}^n A_j = 0$ ensures that ∞ is not a singular point. Let R_S be the algebra of rational functions with possible poles in S, and recall from Proposition 2.9.5 that all finitely generated projective modules over R_S are free. In Example 8.1.8, we interpret $\nabla = d/dz + A(z)$ as an example of a holomorphic connection on a complex vector bundle over the punctured sphere $\mathbf{X} = (\mathbf{C} \cup \{\infty\}) \setminus S$. One usually regards ∇ as a regular singular connection on $\mathbf{C} \cup \{\infty\}$, since all the poles are simple.

We draw a circle around each singular point a_j with small radius $\delta > 0$, and construct a holomorphic solution matrix Ψ on the cut disc $D(a_j, 2\delta) \setminus [a_j, a_j + 2\delta)$. Then as we move around the circle $C(a_j, \delta)$ the solution changes from $\Psi(z)$ to $\Psi(z)M_j$, where $M_j \in GL_m(\mathbf{C})$ is independent of the choice of δ for all sufficiently small $\delta > 0$. Then the matrices M_j for $1 = 1, \dots, n$ give the monodromy data for the differential equation.

Let $\pi_1(\mathbf{X}, \infty)$ be the fundamental group of the space $\mathbf{X}$ relative to the base point. In 5.1 of [99], the authors construct a homomorphism $\pi_1(\mathbf{X}, \infty) \to GL_m(\mathbf{C})$, called the monodromy map, and the image is the monodromy group of ∇.

5.5 Dirac Operators

Results in index theory are usually formulated for complex vector bundles over even-dimensional manifolds, and it is crucially important that the operators have a suitable grading. A particularly important class of examples is given by Dirac operators. In this section, we give a brief account of how these are defined.

Let σ_j be the Pauli matrices of Example 1.3.2. Then the 4×4 matrices

$$\gamma_0 = \begin{bmatrix} 0 & \sigma_0 \\ \sigma_0 & 0 \end{bmatrix}, \quad \gamma_j = \begin{bmatrix} 0 & \sigma_j \\ -\sigma_j & 0 \end{bmatrix} \tag{5.5.1}$$

satisfy

$$\gamma_0^2 = I, \quad \gamma_1^2 = \gamma_2^2 = \gamma_3^2 = -I \tag{5.5.2}$$

and anti-commute so that

$$\gamma_j \gamma_k = -\gamma_k \gamma_j \qquad (j \neq k). \tag{5.5.3}$$

Hence the Dirac operator

$$D_0 = \gamma_0 \frac{\partial}{\partial x_0} + \gamma_1 \frac{\partial}{\partial x_1} + \gamma_2 \frac{\partial}{\partial x_2} + \gamma_3 \frac{\partial}{\partial x_3} \tag{5.5.4}$$

satisfies

$$D_0^2 = I \otimes \left(\frac{\partial^2}{\partial x_0^2} - \left(\frac{\partial^2}{\partial x_1^2} + \frac{\partial^2}{\partial x_2^2} + \frac{\partial^2}{\partial x_3^2} \right) \right). \tag{5.5.5}$$

In the physics literature [40], D_0 is known as Dirac's operator and sometimes referred to as the square root of the d'Alembert operator. In this context, x_0 represents time, and x_1, x_2, x_3 the coordinates of $\mathbf{R}^3$. This is evidently related to the wave equation (5.2.9). We form D_0 by taking the gradient ∇ in (x_0, x_1, x_2, x_3), and then we contract by pairing with $(\gamma_0, \gamma_1, \gamma_2, \gamma_3)$. This construction extends readily to four-dimensional Riemannian manifolds as in [28, p. 77].

Exercise 5.5.1 (Quaternion coefficients) Let $\mathcal{F}$ be the Fourier transform operator defined via $\mathcal{F}f(\xi) = \int_{\mathbf{R}^3} e^{-i\xi\cdot x} f(x)\,dx/(2\pi)^{3/2}$, so that $\mathcal{F}$ gives a unitary operator $L^2(\mathbf{R}^3; \mathbf{C}^4) \to L^2(\mathbf{R}^3; \mathbf{C}^4)$. (See also Exercise 5.8.4.) Consider the differential operator

$$T_0 f = \gamma_0 f + \gamma_1 \frac{\partial f}{i \partial x_1} + \gamma_2 \frac{\partial f}{i \partial x_2} + \gamma_3 \frac{\partial f}{i \partial x_3} \qquad (f \in C_c^\infty(\mathbf{R}^3; \mathbf{C}^4)). \tag{5.5.6}$$

(i) Show that the operator $\mathcal{F}^* T_0 \mathcal{F}$ in $L^2(\mathbf{R}^3; \mathbf{C}^4)$ is given by multiplication by

$$\gamma_1 \xi_1 + \gamma_2 \xi_2 + \gamma_3 \xi_3 + \gamma_0 \tag{5.5.7}$$

which has an inverse given by the operator of multiplication by

$$\frac{\gamma_1 \xi_1 + \gamma_2 \xi_2 + \gamma_3 \xi_3 + \gamma_0}{\xi_1^2 + \xi_2^2 + \xi_3^2 + 1}. \tag{5.5.8}$$

(ii) Deduce that the spectrum of T_0 is $(-\infty, -1] \cup [1, \infty)$, and that T_0 has no eigenvalues. See exercise 9.5 (4) from [41]. The interval $(-1, 1)$ is known as a spectral gap.

Next we consider time-independent Dirac operators with potential q. Let $q\colon \mathbf{R}^3 \setminus \{0\} \to \mathbf{R}$ be continuous, and suppose that

$$|q(x)| \le \frac{\mu}{|x|} \qquad (x \in \mathbf{R}^3 \setminus \{0\}) \tag{5.5.9}$$

where $\mu < \sqrt{3}/2$, and consider the differential operator

$$Tf = \gamma_0 f + \gamma_1 \frac{\partial f}{i\partial x_1} + \gamma_2 \frac{\partial f}{i\partial x_2} + \gamma_3 \frac{\partial f}{i\partial x_3} + qf \qquad (f \in C_c^\infty(\mathbf{R}^3; \mathbf{R}^4)). \tag{5.5.10}$$

The potential is allowed to have a singularity at the origin, like the potential of a hydrogen atom. See [105] for discussion of this example. For general q as above the spectrum can be more complicated, and there can be eigenvalues, but there is still a gap in the essential spectrum.

Proposition 5.5.2 (i) *There exists a unique unitary operator on* $L^2(\mathbf{R}^3; \mathbf{C}^4)$ *such that*

$$U = (iI - T)(iI + T)^{-1}$$

on $C_c^\infty(\mathbf{R}^3; \mathbf{R}^4)$.

(ii) *The spectrum of* T *is contained in* $\mathbf{R}$ *and corresponds to the spectrum of* U *via* $\lambda \mapsto (i - \lambda)(i + \lambda)^{-1}$.

(iii) *The essential spectrum of* U *is contained in a proper closed sub-arc of the unit circle.*

Proof (i) This result is obtained from some delicate analytical estimates. See [108].

This result states that the operator T is essentially self-adjoint on $C_c^\infty(\mathbf{R}^3; \mathbf{C}^4)$.

(ii) The operator T is not bounded below, and the spectrum will intersect both the positive and negative real axes. Write $\lambda = \tan(\theta/2)$ where $\theta \in (-\pi, \pi)$, and observe that

$$e^{i\theta} = \frac{i - \lambda}{i + \lambda}, \tag{5.5.11}$$

gives a bijection between $\sigma(T)$ and $\sigma(U) \setminus \{-1\}$, where $\sigma(U) \subseteq S^1$.

(iii) We can apply the theory of Section 3.8 to U, and consider the essential spectrum of U, which corresponds to the essential spectrum of T by the same correspondence. It is known that the essential spectrum of T is contained in

$(-\infty, -1] \cup [1, \infty)$, so we can take $\theta \in [-\pi, -\pi/2] \cup [\pi/2, \pi]$, so $e^{i\theta}$ lies in the left half plane. Hence $\sigma_e(U)$ is a closed proper subset of S^1. □

The spectrum of T may involve some eigenvalues in $(-1, 1)$, so T may not be invertible. While there are advantages in considering differential operators with matrix coefficients, there are sometimes catches. The operator T^2 superficially resembles some sort of Laplace–Beltrami operator, although there are differences.

5.5.1 Dirac Operators

This basic construction extends from $\mathbf{R}^4$ to $\mathbf{R}^{2n}$ by the results of Section 1.5. As in Corollary 1.5.2, let $(e_j)_{j=1}^{2n}$ be a family of $p \times p$ matrices that satisfy the Clifford relations

$$e_\ell e_m + e_m e_\ell = -2\delta_{m\ell} I_p; \tag{5.5.12}$$

then

$$\left(\sum_{m=1}^{2n} a_m e_m\right)\left(\sum_{m=1}^{2n} b_m e_m\right) + \left(\sum_{m=1}^{2n} b_m e_m\right)\left(\sum_{m=1}^{2n} a_m e_m\right) = -2\left(\sum_{m=1}^{2n} a_m b_m\right) I_p. \tag{5.5.13}$$

The real Clifford algebra Cl_{2n} has complexification $Cl_{2n}\mathbf{C}$, which acts by left multiplication on a p-dimensional complex vector space E. By using the Clifford relations, one checks that $\varepsilon = i^n e_1 \dots e_{2n}$ satisfies $\varepsilon = \varepsilon^*$ and $\varepsilon^2 = I$, so E decomposes as a direct sum $E = E_+ \oplus E_-$, where $E_\pm = \{v \in E : \varepsilon v = \pm v\}$ are the eigenspaces. Then we can consider $D = \sum_{m=1}^{2n} e_m \frac{\partial}{\partial x_m}$ over $\mathbf{R}^{2n}$ as an operator in $L^2(\mathbf{R}^{2n}; E)$, such that $\varepsilon D = -D\varepsilon$, and D interchanges $L^2(M; E_+)$ with $L^2(M; E_-)$. Passing to the Fourier transform, we let $\varepsilon_j \in \{\pm 1\}$ and $\xi = (\xi_j)_{j=1}^{2n} \in \mathbf{R}^{2n}$ have $\|\xi\| = (\sum_{j=1}^{2n} \xi_j^2)^{1/2}$ and

$$F = \sum_{j=1}^{2n} \varepsilon_j e_j \frac{\xi_j}{\|\xi\|},$$

so that $F^2 = I$ and $F^* = F$. Then F is the Fourier multiplier that represents $|D|^{-1} D$ on $L^2(\mathbf{R}^{2n}; E)$. Note that $SO(2n)$ acts on E.

5.5.2 Spin Structure

There is an important class of Riemannian manifolds to which we can generalize this construction. Suppose that M is compact, smooth and connected with dimension $2n$ and is orientable, and consider the fibres T_xM of the tangent bundle, which are isomorphic to $\mathbf{R}^{2n}$. Let M have metric $\sum_{j,k=1}^{2n} g_{jk}dx_jdx_k$ in local coordinates, so $g = [g_{ij}]_{i,j=1}^{2n}$ is positive definite and let $g^{-1} = [g^{ij}]_{i,j=1}^{2n}$ be the inverse matrix, which is also positive definite. Hence there exists a positive square root $q = [q^{ij}]_{i,j=1}^{2n}$. As in Section 1.4, we introduce $\gamma_j = \sum_{k=1}^{2n} q^{jk}e_k$, so by repeatedly using the above identity, we deduce that

$$\gamma_j\gamma_k + \gamma_k\gamma_j = -2g^{jk}I_p, \tag{5.5.14}$$

so we have a Clifford algebra C_q. As before, we introduce in local coordinates the differential operator

$$D = \sum_{j=1}^{2n} \gamma_j \otimes \left(\frac{\partial}{\partial x_j} + \alpha_j(x)\right) \tag{5.5.15}$$

for some smooth matrix functions α_j, so that D satisfies

$$D^2 = -I_p \otimes \sum_{j,k=1}^{2n} g^{ij}\frac{\partial^2}{\partial x_j \partial x_k} + \text{lower order terms.} \tag{5.5.16}$$

Then there is a natural action of the rotation group $SO(2n)$ on T_xM, and $SO(2n)$ is the structure group of the vector bundle TM. We complexify TM and follow the above construction of the graded space $E = E_+ \oplus E_-$. In the case of a spin manifold, the action of $SO(2n)$ extends to an action by $Spin(2n)$ on a vector bundle over M with graded fibres E which gives a spin structure. By results of Sections 5.2 and 5.8, one can show that $(e^{-tD^2})_{t>0}$ defines a semigroup of trace class operator on $L^2(M, \mu; E)$. In the next section, we consider traces and supertraces.

5.6 Theta Summable Fredholm Modules

We begin with the concept of a supertrace. Given isomorphic Hilbert spaces $H_\pm$ and $S_{11}, S_{12}, S_{21}, S_{22} \in \mathcal{L}^1(H_\pm, H_\mp)$, and

$$S = \begin{bmatrix} S_{11} & S_{12} \\ S_{21} & S_{22} \end{bmatrix}, \qquad F = \begin{bmatrix} I & 0 \\ 0 & -I \end{bmatrix} \quad \begin{matrix} H_+ \\ H_- \end{matrix} \tag{5.6.1}$$

we define a supertrace with respect to the grading on the sum of Hilbert spaces by $\tau_F(S) = \operatorname{trace}(S_{11}) - \operatorname{trace}(S_{22})$. For instance, we consider

$$D = \begin{bmatrix} 0 & T_0 \\ T_1 & 0 \end{bmatrix}, \qquad e^{-tD^2} = \begin{bmatrix} e^{-tT_0T_1} & 0 \\ 0 & e^{-tT_1T_0} \end{bmatrix} \quad \begin{matrix} H_+ \\ H_- \end{matrix}. \tag{5.6.2}$$

Proposition 5.6.1 *Suppose that T_0 and T_1 are densely defined with $T_1^* = T_0$ and $T_0^* = T_1$, and that $(e^{-tT_1T_0})_{t>0}$ and $(e^{-tT_0T_1})_{t>0}$ are semigroups of self-adjoint operators such that $e^{-tT_1T_0}$ and $e^{-tT_0T_1}$ are of trace class for all $t > 0$. Then*

$$\tau_F(e^{-tD^2}) = \dim\ker(T_1) - \dim\ker(T_0). \tag{5.6.3}$$

Proof In the literature, this result is often written in terms of $\mathrm{Ind}(T_1) = \dim\ker(T_1) - \dim\ker(T_1^*)$, with the interpretation that T_1 is an elliptic operator; of course, T_1 here is generally not bounded and hence not Fredholm. In view of Exercise 9.2.4, we need to be cautious about some of the spectral inclusions.

For $-A$ the generator of a strongly continuous semigroup $(e^{-tA})_{t>0}$ on H, we have a point spectrum $\sigma_p(A) = \{\lambda \in \mathbf{C} : \exists\, \varphi \in H, \varphi \neq 0, A\varphi = \lambda\varphi\}$, which satisfies the spectral mapping theorem which relates the point spectrum of the semigroup to the generator, namely

$$\{e^{-t\lambda} : \lambda \in \sigma_p(A)\} \subseteq \sigma_p(e^{-tA}) \subseteq \{e^{-t\lambda} : \lambda \in \sigma_p(A)\} \cup \{0\}$$

by [41, p. 64]. We compare these point spectra for the choices of A given by T_1T_0 and T_0T_1.

The operators satisfy $DF + FD = 0$. Since $e^{-tT_1T_0}$ and $e^{-tT_0T_1}$ are trace class, the operators T_1T_0 and T_0T_1 have a discrete spectrum; since the semigroups are self-adjoint, these spectra are real. For $\lambda \neq 0$ an eigenvalue of T_1T_0 with corresponding eigenvector φ_λ, we have $T_1T_0\varphi_\lambda = \lambda\varphi_\lambda$, so $T_0\varphi_\lambda \neq 0$ and $T_0T_1T_0\varphi_\lambda = \lambda T_0\varphi_\lambda$; hence there is an injective map $\varphi \mapsto T_0\varphi$ from the finite-dimensional eigenspace $\{\varphi : T_1T_0\varphi = \lambda\varphi\}$ to the finite-dimensional eigenspace $\{\varphi : T_0T_1\varphi = \lambda\varphi\}$. By symmetry, we deduce that

$$\dim\{\varphi : T_1T_0\varphi = \lambda\varphi\} = \dim\{\varphi : T_0T_1\varphi = \lambda\varphi\} \qquad (\lambda \neq 0). \tag{5.6.4}$$

For $\lambda = 0$, we have finite-dimensional eigenspaces $\{\varphi : T_1T_0\varphi = 0\}$ and $\{\varphi : T_0T_1\varphi = 0\}$, possibly of different dimension, so

$$\mathrm{trace}(e^{-tT_0T_1}) = \sum_{\lambda \in \sigma(T_1T_0)} e^{-t\lambda}\dim\{\varphi : T_1T_0\varphi = \lambda\varphi\}. \tag{5.6.5}$$

Then since $T_1 = T_0^*$, we have a complex

$$H_+ \underset{T_0}{\overset{T_1}{\rightleftarrows}} H_-$$

so that

$$\begin{aligned}
\operatorname{Ind}(D) &= \dim\ker(T_1) - \dim\ker(T_0) \\
&= \dim\ker(T_0T_1) - \dim\ker(T_1T_0) \\
&= \operatorname{trace}(e^{-tT_0T_1}) - \operatorname{trace}(e^{-tT_1T_0}) \\
&= \tau_F(e^{-tD^2}). \qquad (5.6.6)
\end{aligned}$$

□

The most striking feature of (5.6.3) is that the right-hand side is independent of t, while the left-hand side is computable if one knows the kernel $Fe^{-tD^2}(x,y)$ on $x = y$. Given the formula (5.2.13), it reasonable to suppose that for small $t > 0$, the kernel $Fe^{-tD^2}(x,x)$ can be computed in terms of local geometrical quantities. Calculations of this form appear in papers such as [74] and can be extended to an abstract framework, as follows.

Connes [18, 19] considered the basic ingredients of spectral theory to consist of (A, H, D), where H is a Hilbert space, A a $*$-algebra of operators in H, and D a self-adjoint densely defined Dirac operator in H. By the spectral theorem, one can introduce an involution operator F on H by $F = \operatorname{Sign}(D)$, and an orthogonal projection $e = (1 + F)/2$. Suppose that $[F,a]$ determines an operator in $\mathcal{L}^n(H)$ for all $a \in A$; then we can define the cyclic cocycle

$$\tau(a_0,\dots,a_n) = \operatorname{Trace}(a_0[F,a_1]\dots[F,a_n]) \qquad (a_j \in A), \qquad (5.6.7)$$

and one can compare this with the definition in Lemma 3.5.5. Interpreting this formula involves serious technical problems, which are partially addressed in the special case Theorem 9.6.3. Alternatively, we smooth out the cocycle by interposing heat operators between the commutators

$$\tau(a_0,\dots,a_n) = \operatorname{Trace}(a_0e^{-\hbar D^2}[F,a_1]e^{-\hbar D^2}\dots e^{-\hbar D^2}[F,a_n]); \qquad (5.6.8)$$

and then let $\hbar \to 0$. We consider this in Section 11.12.

Example 5.6.2 (Fredholm module) Let M be a compact and connected Riemannian manifold with volume measure μ, and let $A = C^\infty(M;\mathbf{C})$. Let D be a densely defined and self-adjoint first-order differential operator in a Hilbert subspace H of $L^2(M,\mu)$ such that $\{f\colon Df = 0\} = \{0\}$. Then there exists $\lambda_1 > 0$ such that

$$\lambda_1 \int_M |f(x)|^2\mu(dx) \le \int_M |Df(x)|^2\mu(dx). \qquad (5.6.9)$$

The self-adjoint operator D^2 has a positive operator square root $(D^2)^{1/2}$, so we can define $F = D(D^2)^{-1/2}$ such that $F^2 = 1$ and $F^* = F$. In the case of

$\Delta = D^2$, we can use Theorem 5.3.2 and take $H = (I - P)L^2(M)$, where P is the projection onto the constant functions.

A compact operator S on Hilbert space has $(S^*S)^{1/2}$ compact and self-adjoint, with eigenvalues $s_0 \geq s_1 \geq$ and $s_j \to 0$ as $j \to \infty$. The s_j are the singular numbers of S. However, the s_j can decrease arbitrarily slowly, so $\sum_{j=0}^{\infty} e^{-1/s_j^2}$ can diverge. Let J be the ideal in $\mathcal{L}(H)$ given by

$$J = \{S \in \mathcal{K}(H) \colon s_j \leq C/(\log(j+2))^{1/2}; j = 0, 1, 2, \dots \quad \text{for some} \quad C\}. \tag{5.6.10}$$

Let A be a C^*-algebra and (H, F, π, ε) a graded Fredholm module over A. Then (H, F, π, ε) is ϑ-summable if

$$\mathcal{A} = \{a \in A \colon [a, F] \in J\} \tag{5.6.11}$$

is dense in A for the norm topology. The notation ϑ alludes to the heat kernel over the circle, as in Section 5.1. Then Connes [19] proves that there exists a self-adjoint D in H such that:

(a) $D(D^2)^{-1/2} = F$;
(b) $\{a \in A \colon [a, D] \in \mathcal{L}(H)\}$ is dense in A for the norm topology;
(c) the heat operator $e^{-\beta D^2}$ is trace class for some $\beta > 0$.

Up to unitary equivalence, knowledge of the self-adjoint semigroup $(T_t)_{t>0}$ of Proposition 5.2.2 is equivalent to knowledge of the resolvent family $(\lambda I + \Delta)^{-1}$ for all $\lambda \in \mathbf{C} \setminus \sigma(\Delta)$; and for M a smooth compact manifold, this is equivalent to knowledge of the spectrum of Δ with multiplicity. In the 1980s, there was intense interest in the question of how much geometrical information is determined by the partition function $\mathrm{trace}(e^{-t\Delta/2})$ for small $t > 0$. Exercise 5.6.3 indicates how this translates into information about $\mathrm{trace}((\lambda + \Delta)^{-(n+2)/2})$. Connes's result deals with a much more general situation, where $e^{-\beta D^2}$ is trace class for some but not all $\beta > 0$, so we assume much less about the spectral multiplicity of D^2.

Exercise 5.6.3 (Tauberian theorem) The asymptotic expansion of the trace heat kernel [7, theorem E.III] can be expressed in various ways. Let Δ be the Laplace operator on a compact Riemannian manifold of dimension n, and suppose that there exist $R, \beta, M > 0$ such that $g(t) = t^{n/2}\mathrm{trace}(e^{-t\Delta})$ satisfies

$$g(t) \leq Me^{\beta t} \qquad (t > 0) \tag{5.6.12}$$

and $g(t)$ has a convergent power series expansion

$$g(t) = \sum_{k=0}^{\infty} \frac{c_k t^k}{k!} \qquad (0 < t < R). \tag{5.6.13}$$

(i) Show that the Laplace transform of g is given by

$$\int_0^\infty g(t)e^{-\lambda t}\, dt = \Gamma((n+2)/2)\mathrm{trace}\big((\lambda I + \Delta)^{-(n+2)/2}\big) \qquad (\lambda > 0). \tag{5.6.14}$$

(ii) Use Watson's Lemma [100] to deduce that the Laplace transform of g has an asymptotic expansion

$$\int_0^\infty e^{-\lambda t} g(t)\, dt \asymp \sum_{k=0}^{\infty} \frac{c_k}{\lambda^{k+1}} \qquad (\lambda \to \infty). \tag{5.6.15}$$

5.7 Duhamel's Formula

Let M be a compact Riemannian manifold with volume measure μ, and let $A = C^\infty(M; \mathbf{C})$. Suppose that L is a Laplace-type operator that gives rise to a semigroup e^{tL} on $H = L^2(\mu)$ which satisfies (i)–(vii) of Proposition 5.2.2. In this section we show how to perturb L to produce an operator K with similar properties. The main application is to $L = -\Delta$ and $X = K - L$, a differential operator of order zero or one.

For such a semigroup, the resolvent is the family of linear operators on H defined by the Laplace transform

$$(\lambda - L)^{-1} = \int_0^\infty e^{-t\lambda} e^{tL}\, dt \tag{5.7.1}$$

on $\{\lambda \in \mathbf{C} \colon \Re\lambda > \omega_0\}$ for some $\omega_0 \in \mathbf{R}$.

Lemma 5.7.1 *(1) Let $(e^{tL})_{t>0}$ be a strongly continuous semigroup of bounded linear operators on H such that $\|e^{tL}\| \leq Me^{\beta t}$ for all $t > 0$ and some $\beta, M > 0$, and let $X \in \mathcal{L}(H)$. Then one can solve the resolvent equation*

$$(\lambda - L - X)^{-1} = (\lambda - L)^{-1} + (\lambda - L)^{-1} X (\lambda - L - X)^{-1}. \tag{5.7.2}$$

There exists a semigroup $e^{t(L+X)}$ satisfying (i), (ii), (iii) and (vii) of Proposition 5.2.2 such that

$$e^{t(L+X)} = e^{tL} + \int_0^t e^{(t-s)L} X e^{s(L+X)}\, ds. \tag{5.7.3}$$

(2) Suppose that X is a closed linear operator with domain including the domain of L, such that $Xe^{tL} \in \mathcal{L}(H)$ for all $t > 0$ and $\int_0^1 \|Xe^{tL}\|dt$ converges. Then the conclusions of (1) hold.

Proof (1) When $\Re\lambda$ is sufficiently large, one can solve the resolvent equation using geometric series in $(\lambda - L)^{-1}X$. Then one can produce a semigroup $e^{t(L+X)}$ from the resolvents $(\lambda - L - X)^{-1}$ by the Hille–Yoshida generation theorem [41, 31] via $e^{t(L+X)} = \lim_{n\to\infty}(1 - t(L + X)/n)^{-n}$. The resolvent equation is the Laplace transform of the semigroup equation (5.7.3), so the expressions (5.7.2) and (5.7.3) are equivalent. By repeatedly substituting the left-hand side into the integrand on the right-hand side, we obtain

$$e^{t(L+X)} = e^{tL} + \sum_{n=1}^{\infty} \int_{0\leq t_j;\, t_1+\dots+t_n<t} e^{(t-t_1-\dots-t_n)L} X e^{t_1 L} X e^{t_2 L} \dots X e^{t_n L} dt_1 \dots dt_n, \tag{5.7.4}$$

where we can view the nth integral as being an n-fold Laplace convolution, taken over a simplex in the positive cone of $\mathbf{R}^n$. The volume of this simplex is $t^n/n!$, so with $\|X\| \leq M_1$, the series is dominated by the convergent series $\sum_n M_1^n M^{n+1} e^{\beta t} t^n/n!$, hence converges in operator norm.

(2) The proof of this is similar and given in theorem VIII.1.19 of [31]. □

Example 5.7.2 (Perturbed Laplace operator) Lemma 5.7.1(2) applies to the case of $H = L^2(M)$ with M a compact Riemannian manifold, $L = -\Delta$ the Laplace operator and X a first-order differential operator with smooth coefficients. By (5.2.20), we have

$$\|\nabla T_t f\| \leq \|\Delta T_t f\|^{1/2} \|\Delta T_t f\|^{1/2}, \tag{5.7.5}$$

where by the spectral theorem applied to $T_t = e^{-t\Delta/2}$, we have

$$\begin{aligned} \|\Delta T_t f\|^2 &= \int_0^\infty \lambda^2 e^{-\lambda t} \langle e(d\lambda) f, f\rangle \\ &\leq \frac{2}{t^2} \int_0^\infty \langle e(d\lambda) f, f\rangle \\ &= \frac{2}{t^2} \|f\|^2 \end{aligned} \tag{5.7.6}$$

so $\|\nabla T_t f\| \leq 2^{1/4}\|f\|/\sqrt{t}$, and Lemma 5.7.1(2) applies.

Proposition 5.7.3 (Duhamel's formula) *Suppose that $K(\hbar) = L + X(\hbar)$ for $\hbar \geq 0$ is a one-parameter family of operators in H such that*

(i) $(e^{tK(\hbar)})_{t>0}$ *are semigroups where L and $X(\hbar)$ satisfy Lemma 5.7.1;*

(ii) $e^{tK(0)} \in \mathcal{L}^1(H)$ *for all* $t > 0$;
(iii) $K(\hbar) \to K(0)$ *as* $\hbar \to 0+$, *with*

$$\delta K = \lim_{\hbar \to 0+} \frac{K(\hbar) - K(0)}{\hbar} \in \mathcal{L}(H).$$

Then the usual trace $\tau : \mathcal{L}^1(H) \to \mathbf{C}$ *and right-hand derivative satisfy*

$$\left(\frac{d}{d\hbar}\right)_{\hbar=0} \tau(e^K) = \tau\big(e^{K(0)}\delta(K)\big). \tag{5.7.7}$$

Proof We take the limit of

$$\frac{e^{K(\hbar)} - e^{K(0)}}{\hbar} = \int_0^1 e^{(1-t)K(\hbar)} \frac{K(\hbar) - K(0)}{\hbar} e^{tK(0)}\, dt \tag{5.7.8}$$

as $\hbar \to 0+$, to obtain

$$\delta e^K = \int_0^1 e^{(1-t)K(0)} \delta(K(0)) e^{tK(0)}\, dt, \tag{5.7.9}$$

where $e^{(1-t)K(0)}$ and $e^{tK(0)}$ belong to $\mathcal{L}^1(H)$, so we can rearrange

$$\begin{aligned} \tau(\delta e^K) &= \tau\left(\int_0^1 e^{(1-t)K(0)} \delta(K) e^{tK(0)}\, dt\right) \\ &= \tau(e^{K(0)} \delta K). \end{aligned} \tag{5.7.10}$$

□

Exercise 5.7.4 (Zeta functions) (i) Let $H = L^2(S^1, d\theta/(2\pi))$ and

$$P^\perp H = \left\{ f \in L^2(S^1, d\theta/(2\pi)) : \int f(\theta)\, d\theta = 0 \right\}$$

and let Δ be the self-adjoint operator in $P^\perp H$ given by $\Delta f = -f''$. Find the spectrum of Δ, and hence find

$$\zeta_\Delta(s) = \frac{1}{\Gamma(s)} \int_0^\infty t^{s-1} \operatorname{trace}\big(e^{-t\Delta}\big) dt \qquad (\Re s > 1/2). \tag{5.7.11}$$

(ii) Let M be a compact and connected Riemannian manifold without boundary, let $H = L^2(M; \mu)$ and let $P^\perp H$ be as in Theorem 5.3.2. With Δ the Laplace operator restricted to $P^\perp H$, show that there exists s_0 such that the formula (5.7.11) defines a holomorphic function of s for $\Re s > s_0$.

(iii) Let $X \in \mathcal{L}(P^{\perp}H)$ and define $(e^{-t(\Delta+\hbar X)})_{t>0}$ as in Lemma 5.7.1. Show that

$$\left(\frac{d}{d\hbar}\right)_{\hbar=0} \zeta_{\Delta+\hbar X}(s) = \frac{-1}{\Gamma(s)}\int_0^\infty t^s \operatorname{trace}\big(Xe^{-t\Delta}\big)dt \qquad (\Re s > s_0). \tag{5.7.12}$$

This zeta function has been used to compute determinants associated with Δ. In particular, this calculation was used by Quillen in his computation of the curvature of the determinant line bundle; see [28, 86].

5.8 Quantum Harmonic Oscillator

The quantum harmonic operator has various roles in cyclic theory. The number operator with spectrum $\{0, 1, 2, \dots\}$ generates the Ornstein–Uhlenbeck operator semigroup, although it can be expressed as a second-order differential operator on $\mathbf{R}$ of limit point type.

Example 5.8.1 (Hermite polynomials) We express the usual grading on $\mathbf{C}[x]$ in a style suitable for the quantum harmonic oscillator. The standard grading is expressed by the operator $H_0 : \mathbf{C}[x] \to \mathbf{C}[x]$, and given by

$$H_0 \sum_{j=0}^{\infty} a_j x^j = \sum_{j=0}^{\infty} j a_j x^j. \tag{5.8.1}$$

We introduce $\partial = \frac{\partial}{\partial x}$ and $\partial^* = -\partial + x$, so $\partial^* f = -e^{x^2/2}\partial(e^{-x^2/2} f)$. Also, the anti-commutation relation for the quantum harmonic oscillator may be expressed as

$$[\partial, \partial^*] = 1. \tag{5.8.2}$$

We define the (probabilists') Hermite polynomials by the recurrence relation

$$\begin{aligned} He_0 &= 1, \\ He_n &= \partial^* He_{n-1}, \end{aligned} \tag{5.8.3}$$

so He_n is monic and of degree n. Then by induction one proves that

$$\begin{aligned} \partial He_n &= n He_{n-1}, \\ \partial^* \partial He_n &= n He_n. \end{aligned} \tag{5.8.4}$$

Proposition 5.8.2 (Mehler's formula) (i) *The operator*

$$L = -\hbar^2 \frac{d^2}{dx^2} + x^2 \tag{5.8.5}$$

in $L^2(e^{-x^2/2\hbar} dx/\sqrt{2\pi\hbar})$ *gives rise to a semigroup* e^{-tL} *satisfying Proposition 5.2.2(i)–(vi);*

(ii) *the kernel of the corresponding integral operator on* $L^2(\mathbf{R})$ *corresponding to* $e^{-L/2}$ *is*

$$M(x, y; \hbar/2) = \frac{1}{\sqrt{1 - e^{-\hbar}}} \exp\left(-\frac{e^{-\hbar/2}y^2 - 2xy + e^{-\hbar/2}x^2}{4\hbar \sinh(\hbar/2)} - \frac{x^2 + y^2}{\hbar}\right) \quad (x, y \in \mathbf{R}); \tag{5.8.6}$$

(iii)

$$e^{-tH} f \to \int_{-\infty}^{\infty} f(x) \frac{e^{-x^2/2\hbar} dx}{\sqrt{2\pi\hbar}} \qquad (t \to \infty). \tag{5.8.7}$$

Proof (i) We introduce

$$\varphi_n(x) = \frac{He_n(\sqrt{2}x)e^{-x^2/2}}{\pi^{1/4}\sqrt{n!}} \tag{5.8.8}$$

so that the raising and lowering operations of (5.8.3) and (5.8.4) become

$$\begin{aligned} \left(-\frac{d}{dx} + x\right)\varphi_n(x) &= \sqrt{2}\sqrt{n+1}\varphi_{n+1}(x), \\ \left(\frac{d}{dx} + x\right)\varphi_n(x) &= \sqrt{2n}\varphi_{n-1}(x). \end{aligned} \tag{5.8.9}$$

Taking the Fourier transform, we find

$$\hat{\varphi}_0(x) = \sqrt{2\pi}\,\varphi_0(x) \tag{5.8.10}$$

and the raising and lowering operations are unchanged, up to constants

$$\begin{aligned} \left(i\frac{d}{d\xi} - i\xi\right)\hat{\varphi}_n(\xi) &= \sqrt{2}\sqrt{n+1}\hat{\varphi}_{n+1}(\xi), \\ \left(i\frac{d}{d\xi} + i\xi\right)\hat{\varphi}_n(\xi) &= \sqrt{2n}\hat{\varphi}_{n-1}(x); \end{aligned} \tag{5.8.11}$$

so that we can solve the recurrence relations and prove

$$\hat{\varphi}_n(\xi) = (-i)^n \sqrt{2\pi}\,\varphi_n(\xi). \tag{5.8.12}$$

By Fourier analysis, one shows that the $(\varphi_n)_{n=0}^{\infty}$ give a complete orthogonal basis for $L^2(\mathbf{R})$, or equivalently that the h_n give a complete orthogonal basis for $L^2(e^{-x^2/2}dx/\sqrt{2\pi})$. Now $L = \partial^*\partial$ is known as the number operator or Ornstein–Uhlenbeck operator, so H determines a self-adjoint operator in $L^2(\mathbf{R}; e^{-x^2/2})$ with spectrum $\{0, 1, 2, \dots\}$, hence is unitarily equivalent to a diagonal matrix with diagonal entries $0, 1, 2, \dots$. So by rescaling $x \mapsto x/\sqrt{\hbar}$, we deduce that $\rho_t = e^{-tL}$ is the Ornstein–Uhlenbeck semigroup, which satisfies Proposition 5.2.2(i)–(vi) on $L^2(\mathbf{R}; e^{-x^2/2\hbar})$.

(ii) We have

$$LHe_n(x/\sqrt{\hbar}) = n\hbar He_n(x/\sqrt{\hbar}) \tag{5.8.13}$$

and

$$\int_{-\infty}^{\infty} He_n(x)He_m(y)e^{-x^2/2\hbar}\frac{dx}{\sqrt{2\pi\hbar}} = \delta_{nm}n!, \tag{5.8.14}$$

so

$$e^{-tL/2} \leftrightarrow \sum_{n=0}^{\infty}\frac{e^{-tn\hbar/2}}{n!}He_n(y/\sqrt{\hbar})He_n(x/\sqrt{\hbar}). \tag{5.8.15}$$

By Mehler's formula we have for $0 < \theta < 1$,

$$\sum_{n=0}^{\infty}\frac{\theta^n}{n!}He_n(y)He_n(x) = \frac{1}{\sqrt{1-\theta^2}}\exp\left(-\frac{\theta^2y^2 - 2\theta xy + \theta^2x^2}{2(1-\theta^2)}\right). \tag{5.8.16}$$

For $0 < \theta < 1$ and $\sqrt{(1+\theta)/\theta} > \beta > 0$, we observe that $\varphi_n(\beta x)$ satisfy

$$-\frac{1}{\beta^2}\frac{d^2}{dx^2}\varphi_n(\beta x) + \beta^2x^2\varphi_n(\beta x) = (2n+1)\varphi_n(\beta x)$$

and Mehler's formula suggests we consider

$$\begin{aligned}
M(x, y; t) &= \beta\sum_{n=0}^{\infty}\theta^n\varphi_n(\beta x)\varphi_n(\beta y)\\
&= \frac{\beta}{\sqrt{\pi}}\sum_{n=0}^{\infty}\frac{\theta^n}{n!}He_n(\sqrt{2}\beta x)He_n(\sqrt{2}\beta y)e^{-\beta^2(x^2+y^2)/2}\\
&= \frac{\beta}{\sqrt{\pi(1-\theta^2)}}\exp\left(-\frac{\beta^2\theta^2y^2 - 2\beta^2\theta xy + \beta^2\theta^2x^2}{1-\theta^2}\right.\\
&\qquad\left.-\frac{\beta^2x^2 + \beta^2y^2}{2}\right).
\end{aligned} \tag{5.8.17}$$

The quadratic form in the last expression involves the real symmetric matrix

$$\frac{1}{2}\frac{\beta^2}{1-\theta^2}\begin{bmatrix}1+\theta^2 & -2\theta\\ -2\theta & 1+\theta^2\end{bmatrix} = \frac{1}{2t\sinh t}\begin{bmatrix}\cosh t & -1\\ -1 & \cosh t\end{bmatrix}, \tag{5.8.18}$$

when we take $\beta = 1/\sqrt{t}$ and $\theta = e^{-t}$. The determinant of the matrix is $1/(4t^2)$, hence (as in (5.9.14)) the kernel has integral

$$\iint_{\mathbf{R}^2} M(x, y; t)dxdy = \frac{\beta t\sqrt{\pi}}{\sqrt{1-\theta^2}} = \frac{\sqrt{\pi}e^{t/2}}{\sqrt{2}}\sqrt{\frac{t}{\sinh t}}. \tag{5.8.19}$$

Finally, let $t/2 = \hbar$.

(iii) It is evident from the expansion (5.8.15) that

$$e^{-tL/2}f(y) = \sum_{n=0}^{\infty} \frac{e^{-nt\hbar/2}}{n!} He_n(y/\sqrt{\hbar}) \int_{-\infty}^{\infty} f(x)He_n(x/\sqrt{\hbar})e^{-x^2/2\hbar}\frac{dx}{\sqrt{2\pi\hbar}}$$
$$\to \int_{-\infty}^{\infty} f(x)e^{-x^2/2\hbar}\frac{dx}{\sqrt{2\pi\hbar}} \tag{5.8.20}$$

as $t \to \infty$ since $He_0(x/\sqrt{\hbar}) = 1$. □

Now we consider the corresponding statements for even and odd polynomials. Let $Q = \mathbf{C}[x]$ and $R = \mathbf{C}[x^2]$ be the algebra of even complex polynomials, so $Q = R \oplus xR$ is an R module. Hence the summands in $Q = R \oplus xR$ are interchanged by $\partial^*\colon R \to xR$ and $\partial\colon xR \to R$, so $\partial^*\partial\colon xR \to xR$ is invertible.

Let $I = (x^2) = x^2R$, and observe that R/I^{n+1} is the space of even polynomials of degree $\leq 2n$, which is spanned by $He_0, He_2, \ldots, He_{2n}$ mod I^{n+1}.

Proposition 5.8.3 (i) $\rho_t = e^{-tL}$ *determines semigroups*

$$\rho_t : Q \to Q; \quad \rho_t : R \to R; \quad \rho_t : R/I^{n+1} \to R/I^{n+1}; \tag{5.8.21}$$

(ii) $\rho_t \to \rho_\infty$ *as* $t \to \infty$, *where* $\rho_\infty\colon \mathbf{C}[x] \to \mathbf{C}1$ *is the projection onto the constant polynomial;*

(iii) *The map* $U\colon \mathbf{C}[x] \to Q$

$$\sum_{j=0}^{\infty} a_j x^j \to \sum_{j=0}^{\infty} a_j He_j \tag{5.8.22}$$

is invertible and satisfies $H_0 = U^{-1}HU$.

Proof This follows from Proposition 5.8.2. One can express U and HU in terms of an upper triangular matrix. □

Exercise 5.8.4 (Fourier transform and Hermite functions) (i) Show that the Fourier transform $\mathcal{F}f = \hat{f}/\sqrt{2\pi}$ on $L^2(\mathbf{R})$ is a unitary operator with $\mathcal{F}^4 = I$.

(ii) Let φ_n be as in (5.8.8). Show that

$$\varphi_{2n}(\sqrt{\Delta}) = \frac{(-1)^n}{\sqrt{2\pi}} \int_{-\infty}^{\infty} \varphi_{2n}(s) \cos(s\sqrt{\Delta})\, ds. \qquad (5.8.23)$$

(iii) Let $\phi_{2n}(x) = x^{2n}e^{-tx^2/2}$ for $t > 0$, $x \in \mathbf{R}$. Show that $\hat{\phi}_{2n}(s) = p_{2n}(s)e^{-s^2/(2t)}$, where p_{2n} is a real even polynomial of degree $2n$.

5.9 Chern Polynomials and Generating Functions

In Section 8.2, we define the Chern character in the context of differential geometry for connections over manifolds. In this section, we introduce polynomials and generating functions which are used in the subsequent definitions. We can pass between traces and determinants by working in $\mathbf{C}[[t]]$ and using the formula $\log\det(I + tA) = \operatorname{trace}\log(I + tA)$ or equivalently

$$\det(I + tA) = \exp\big(t\operatorname{trace}A - t^2\operatorname{trace}A^2/2 + t^3\operatorname{trace}A^3/3 - \cdots\big).$$

5.9.1 Chern Polynomials

(i) We start with $1 + zt$. Let $A, B \in M_n(\mathbf{C})$ and form the Chern polynomial

$$c(t; A) = \det(I + tA) = \sum_{j=0}^{n} t^j \sigma_j(A) \in \mathbf{C}[t] \qquad (5.9.1)$$

where σ_j is the jth elementary symmetric polynomial in the eigenvalues of A. Note that

$$c(t; A \oplus B) = \det\left(I_{2n} + t\begin{bmatrix} A & 0 \\ 0 & B\end{bmatrix}\right) = c(t; A)c(t; B), \qquad (5.9.2)$$

which is used to define the Chern polynomial of the sum of two vector bundles. Now suppose that A has eigenvalues ξ_j with corresponding eigenvectors X_j and B has eigenvalues η_k with corresponding eigenvectors Y_k; then $A \otimes I + I \otimes B$ has eigenvalues $\xi_j + \eta_k$ with corresponding eigenvectors $X_j \otimes Y_k$; so we introduce

$$\begin{aligned} c(t; A \otimes I + I \otimes B) &= \det\big(I_n \otimes I_n + tA \otimes I + tI \otimes B\big) \\ &= \prod_{j,k=1}^{n} \big(1 + t(\xi_j + \eta_k)\big) \end{aligned} \qquad (5.9.3)$$

to define the Chern polynomial of the tensor product of two vector bundles.

(ii) Now consider e^z. Let

$$\mathrm{ch}\,(A) = \mathrm{trace}(e^A) = n + \sigma_1(A) + \frac{\sigma_1(A)^2 - 2\sigma_1(A)}{2!} + \cdots \tag{5.9.4}$$

be the Chern series of A, which is associated with the Chern character of a vector bundle, as in Section 8.2. We have the addition rule

$$\mathrm{ch}\,(A \oplus B) = \mathrm{trace}\left(\begin{bmatrix} e^A & 0 \\ 0 & e^B \end{bmatrix}\right) = \mathrm{ch}\,(A) + \mathrm{ch}\,(B), \tag{5.9.5}$$

and the multiplication rule

$$\mathrm{ch}\,(A)\mathrm{ch}\,(B) = \mathrm{trace}\big(e^A \otimes e^B\big) \tag{5.9.6}$$

where $e^A \otimes e^B$ is generated by $A \otimes I + I \otimes B$. (This coincides with the addition rule for generators of commuting semigroups.) For reasons discussed in Normalization 8.2.2, the definition of the Chern character used for complex vector bundles involves an extra constant, namely $-A/(2\pi i)$ instead of A.

Definition 5.9.2 (Generating functions) Let

$$f_t(z) = \frac{ze^{tz}}{e^z - 1} \tag{5.9.7}$$

so that f_t defines a meromorphic function of z with simple poles at $z = 2\pi ni$ for $n = \pm 1, \pm 2, \ldots$. There is a convergent power series expansion

$$f_t(z) = \sum_{n=0}^{\infty} \frac{P_n(t)}{n!} z^n \qquad (|z| < 2\pi) \tag{5.9.8}$$

where the $P_n(t)$ turn out to be rational polynomials of degree $\leq n$, known as the Bernoulli polynomials. The Bernoulli numbers are $B_n = P_n(0)$. The following related expansions appear in index theory:

(i) Todd genus

$$f_1(z) = \frac{z}{1 - e^{-z}} = 1 + (1/2)z + \sum_{j=0}^{\infty} \frac{B_{2j} z^{2j}}{(2j)!}; \tag{5.9.9}$$

(ii) Hirzebruch genus

$$(1/2)\big(f_1(2\sqrt{z}) + f_0(2\sqrt{z})\big) = \frac{\sqrt{z}}{\tanh\sqrt{z}} = \sum_{j=0}^{\infty} \frac{2^{2j} B_{2j} z^j}{(2j)!}; \tag{5.9.10}$$

(iii) The $\hat{A}$ genus is related to the supertrace via some determinant formulas. The $\hat{A}$ genus has generating function

$$f_{1/2}(\sqrt{z}) = \frac{\sqrt{z}/2}{\sinh\sqrt{z}/2}. \tag{5.9.11}$$

Remarks 5.9.3 In (5.8.19) we encountered the function

$$f_{1/2}(z) = \frac{z/2}{\sinh(z/2)}, \tag{5.9.12}$$

which is meromorphic with simple poles at $z = 2\pi i n$ for $n \in \mathbf{Z} \setminus \{0\}$), and even, so $f_{1/2}(\sqrt{z})$ is also meromorphic. Also, $1/f_{1/2}(z)$ and $1/f_{1/2}(\sqrt{z})$ are entire. By contour integration, one shows that it satisfies

$$f_{1/2}(x) = \frac{\pi}{2} \int_{-\infty}^{\infty} \frac{e^{ixt}\,dt}{\cosh^2 \pi t} \qquad (x \in \mathbf{R}), \tag{5.9.13}$$

hence the matrix $[f_{1/2}(x_j - x_k)]$ is positive definite.

Finally, we introduce a notion of the square root of a determinant, which arises in various contexts in physics.

Proposition 5.9.4 (Gaussian integrals) *Let A be a real symmetric $n \times n$ matrix with all eigenvalues > 0. Then A is invertible, with real symmetric inverse, and there is real quadratic form $x^t A^{-1} x$ such that*

$$\int_{\mathbf{R}^n} e^{-(1/2)x^t A^{-1} x} dx_1 \dots dx_n = (2\pi)^{n/2} \big(\det A\big)^{1/2}. \tag{5.9.14}$$

Proof To see this, we write $A = U^t D U$ where U is a real orthogonal matrix and D is a diagonal matrix with diagonal entries given by the eigenvalues of A. The volume measure does not change under the transformation $x \mapsto U^t x$, so the integral reduces to a product of one-dimensional Gaussian integrals, where $\int_{-\infty}^{\infty} e^{-x^2/2\lambda} dx = \sqrt{2\pi\lambda}$. □

6
The Algebra of Noncommutative Differential Forms

Noncommutative geometry involves extending concepts from classical geometry, described by commutative algebras, to geometric spaces which are described by noncommutative coordinate algebras. Connes [20] described the system of analogies between classical geometry and noncommutative differential geometry as follows.

Classical	**Noncommutative**
Space	Coordinate algebra
Vector bundle	Finite projective module
Differential form	Class of Hochschild cocycle
De Rham current	Cyclic cohomology
Chern–Weil theory of connections	Pairing of K-theory and cyclic homology

In Sections 6.1 and 6.2 we recall some results about the Kähler differentials of a commutative algebra, with particular emphasis on the coordinate ring of meromorphic functions on a compact Riemann surface, and introduce the notion of a smooth variety and connections in this context. The coordinate functions on a non-singular variety have an infinitesimal lifting property as in [49, p. 188] which can be extended to noncommutative algebras. The 'non' in non-singular refers to a specific property and means something rather different from the 'non' in noncommutative. Nevertheless, we introduce noncommutative differential forms ΩA in Section 6.3 and develop the analogy between smooth algebras and quasi-free algebras. The prototype of a quasi-free algebra is the algebra of coordinate functions on a compact Riemann surface, and we describe this example in some detail. In Section 6.5, we introduce ΩA, based upon the Fedosov product, and in Sections 6.6 and 6.7 consider the Cuntz algebra and the tensor algebra with the Fedosov product. In Chapter 7, we introduce cyclic homology, and in Chapter 8 develop the Chern–Weil theory of connections.

6.1 Kähler Differentials on an Algebraic Curve

An algebraic variety $\mathbf{X}$ over $\mathbf{C}$ may be viewed as a topological space, a differentiable manifold and as the locus defined by polynomial equations. In this section, we discuss how to define the Kähler differentials on $\mathbf{X}$ by purely algebraic means. Whereas the Kähler differentials give an algebraic cotangent space on $\mathbf{X}$, we do not use the differentiable structure or tangent spaces to define them. Instead, we follow the approach used historically to define abelian integrals on compact algebraic curves. The Kähler differentials have certain universal properties which are intuitively natural and helpful in computations. Further, the Kähler differentials point the way towards noncommutative generalizations.

In this subsection we describe the basic example which arises in classical commutative algebra. Suppose that A is a finitely generated and commutative unital algebra over $\mathbf{C}$, so $A = \mathbf{C}[f_1, \dots, f_n]$. The nil radical of A is the ideal generated by the nilpotent elements. Then there is an exact sequence

$$0 \longrightarrow J \longrightarrow \mathbf{C}[X_1, \dots, X_n] \longrightarrow A \longrightarrow 0, \tag{6.1.1}$$

where J is an ideal of the complex polynomial algebra $\mathbf{C}[X_1, \dots, X_n]$, and $X_j \mapsto f_j$ for $j = 1, \dots, n$ determines the algebra homomorphism. The ideal J has a finite basis $(F_1, \dots, F_m)$ by Hilbert's basis theorem; see [8].

We introduce the maximal ideal space $\mathbf{X} = \mathrm{Hom}_{alg}(A, \mathbf{C})$, and observe that this may be identified with the affine variety

$$\mathbf{X} = \big\{(t_1, \dots, t_n) \in \mathbf{C}^n : f(t_1, \dots, t_n) = 0, \quad \text{for all} \quad f \in J\big\}. \tag{6.1.2}$$

We introduce the algebra of polynomials vanishing on $\mathbf{X}$ by

$$\mathcal{A}_{\mathbf{X}} = \big\{g \in \mathbf{C}[X_1, \dots, X_n] : g(t_1, \dots, t_n) = 0 \quad \text{for all} \quad (t_1, \dots, t_n) \in \mathbf{X}\big\}. \tag{6.1.3}$$

The radical of J is defined to be $\sqrt{J} = \{f \in \mathbf{C}[X_1, \dots, X_n]\colon f^m \in J \ \text{ for some } \ m\}$. By Nullstellensatz [8, p. 85], the radial of J satisfies $\sqrt{J} = \mathcal{A}_{\mathbf{X}}$.

Lemma 6.1.1 *The finitely generated algebra A over $\mathbf{C}$ is isomorphic to the algebra $\mathbf{C}[\mathbf{X}]$ of coordinate functions on a closed irreducible algebraic set $\mathbf{X}$, if and only if A is an integral domain.*

Proof Suppose that A is an integral domain, so J is a prime ideal. Let $g \in \mathcal{A}_{\mathbf{X}}$, so $g \in \sqrt{J}$ and $g^r \in J$ for some r, hence $g \in J$ since J is prime. Hence $\mathcal{A}_{\mathbf{X}} \subseteq J$. But $J \subseteq \mathcal{A}_{\mathbf{X}}$, and we have $J = \mathcal{A}_{\mathbf{X}}$, so $A = \mathbf{C}[\mathbf{X}]/\mathcal{A}_{\mathbf{X}}$ is the ring of coordinate functions on $\mathbf{X}$. Conversely, the algebra of coordinate functions on $\mathbf{X}$ is an integral domain. □

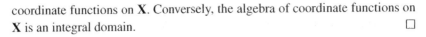

Proposition 6.1.2 *An extension field* $\mathbf{K}$ *of* $\mathbf{C}$ *is isomorphic to the field of rational functions on an irreducible closed algebraic set if and only if* $\mathbf{K}$ *is finitely generated.*

Proof If $\mathbf{K}$ is generated over $\mathbf{C}$ by $f_1, \dots, f_n$, then we can introduce $A = \mathbf{C}[f_1, \dots, f_n]$, where A is an integral domain since $A \subset \mathbf{K}$. Then $A = \mathbf{C}[\mathbf{X}]$, as above, and $\mathbf{K}$ is the field of fractions of A. That is, we can introduce $\mathbf{K} = \mathbf{C}(\mathbf{X})$ as the field of fractions on $\mathbf{X}$. □

We now continue to introduce derivations on fields such as $\mathbf{K}$. Let $\mathbf{k}$ be a field that contains $\mathbf{R}$, and let $\mathbf{k}[z_1, \dots, z_n]$ be the polynomials with coefficients in $\mathbf{k}$, and $R = \mathbf{k}(z_1, \dots, z_n)$ the quotient field of $\mathbf{k}[z_1, \dots, z_n]$, known as the field of rational functions in the indeterminates $z_1, \dots, z_n$. Then R has derivations $\frac{\partial}{\partial z_j}$ such that the $\partial k/\partial z_j = 0$ for all $k \in \mathbf{k}$; indeed, one calls $\mathbf{k}$ the field of constant elements.

Now let a be an algebraic element over R so that $p(a) = 0$ for some monic irreducible polynomial p with coefficients in $\mathbf{K}$. Also let $\partial : R \to R$ be any derivation on R such that $\partial k = 0$ for all $k \in \mathbf{k}$. Then $R[a]$ is a field, and there exists a unique extension of ∂ to a derivation $\partial : R[a] \to R[a]$. To see this, one applies ∂ to the formula $p(a) = 0$, and obtains an expression for $\partial(a)$.

The finite algebraic extensions of $\mathbf{C}(x)$ are described in the following result.

Theorem 6.1.3 (i) *Let* $p(z) = z^n + \sum_{j=0}^{n-1} a_j(x) z^j$ *be a monic irreducible polynomial with coefficients in* $\mathbf{C}(x)$. *Then there exists a compact Riemann surface* $\mathbf{X}$ *and a branched* n*-sheeted holomorphic covering map* $\pi : \mathbf{X} \to S^2$ *to the Riemann sphere, and a meromorphic function* Z *on* $\mathbf{X}$ *such that* $p(Z) = 0$.

(ii) *Conversely, the field of meromorphic functions on a compact Riemann surface is isomorphic to a finite algebraic extension of the field* $\mathbf{C}(x)$.

Proof See Forster [36, 8.9] and [34]. □

The theory of compact Riemann surfaces $\mathbf{X}$ can be approached by two different routes: as in Theorem 6.1.3, one can view $\mathbf{X}$ as a complex curve and study the function field defined on $\mathbf{X}$. Alternatively, one can follow the approach made precise by Weyl, introducing a two-dimensional smooth manifold, introducing analytic forms on open neighbourhoods in $\mathbf{X}$ and establishing the existence and properties of analytic differentials on $\mathbf{X}$ by potential theory. Ultimately the two view points are reconciled via abelian integrals and the theory of uniformization. One needs to prove the existence of a non-constant meromorphic function Z on $\mathbf{X}$.

The differential of a meromorphic function on a Riemann surface is a meromorphic differential, not really a meromorphic function. Kirwan [62, 6.5] resolves this issue. Given a compact Riemann surface $\mathbf{X}$ with a non-constant meromorphic function $Z\colon \mathbf{X} \to S^2$, one can produce a triangulation of $\mathbf{X}$, as follows. Given any finite set $\{b_j : j = 1\dots,m\}$ of points in the Riemann sphere S^2, one can choose a triangulation of S^2 with faces Δ_k such that the vertices include all the b_j. We can then use the inverse images $Z^{-1}(\Delta_k)$ to triangulate $\mathbf{X}$; see [34]. With this triangulation, one can compute the genus g, Euler characteristic and other topological invariants of $\mathbf{X}$. This shortcut from algebra to topology leaves us with the problem of showing that the genus, etc. are true topological invariants, and do not depend upon the choice of triangulation. Kirwan [62, p. 98] discusses the triangulation of a complex curve, and proves the uniqueness of the invariants.

Example 6.1.4 (Elliptic curve) We can take $\mathbf{k} = \mathbf{C}$, then $R = \mathbf{C}(X)$ has derivation $\frac{\partial}{\partial X}$, and $\mathbf{K} = R[Z]$, where Z satisfies the polynomial equation

$$Z^2 = (X-a)(X-b)(X-c) \tag{6.1.4}$$

for distinct $a,b,c \in \mathbf{C}$. This determines an elliptic curve $\mathbf{X}$. Also, $\mathbf{K}$ is known as the field of rational functions on $\mathbf{X}$ and $\mathbf{K}$ has transcendence degree one over $\mathbf{C}$, since Z is algebraic over R. The classical way of introducing differentials is to write $Z = dX/dY$ and

$$Y = \frac{1}{2}\int \frac{dX}{\sqrt{(X-a)(X-b)(X-c)}}. \tag{6.1.5}$$

Then one introduces Weierstrass's function $\wp$, which is meromorphic on $\mathbf{X}$ and gives a 2:1 map onto S^2, so that $Z = \wp'(Y)/2$ and $X = \wp(Y)$. A typical element of $\mathbf{K}$ has the form $g(X) + h(X)Z$, so a typical difference quotient has the form

$$\begin{aligned}&\frac{g(X_1)+h(X_1)Z_1 - g(X_2) - h(X_2)Z_2}{X_1 - X_2}\\ &= \frac{g(X_1)-g(X_2)}{X_1-X_2} + \frac{h(X_1)-h(X_2)}{X_1-X_2}Z_1 + h(X_2)\frac{Z_1-Z_2}{X_1-X_2},\end{aligned} \tag{6.1.6}$$

where the difference quotients in g and h belong to $\mathbf{C}(X_1, X_2)$. For a suitable choice of parameters a,b,c we can use the functional equation of $\wp$ to write the final term as

$$\frac{Z_1 - Z_2}{X_1 - X_2} = \frac{\wp'(y_1) - \wp'(y_2)}{2(\wp(y_1) - \wp(y_2))}$$
$$= \int_{1/2}^{y_1} \wp(t)\,dt + \int_{1/2}^{y_2} \wp(t)\,dt - \int_{1/2}^{y_1+y_2} \wp(t)\,dt - \kappa \qquad (6.1.7)$$

for some constant κ. The functional relation for $\wp$ expresses the group law on the elliptic curve, as discussed in [75]. The elliptic curve has the important property that the tangent bundle is trivial as in [75, p. 120], and one can express the points on the torus by a simple system of coordinates. For these reasons, $\mathbf{X}$ is used as a model example in index theory [7, p. 311].

The notion of a connection extends to vector fields in this context. With $\mathbf{K} = \mathbf{C}(\mathbf{X})$, the space $M = \mathbf{K}^{n\times 1}$ of $n \times 1$ columns is a $\mathbf{K}$-module, and we write $\Omega_{\mathbf{K}/\mathbf{k}} = \mathbf{K}dX$ where X is the differential on $\mathbf{X}$. Given $\theta \in M_n(\mathbf{K})$, so θdX is a matrix-valued 1-form, we introduce $\nabla\colon M \to \Omega_{\mathbf{K}/\mathbf{k}} \otimes M$ by $\nabla\xi = \frac{\partial\xi}{\partial X}dX + \theta\xi dX$, where $\theta\xi$ is the matrix product on the vector and $\frac{\partial\xi}{\partial X}$ is componentwise differentiation. Then

$$\nabla(f\xi) = (df)\xi + f\nabla\xi \qquad (f \in \mathbf{K}, \xi \in M). \qquad (6.1.8)$$

The properties of this connection are discussed in the classical theory of differential equations.

To match with notation in Section 6.2, we introduce

$$F(X_1, X_2) = X_1^2 - (X_2 - a)(X_2 - b)(X_2 - c)$$

so $\mathbf{C}[\mathbf{X}] = \mathbf{C}[X_1, X_2]/(F)$ and we compute the matrix of first-order partial derivatives

$$DF = \begin{bmatrix} \frac{\partial F}{\partial X_1} \\ \frac{\partial F}{\partial X_2} \end{bmatrix} = \begin{bmatrix} 2X_1 \\ -3X_2^2 + 2(a+b+c)X_2 - (ab+ac+bc) \end{bmatrix}. \qquad (6.1.9)$$

Using the condition that the a, b, c are distinct, one can easily check that DF has rank one at all points on $\mathbf{X}$. When suitably compactified, $\mathbf{X}$ can be realized as a torus; see [75]. More generally, a complex polynomial p of degree $2g+1$ is associated with curve $\mathbf{P}$: $Z^2 = p(X)$, which is typically hyperelliptic. The differentials are easy to interpret in this setting.

We resume the previous discussion and, more generally, let $\mathbf{K}$ be a finite algebraic extension of R, and $\Omega_{\mathbf{K}/\mathbf{k}}$ be the vector space over $\mathbf{K}$ with basis $dz_1, \dots, dz_n$ so $\Omega_{\mathbf{K}/\mathbf{k}} = \{\sum_{j=1}^n f_j dz_j\colon f_j \in \mathbf{K}\}$ and $df = \sum_{j=1}^n \frac{\partial f}{\partial z_j} dz_j$ gives a derivation $d\colon \mathbf{K} \to \Omega_{\mathbf{K}/\mathbf{k}}$. Then $\Omega_{\mathbf{K}/\mathbf{k}}$ has the following universal property.

Proposition 6.1.5 *Let M be a $\mathbf{K}$ module, and $D\colon \mathbf{K} \to M$ any derivation such that $Dk = 0$ for all $k \in \mathbf{k}$. Then there exists a unique $\mathbf{K}$-linear map $L\colon \Omega_{\mathbf{K}/\mathbf{k}} \to M$ such that $D = L \circ d$.*

Proof We have

$$L\left(\sum_{j=1}^{n} f_j dz_j\right) = \sum_{j=1}^{n} f_j D(z_j), \tag{6.1.10}$$

so L is unique. Also, $D - L \circ d\colon \mathbf{K} \to M$ is a derivation such that $(D - L \circ d)| R = 0$; but by uniqueness of extensions, $D - L \circ d$ must be trivial on $\mathbf{K}$, so $D = L \circ d$. □

We can thus regard $d\colon \mathbf{K} \to \Omega_{\mathbf{K}/\mathbf{k}}$ as universal for derivations on $\mathbf{K}$ that are zero on $\mathbf{k}$. The map $L \mapsto L \circ d = \partial$ gives a one-to-one correspondence

$$\mathrm{Hom}_{\mathbf{K}}(\Omega_{\mathbf{K}/\mathbf{k}}, \mathbf{K}) \leftrightarrow \mathrm{Der}_{\mathbf{k}}(\mathbf{K}, \mathbf{K}) \tag{6.1.11}$$

where $\partial(af + bg) = a\partial f + b\partial g$ for all $a, b \in \mathbf{k}$ and $f, g \in \mathbf{K}$.

Example 6.1.6 (Algebraic derivations) In the context of Proposition 6.1.5, let $D\colon \mathbf{K} \to \mathbf{K}$ be the derivation $Df = \sum_{j=1}^{n} h_j \frac{\partial f}{\partial x_j}$ where $h_j \in \mathbf{K}$, so D is the derivation in the direction of $(h_1, \dots, h_n)$. Then we can take $L(dx_j) = h_j$, so $D = L \circ d$.

Definition 6.1.7 (Connection) Given a derivation $d\colon \mathbf{K} \to \mathbf{K}$ such that $dk = 0$ for all $k \in \mathbf{k}$, a connection for $\mathbf{K}/\mathbf{k}$ is a $\mathbf{k}$-linear map $\nabla\colon M \to \Omega_{\mathbf{K}/\mathbf{k}} \otimes_{\mathbf{K}} M$ where

(i) M is a finitely generated $\mathbf{K}$-module;
(ii) $\nabla(fm) = df \otimes m + f\nabla m$ for all $f \in \mathbf{K}$, and $m \in M$.

Given D and ∇, we can introduce L such that $D = L \circ d$ and introduce $\nabla_D = (L \otimes id)\nabla\colon M \to M$ given by

$$M \overset{\nabla}{\longrightarrow} \Omega_{\mathbf{K}/\mathbf{k}} \otimes_{\mathbf{K}} M \overset{L \otimes id}{\longrightarrow} \mathbf{K} \otimes M \longrightarrow M \tag{6.1.12}$$

such that ∇_D is a derivation on the $\mathbf{K}$-module M, as in

$$\nabla_D(fm) = (Df)m + f\nabla m \qquad (f \in \mathbf{K}, m \in M). \tag{6.1.13}$$

With the connection ∇, we can differentiate with ∇_D along the direction of D.

There are various ways of extending this construction. However, when one replaces $\mathbf{C}[\mathbf{X}]$ by other algebras, many subtle phenomena emerge. Theorem

6.1.3 depends crucially on the fact that **K** has finite transcendence degree over **k**, and Proposition 6.1.5 needs adjusting to make it valid in closely related contexts, as discussed on p. 158 of [99]. In Section 6.2, we replace the integral domain A by a commutative algebra, and then restrict attention to the cases in which A is finitely generated and has no nilpotent elements other than zero.

Exercise 6.1.8 (Transcendental derivatives) Let A be the algebra of holomorphic functions on the open unit disc, so that A is a transcendental extension of **C**. There is a differential $d: A \to Adz: f \mapsto f'(z)dz$, where $f'(z)$ is the usual complex derivative of f. This derivation does not have the universal property of Proposition 6.1.5. See [99, p. 159].

Exercise 6.1.9 (Hyperelliptic curves) Let $p(x)$ be an irreducible complex polynomial of degree $2g+1$, and let $\mathbf{X}$ be the curve $\mathbf{X}: Z^2 = p(X)$. There is a non-constant rational function $\wp: \mathbf{X} \to S^2$, so we can define smooth curves on S^2, and use $\wp^{-1}$ to pull them back to $\mathbf{X}$; thus we can introduce a formal integration theory. We introduce the differential $X^{j-1}dX/Z$ on $\mathbf{X}$ for $j = 1, \dots, g$, then for fixed $q_j \in \mathbf{X}$ and variable $p_j \in \mathbf{X}$ we let $F: \mathbf{X}^g \to \mathbf{C}^g$

$$F(p_1, \dots, p_g) = \left(\sum_{j=1}^{g} \int_{q_j}^{p_j} X^{k-1} dX/Z \right)_{k=1}^{g}, \tag{6.1.14}$$

where the integrals are taken in $\mathbf{X}$ along some path from q_j to p_j; see [62, p. 148] and [78, p. 135]. This is a symmetric function in $(p_1, \dots, p_g)$, known as the Abel map.

(i) Then the Jacobian matrix of partial derivatives is

$$\left[\frac{\partial F}{\partial p_k} \right] = \left[\frac{X^{j-1}(p_k)}{Z(p_k)} \right]_{j,k=1}^{n}. \tag{6.1.15}$$

Compute the determinant of this.

(ii) Choose $x_0 \in \mathbf{X}$ and consider loops in $\mathbf{X}$ that are based at x_0. Show that there is a group homomorphism $H_1(\mathbf{X}, \mathbf{Z}) \to \mathbf{C}^g$ given by

$$\Phi_A : \gamma \mapsto \left(\int_\gamma X^{k-1} dX/Z \right)_{k=1}^{g}. \tag{6.1.16}$$

The image of this map is a lattice Λ in $\mathbf{C}^g$, and $\mathbf{C}^g/\Lambda$ is a compact complex torus, known as the Jacobian variety. To determine the range of Φ_A, one needs much more theory. Mumford [79, 3.28] gives an algebraic construction of the Jacobian of a hyperelliptic curve. See [36] for further discussion of the analysis, and [96] for the algebra.

(iii) Let $\Omega^1_{\mathbf{X}}$ be the complex vector space of holomorphic differentials on $\mathbf{X}$. Prove that

$$\dim_{\mathbf{C}} \Omega^1_{\mathbf{X}} = g,$$

which is a special case of the Riemann–Roch theorem. See [43, p. 103], and [96].

Example 6.1.10 (Homology of elliptic curves) Deligne [27, pp. 12–25] discusses how the various $H_\bullet$ groups are related for the elliptic curve $y^2z = 4x^3 - g_2xz^2 - g_3z^3$. This is the projective version of Example 6.1.4.

6.2 Homology of Kähler Differential Forms

The de Rham complex can be constructed for any unital commutative algebra A over $\mathbf{C}$, as follows. We wish to have an A-bimodule Ω^1_A and a $\mathbf{C}$-linear map $d\colon A \to \Omega^1_A$ such that

$$d1 = 0, \qquad d(ab) = (da)b + a(db) \qquad (a, b \in A). \tag{6.2.1}$$

Lemma 6.2.1 *Let A be a commutative unital algebra with multiplication m. Then there exists an ideal I of $A \otimes A$ such that*

$$0 \longrightarrow I \longrightarrow A \otimes A \overset{m}{\longrightarrow} A \longrightarrow 0 \tag{6.2.2}$$

is an exact sequence of A-bimodules and complex algebras.

Proof The multiplication $m(a \otimes b) = ab$ gives an A-bimodule map and an algebra homomorphism, so $I = \mathrm{Ker}(m\colon A \otimes A \to A)$ is an ideal in $A \otimes A$ and an A-bimodule. Observe that a typical element of $A \otimes A$ is

$$\sum_{j=1}^n a_j \otimes b_j = \sum_{j=1}^n a_j(1 \otimes b_j - b_j \otimes 1) + \sum_{j=1}^n a_jb_j \otimes 1, \tag{6.2.3}$$

where $\sum_{j=1}^n a_j(1 \otimes b_j - b_j \otimes 1) \in I$ and $\sum_{j=1}^n a_jb_j = m(\sum_{j=1}^n a_j \otimes b_j)$. Note also that

$$\sum_{j=1}^n \sum_{k=1}^p a_jc_k(1 \otimes b_jd_k - b_jd_k \otimes 1) = \sum_{j=1}^n \sum_{k=1}^p a_jc_k(1 \otimes b_j - b_j \otimes 1)d_k + \sum_{j=1}^n \sum_{k=1}^p a_jb_jc_k(1 \otimes d_k - d_k \otimes 1). \tag{6.2.4}$$

□

Note that $[A \otimes A, A] \subseteq I$, and let the space of Kähler differentials be $\Omega^1_A = I/I^2$; then let $d\colon A \to \Omega^1_A$ by $da = a \otimes 1 - 1 \otimes a$ modulo I^2. Then Ω^1_A is an $(A \otimes A)/I$ module. Then the exterior tensor power of Ω^1_A as an A module is $\Omega^i_A = \wedge^i_A \Omega^1_A$, where $\wedge_A$ denotes the exterior tensor power as an A-module. In Section 6.3, we will consider $I = \Omega^1 A$, as opposed to $I/I^2 = \Omega^1_A$.

Definition 6.2.2 (Kähler differentials) This is the complex $A = \Omega^0_A$ and

$$\Omega^0_A \overset{d}{\longrightarrow} \Omega^1_A \overset{d}{\longrightarrow} \Omega^2_A \overset{d}{\longrightarrow} \cdots. \tag{6.2.5}$$

Proposition 6.2.3 *The space Ω^1_A of Kähler differentials has the universal property that given any A-module M and derivation $\partial\colon A \to M$ such that*

$$\partial(\lambda a + \mu b) = \lambda\partial(a) + \mu\partial(b), \tag{6.2.6}$$

$$\partial(ab) = b(\partial a) + a\partial(b) \qquad (\lambda, \mu \in \mathbf{C}; a, b \in A), \tag{6.2.7}$$

there exists an A-module map $L\colon \Omega^1_A \to M$ such that $L(df) = \partial(f)$.

Proof (See [33].) We introduce the algebra $A \oplus \Omega^1_A$ with the multiplication

$$(a, u) \cdot (b, v) = (ab, av + bu), \tag{6.2.8}$$

so that

$$0 \longrightarrow \Omega^1_A \longrightarrow A \oplus \Omega^1_A \longrightarrow A \longrightarrow 0 \tag{6.2.9}$$

is an extension of A by the ideal Ω^1_A with square zero, and $\psi\colon A \otimes A \to A \oplus \Omega^1_A$ by $\psi(a \otimes b) = (ab, adb)$. Then $\psi(1 \otimes b - b \otimes 1) = (0, db)$ and $c\psi(a \otimes b) = \psi(ca \otimes b)$, so ψ is an A-module map. Also, ψ is an algebra homomorphism since

$$\begin{aligned}
\psi((a_1 \otimes b_1)(a_2 \otimes b_2)) &= \psi(a_1 a_1, b_1 b_2)\\
&= (a_1 a_2 b_1 b_2, a_1 a_2 d(b_1 b_2))\\
&= (a_1 a_2 b_1 b_2, a_1 a_2 b_1 db_2) + (a_1 a_2 b_1 b_2, a_1 a_2 b_2 db_1)\\
&= (a_1 b_1, a_1 db_1) \cdot (a_2 b_2, a_2 db_2)\\
&= \psi(a_1 \otimes b_1)\psi(a_2 \otimes b_2).
\end{aligned} \tag{6.2.10}$$

We introduce the homomorphisms $\varphi_1, \varphi_2\colon A \to A \otimes A$ by $\varphi_1(a) = a \otimes 1$ and $\varphi_2(a) = 1 \otimes a$; then let $\delta = \varphi_1 - \varphi_2$. Observe that $m \circ \varphi_1 = id = m \circ \varphi_2$, so we have split the sequence (6.2.9). We have $\delta(a) = a \otimes 1 - 1 \otimes a \in I$, so $\varphi(a)\varphi(b) \in I^2$. We next check that δ is a derivation (of course, $\delta(a) = da$, but we don't want to prejudge the issue). Now the identity $\varphi_1(ab) = \varphi_1(a)\varphi_1(b)$ leads to

$$(\varphi_2(ab) + \delta(ab)) = (\varphi_2(a) + \delta(a))(\varphi_2(b) + \delta(b) \tag{6.2.11}$$

so that

$$\delta(ab) = \varphi_2(a)\delta(b) + \delta(a)\varphi_2(b) \qquad (mod \quad I^2). \tag{6.2.12}$$

This shows that φ is a derivation; also $\psi \circ \delta(b) = (0, db)$.

Given any derivation $\partial\colon A \to M$ as above, we define $L(db) = \partial b$, so $L \leftrightarrow \partial$ gives the correspondence

$$\mathrm{Hom}_A(\Omega_{A/\mathbf{k}}, M) \leftrightarrow \{\text{derivations } \partial\colon A \to M\colon \partial k = 0, \forall k \in \mathbf{k}\}. \tag{6.2.13}$$

□

Example 6.2.4 (Differentials on polynomials) (i) Suppose that $A = \mathbf{C}[X_1, \dots, X_n]$ is the complex polynomial algebra. Then Ω^1_A is the free A-module of rank n that is generated by $dX_1, \dots, dX_n$; see [49, p. 173]. Also, for $F(X_1, \dots, X_n) \in A$, the usual partial derivatives give the differential via

$$dF = \sum_{j=1}^{n} \frac{\partial F}{\partial X_j} dX_j \tag{6.2.14}$$

[45, chapitre 0, example 20.5.13]. Hence $\Omega^k_A = 0$ for all $k > n$.

(ii) Let $C = \mathbf{C}[\mathbf{X}]$ be the coordinate ring on a non-singular affine variety $\mathbf{X}$. Then $\mathbf{C}[\mathbf{X}] \otimes \mathbf{C}[\mathbf{X}] \cong \mathbf{C}[\mathbf{X} \times \mathbf{X}]$ and $m\colon \mathbf{C}[\mathbf{X} \times \mathbf{X}] \to \mathbf{C}[\mathbf{X}]$ is a restriction to the diagonal $\mathbf{X} \to \mathbf{X} \times \mathbf{X}$, as in $f(x,y) \mapsto f(x,x)$. We recall from Section 6.1 that $C = \mathbf{C}[\mathbf{X}]$ arises when $C = A/J$, where J is a prime ideal in $A = \mathbf{C}[X_1, \dots, X_n]$, so that C is an integral domain, with field of fractions $\mathbf{K}$, so differentials are defined as in Proposition 6.1.5. By [49, p. 173], we have an exact sequence

$$0 \longrightarrow J/J^2 \overset{\delta}{\longrightarrow} \Omega^1_A \otimes_A C \longrightarrow \Omega^1_C \longrightarrow 0, \tag{6.2.15}$$

where $\delta a = da \otimes 1$ and Ω^1_A is as in (i). The image of δ gives the relations that the differentials in Ω^1_C satisfy.

(iii) Let $R = \mathbf{R}[X,Y,Z]/(X^2 + Y^2 + Z^2 - 1)$ be the coordinate ring on the sphere S^2 in $\mathbf{R}^3$. There exists an R-module that is projective, but not free; see [64]. This relates to the topological result that the tangent bundle to S^2 is non-trivial, and there does not exist a continuous function $v\colon S^2 \to \mathbf{R}^3$ such that $v(x) \cdot x = 0$ and $v(x) \neq 0$ for all $x \in S^2$; see [42, p. 95].

Definition 6.2.5 (Smoothness) More generally, we can consider the case in which A is finitely generated and has no non-zero nilpotent elements, so A has zero nil radical. That is, we assume $A = \mathbf{C}[X_1, \dots, X_n]/J$ where $J = \sqrt{J}$ and $J = \mathbf{C}[F_1, \dots, F_m]$. See [96] for further discussion of the notion of a non-singular variety. As before, we introduce $\mathbf{X} = \mathrm{Hom}_{alg}(A, \mathbf{C})$, and say that $\mathbf{X}$

is smooth at $t = (t_1, \dots, t_n)$ if $\dim_t \mathbf{X} = \dim_t T\mathbf{X}$. Now we have $\dim_t \mathbf{X} \leq \dim_t T\mathbf{X}$, and $(T\mathbf{X})_t$ is determined by the system of linear equations

$$\sum_{k=1}^{n} \left(\frac{\partial F_i}{\partial X_k}\right)_t (X_k - t_k) = 0 \qquad (i = 1, \dots, m) \tag{6.2.16}$$

in which the matrix of first-order partial derivatives $\frac{\partial F_i}{\partial X_k}$ is evaluated by $(X_1, \dots, X_n) \mapsto (t_1, \dots, t_n)$. So by the rank-plus-nullity theorem, $\mathbf{X}(A)$ is smooth at t if

$$\mathrm{rank}\Big[\left(\frac{\partial F_i}{\partial X_k}\right)_t\Big]_{i=1,\dots,m;\,k=1,\dots,n} + \dim_t \mathbf{X} = n. \tag{6.2.17}$$

Then $\mathbf{X}$ is non-singular if $\mathbf{X}$ is smooth at all points; see [64, p. 166] and [96]. Exercise 6.2.11 gives a criterion for smoothness in the case of varieties in $\mathbf{C}^2$.

Lefschetz proved that every n-dimensional complex algebraic variety is an orientable C^∞ real differentiable manifold of dimension $2n$. Also, every algebraic variety can be triangulated; see [49, p. 447]. Hence $\mathbf{X}$ has interesting homology as a topological space, a complex differentiable manifold and an algebraic variety, which can be computed in terms of suitable spaces of differentials. There are definitions of homology and cohomology groups within each theory, which raises the unattractive prospect that the various groups could be different and hard to interpret. Fortunately, the theories resolve and produce consistent results which we summarize here.

First, let γ be a differentiable ℓ-simplex in $\mathbf{X}$, and $d\gamma$ the alternating sum of the faces of γ. Then we form the homology group $H^\ell(\mathbf{X}, \mathbf{R})$ as the real vector space spanned by the differentiable ℓ simplices modulo the boundaries. Also, let $\Omega^\ell_{\mathbf{R}}(\mathbf{X})$ be the space of C^∞ real differential forms on $\mathbf{X}$ with de Rham differentials

$$\Omega^0_{\mathbf{R}}(\mathbf{X}) \xrightarrow{d} \Omega^1_{\mathbf{R}}(\mathbf{X}) \xrightarrow{d} \Omega^2_{\mathbf{R}}(\mathbf{X}) \xrightarrow{d} \dots . \tag{6.2.18}$$

A systematic account of this differential analysis is contained in [107]. Then we let

$$H^\ell_{DR}(\mathbf{X}) = \{\text{closed } \ell - \text{forms}\}/\{\text{exact } \ell - \text{forms}\}. \tag{6.2.19}$$

By Stokes's theorem, the boundary operation and the de Rham differential are related by

$$\int_\gamma d\omega = \int_{\partial\gamma} \omega. \tag{6.2.20}$$

We note that $H^1_{DR}(\mathbf{X})$ is the quotient group of all closed C^∞ differential 1-forms on $\mathbf{X}$ by the subgroup of exact differential 1-forms on $\mathbf{X}$.

Proposition 6.2.6 *There is a group homomorphism*

$$H^1_{DR}(\mathbf{X}) \times H_1(\mathbf{X};\mathbf{Z}) \to \mathbf{C}\colon ([\omega],[\gamma]) \mapsto \int_\gamma \omega, \tag{6.2.21}$$

where $\gamma\colon S^1 \to \mathbf{X}$ is a closed contour on $\mathbf{X}$ and ω a 1-form.

Proof We show that $\int_\gamma \omega$ depends only upon the de Rham class of ω, and the homology class of γ. If ω is exact, so $\omega = df$ for some $f \in C^\infty(\mathbf{X};\mathbf{C})$, then $\int_\gamma \omega = \int_\gamma df = 0$ since γ is closed. Conversely, if $\int_\gamma \omega = 0$ for all γ, then $\omega = df$ for some $f \in C^\infty(\mathbf{X};\mathbf{C})$, essentially by the fundamental theorem of calculus. Now consider a closed contour γ' that is homologous to γ, so $\gamma - \gamma' = \partial\Omega$ for some region Ω. Then by Stokes's theorem

$$\int_\gamma \omega - \int_{\gamma'} \omega = \int_{\partial\Omega} \omega = \iint_\Omega d\omega = 0. \tag{6.2.22}$$

Hence the integral $\int_\gamma \omega$ depends depends only on $([\omega],[\gamma])$ so we have a pairing which clearly gives a group homomorphism of additive groups. □

Recalling from Section 6.1 that $\mathbf{X}$ is a complex manifold, we introduce the sheaf $\Omega^\ell(\mathbf{X})$ of holomorphic differentials $\omega = \sum_{i_1,\cdots,i_\ell} \alpha_{i_1,\dots,i_\ell} dz_{i_1} \wedge \cdots \wedge dz_{i_\ell}$, where $\alpha_{i_1,\dots,i_\ell}$ is a holomorphic function of $(z_1,\dots,z_n)$, defined on a local coordinate chart; see [107]. There is a holomorphic differential $d\colon \Omega^\ell(\mathbf{X}) \to \Omega^{\ell+1}(\mathbf{X})$ such that $d^2 = 0$, and hence a complex of holomorphic differentials $\Omega_{\mathbf{C}}(\mathbf{X})$ with analytic homology $H_a(\mathbf{X};\mathbf{C})$.

Definition 6.2.7 (Smooth) Say that a commutative, finitely generated and unital algebra A over $\mathbf{C}$ is smooth if A has no nilpotent elements other than 0 and that $\mathbf{X}$ is non-singular.

Theorem 6.2.8 *Suppose that A is smooth.*

(i) *Then there is a natural isomorphism between $H^\ell(\mathbf{X},\mathbf{R})$ and $H^\ell_{DR}(\mathbf{X})$ given by the pairing $(\omega,\gamma) \mapsto \int_\gamma \omega$, in which a differential form ω is integrated over a simplex γ.*

(ii) *There is a natural isomorphism between $H^\ell(\mathbf{X},\mathbf{R}) \otimes \mathbf{C}$ and $H^\ell_a(\mathbf{X};\mathbf{C})$ and there is a decomposition*

$$H^\ell_a(\mathbf{X};\mathbf{C}) \cong \oplus_{p+q=\ell} H^p(\mathbf{X},\Omega^q). \tag{6.2.23}$$

Proof This result depends upon a reduction to local calculations based on the complex Poincaré lemma; see [107, p. 60], and shows that the topological homology theory coincides with the analytical, where the latter involves the smooth structure of complex manifolds. □

A more profound analysis introduces the differential forms on $\mathbf{X}$ with algebraic coefficients, algebraic differentials of the first kind, with a corresponding homology. For non-singular algebraic varieties $\mathbf{X}$, it turns out that the algebraic and analytic homology groups are isomorphic. Hartshorne develops an algebraic de Rham theory in [48], and see [27] for a discussion using Hodge theory. Mumford discusses the connection between the various results in [77]. In appendix E to [68], Ronco discusses various notions of smoothness for a commutative algebra.

Theorem 6.2.9 (Lefschetz, Atiyah–Hodge, Grothendieck) (i) *Assume that A is smooth. Then*

$$H^j(\Omega_A) \cong H^j(\mathbf{X}, \mathbf{C}). \tag{6.2.24}$$

(ii) *In particular, if $\mathbf{X}$ is a non-singular affine algebraic variety, then $H^j(\mathbf{X};\mathbf{C}) \cong \{$ closed algebraic j-forms $\}/\{$ exact algebraic j-forms $\}$.*

See [6] for further discussion. We can also deal with a more general situation by resolving singularities. In Section 6.8, we discuss inverse limits and completions, as in (ii) of the following result.

Theorem 6.2.10 (Zariski–Grothendieck) (i) *A is smooth if and only if A has the lifting property with respect to nilpotent extensions; that is, given an extension by commutative algebras*

$$0 \longrightarrow I \longrightarrow R \overset{\pi}{\longrightarrow} A \longrightarrow 0, \tag{6.2.25}$$

where $I^n = 0$ for some n, there exists a lifting homomorphism $\rho\colon A \to R$ such that $\pi \circ \rho = id$.

(ii) *Suppose that $A = R/I$ where R is smooth, for example when $R = \mathbf{C}[x_1, \dots, x_n]$. Then*

$$H^i(\mathbf{X};\mathbf{C}) = H^i(\lim_{\leftarrow k} \Omega_R/I^k\Omega_R). \tag{6.2.26}$$

Exercise 6.2.11 (Smoothness criterion) (After [8, ex. 11.1]) Let $F \in \mathbf{C}[X_1, X_2]$ be an irreducible complex polynomial, and let $A = \mathbf{C}[X_1, X_2]/(F)$, so $\mathbf{X}$ is the algebraic variety $F = 0$. Let I be the maximal ideal in A that corresponds to $(t_1, t_2) \in \mathbf{X}$ and consider

$$DF(t_1,t_2) = \begin{bmatrix} \frac{\partial F}{\partial X_1}(t_1,t_2) \\ \frac{\partial F}{\partial X_2}(t_1,t_2) \end{bmatrix}.$$

(i) Using the Taylor expansion of F show that $F \in I^2$, if and only if the rank of $DF(t_1,t_2)$ equals zero.

(ii) Deduce that the rank of $DF(t_1, t_2)$ is equal to the dimension of

$$I/I^2 = \big(X_1 - t_1, X_2 - t_2\big)/\big((X_1 - t_1, X_2 - t_2)^2 + (F)\big).$$

(iii) Let $g_I(A) = \oplus_{n=1}^{\infty} I^n/I^{n+1}$. Deduce that $g_I(A)$ is isomorphic either to $\mathbf{k}[t]$ or to $\mathbf{k}$. (See theorem 11.22 of [8].)

(iv) Show that the set $\{(t_1, t_2) \in \mathbf{C}^2 \colon t_1 t_2 = 0\}$ is connected but not irreducible.

Exercise 6.2.12 (Generators of algebra) Let A be a unital commutative Banach algebra with n generators. Prove that the maximal ideal space X of A is homeomorphic to a compact subset of $\mathbf{C}^n$.

Example 6.2.13 (i) *(Not smooth algebras)* The algebras $\mathbf{C}[X]/(X^2)$ and $\mathbf{C}[X_1, X_2]/(X_1X_2)$ are not smooth.

(ii) The following are smooth: the polynomial ring $\mathbf{C}[X_1, \dots, X_n]$, the finite Laurent polynomials $\mathbf{C}[X, 1/X]$ and $\mathbf{C}[x, y, s, t]/(xt - ys - 1)$, which we regard as the coordinate algebra of the matrices $\left[\begin{smallmatrix} x & y \\ s & t \end{smallmatrix}\right]$ with determinant one.

(iii) *(Infinitesimal lifting property)* Exercise 8.6 on p. 188 of [49] considers the coordinate ring $C[\mathbf{X}]$ on a non-singular variety, so there is an exact sequence of Kähler differentials as in (6.2.15). Given an exact sequence of complex commutative algebras $0 \to I \to B' \to B \to 0$, such that $I^2 = 0$ and given an homomorphism $\rho \colon C[\mathbf{X}] \to B$, it is shown that there exists a lifting homomorphism $\rho' \colon C[\mathbf{X}] \to B'$. This example extends to the context of noncommutative differentials forms, and points towards the appropriate generalization of smooth algebras.

In Section 6.3, we introduce the concept of noncommutative differential forms. The correct analogue of smooth algebras in the noncommutative setting turns out to be quasi-free algebras, as we introduce in Section 6.4.

6.3 Noncommutative Differential Forms ΩA

The main idea of this approach to cyclic theory is to calculate with chains in $A^{\otimes n}$ and use universal properties to characterize algebras such as the universal tensor algebra $T(A)$ and the universal algebra of noncommutative differential forms ΩA. In this section, we define ΩA as a differential graded algebra and then study the odd and even parts of the grading. We reserve the notation Ω_A for Kähler differentials, as in Section 6.2, whereas in the literature [88], the notation Ω_A is sometimes used for noncommutative differentials. Later we compare these modules.

Let A be a unital $\mathbf{k}$-algebra and quotient A by $\{t1 : t \in \mathbf{k}\}$ to form the quotient vector space $\bar{A} = A/\mathbf{k}$, and let $\bar{a}$ be the image of $a \in A$ under the quotient map. We write $R = \oplus_{n=-\infty}^{\infty} R_n$ for a differential graded algebra with differentials $d_n : R_n \to R_{n+1}$ such that $d_{n+1}d_n = 0$. The main result is an analogue of Proposition 6.2.3.

Theorem 6.3.1 (i) *There exists a differential graded algebra ΩA, which is unique up to canonical isomorphism, such that $\Omega^n A = 0$ for $n < 0$, $\Omega^0 A = A$, and*

$$A \otimes (\bar{A})^{\otimes n} \to \Omega^n A : (a_0, a_1, \dots, a_n) \mapsto a_0 da_1 \dots da_n \tag{6.3.1}$$

extends to a $\mathbf{k}$-vector space isomorphism for all $n > 0$.

(ii) *Given any differential graded algebra R and homomorphism $u_0 : A \to R_0$, there exists a unique homomorphism of differential graded algebras $u : \Omega A \to R$ that extends u_0.*

Proof The basic idea is that

$$(a_0, \bar{a}_1, \dots, \bar{a}_n) \longleftrightarrow a_0 da_1 \dots da_n \tag{6.3.2}$$

and that multiplication of differential forms should satisfy

$$\begin{aligned}(a_0 da_1 \dots da_n)\alpha = {} & a_0 da_1 \dots d(a_n\alpha) \\ & - a_0 da_1 \dots d(a_{n-1}a_n)\alpha \\ & + a_0 da_1 \dots d(a_{n-2}a_{n-1})da_n d\alpha \\ & + \cdots \\ & + (-1)^{n-1} a_0 d(a_1a_2) da_3 \dots da_n d\alpha \\ & + (-1)^n a_0 a_1 da_2 \dots da_n d\alpha,\end{aligned} \tag{6.3.3}$$

so as to satisfy Leibniz's rule. The proof involves checking that all this works. First, we establish existence by producing a specific model for ΩA. Let K be the complex of $\mathbf{k}$-vector spaces that has components

$$K_n = \begin{cases} A \otimes (\bar{A})^{\otimes n}, & \text{for } n \geq 1; \\ A, & \text{for } n = 0; \\ 0, & \text{for } n < 0, \end{cases} \tag{6.3.4}$$

and with differential $\partial : K_n \to K_{n+1}$

$$\partial(a_0, \bar{a}_1, \dots, \bar{a}_n) = (1, \bar{a}_0, \bar{a}_1, \dots, \bar{a}_n). \tag{6.3.5}$$

□

Lemma 6.3.2 *Let R be a differential graded algebra and $u\colon A \to R_0$ a homomorphism; let $\tilde{u}\colon K \to R$ be*

$$\tilde{u}(a_0, \bar{a}_1, \dots, \bar{a}_n) = u(a_0)du(a_1)\dots du(a_n). \tag{6.3.6}$$

If L is the smallest differential graded subalgebra of R that contains $u(A)$, then $L = R$ and $\tilde{u}$ is surjective onto R.

Proof Let $L = \mathrm{Im}(\tilde{u})$, so L is a subcomplex of R since $\tilde{u}$ is a map of complexes. Now consider $J = \{x \in R\colon xL \subseteq L\}$; then by Leibniz's rule, J is a differential graded subalgebra of R, and J contains $u(A)$ since u is a homomorphism; hence $J = R$. We deduce that L is a left ideal of R, and $1 \in L$, so $L = R$. □

Proof of 6.3.1(i) Now we prove the existence of ΩA. The $\mathbf{k}$-linear operators $K \to K$ form a differential graded algebra $\mathrm{End}(K)$ with differential

$$d\omega = \partial \circ -(-1)^{deg(\omega)}\omega \circ \partial \tag{6.3.7}$$

for ∂ as above, and $\iota\colon A \to \mathrm{End}(K)$

$$\iota(a)\colon (a_0, \bar{a}_1, \dots, \bar{a}_n) \mapsto (aa_0, \bar{a}_1, \dots, \bar{a}_n). \tag{6.3.8}$$

Now let Ω be the smallest differential graded algebra that is a subalgebra of $\mathrm{End}(K)$ and contains $\iota(A)$.

We have $\Psi = \tilde{\iota}\colon K \to \Omega$ given by

$$\Psi(a_0, \bar{a}_1, \dots, \bar{a}_n) = \iota(a_0)d\iota(a_1)\dots d\iota(a_n) \tag{6.3.9}$$

which is surjective by the Lemma 6.3.2. Also, K is a left Ω module since $\Omega \subseteq \mathrm{End}(K)$, and there is a map $\Phi\colon \Omega \to K\colon \omega \mapsto \omega \circ 1$ which applies the endomorphisms in Ω to $1 \in A$. Then

$$[\partial, \iota(a_n)] \circ 1 = \partial a_n = (1, \bar{a}_n), \tag{6.3.10}$$
$$[\partial, \iota(a_{n-1})](1, \bar{a}_n) = \partial \iota a_{n-1}(1, \bar{a}_n) = \partial(a_{n-1}, \bar{a}_n) = (1, \bar{a}_{n-1}, \bar{a}_n); \tag{6.3.11}$$

and after repeating this computation n times, we obtain

$$\begin{aligned}\Phi(\Psi(a_0, \bar{a}_1, \dots, \bar{a}_n)) &= \Psi(a_0\bar{a}_1, \dots, \bar{a}_n) \circ 1 \\ &= \iota(a_0)d\iota(a_1)\dots d\iota(a_n) \\ &= a_0[\partial, \iota(a_1)]\dots[\partial, \iota(a_n)] \\ &= (a_0, \bar{a}_1, \dots, \bar{a}_n).\end{aligned} \tag{6.3.12}$$

Hence $\Phi\Psi$ is the identity on Ω and we deduce that ϕ and Ψ are isomorphisms. Thus Ω with $\iota(A) \cong \Omega^0 A$ is the required differential graded algebra. □

The next result addresses uniqueness, as required for Theorem 6.3.1(ii).

Lemma 6.3.3 *Let (R,u) be a differential graded algebra and a homomorphism $u\colon A \to R_0$ such that u extends to an isomorphism $\tilde{u}\colon K \to R$. Then given any (R',u') with the same property, there exists a unique homomorphism of differential graded algebras $v\colon R \to R'$ such that $vu = u'$.*

Proof Introduce $R \times R'$ and $u_0 = (u,u')\colon A \to R_0 \times R'_0$; then let S be the differential graded subalgebra of $R \times R'$ that is generated by the image of u_0. The projections $R \times R' \to R$ and $R \times R' \to R'$ induce differential graded homomorphisms $p_1\colon S \to R$ and $p_2\colon S \to R'$. The maps $\tilde{u}_0\colon K \to S$ and $\tilde{u}\colon K \to R$ are both surjective by Lemma 6.3.2, and $\tilde{u}$ is an isomorphism. Hence p_1 is also an isomorphism, as manifested by the following diagram:

$$\begin{array}{ccc} K & & \\ \tilde{u}_0 \downarrow & \searrow \tilde{u} & \\ S & \underset{p_1}{\longrightarrow} & R. \end{array} \tag{6.3.13}$$

So we take $v = p_2 p_1^{-1}$ to give $v\colon R \to R'$ such that $vu = u'$ and conclude the proof of the Theorem 6.3.1. □

Example 6.3.4 (Noncommutative differentials) (i) Let $A = C^\infty(M)$ where M is a manifold, and E and A-bimodule. We can form $T_A(E)$, the tensor algebra of E over A, with the nth summand $E \otimes_A E \otimes_A \cdots \otimes_A E$ with n factors. This is projective as an A-bimodule. In particular, we can take $E = \Omega^1(A)$ and form

$$\Omega^1 A \otimes_A \Omega^1 A \otimes_A \cdots \otimes_A \Omega^1 A = (A \otimes \bar{A}) \otimes_A \cdots \otimes_A (A \otimes \bar{A}) \cong \Omega^n A. \tag{6.3.14}$$

Then $\Omega A = T_A(\Omega^1 A)$ is the space of noncommutative differential forms over the commutative algebra A. Then $\Omega^1 A$ is the noncommutative analogue of the tangent bundle over M, and ΩA is the noncommutative analogue of the differential forms over M.

(ii) In the case of Kähler differentials, there are two algebras attached to an A-module E namely the symmetric algebra $S_A(E)$ and the alternating algebra $\wedge_A E$. When $E = \Omega^1_A$, the tangent bundle, these give

$$\begin{aligned} S_A(\Omega^1_A) &= \text{algebra of polynomial functions on the tangent bundle,} \\ \wedge \Omega^1_A &= \Omega_A = \text{de Rham complex of differential forms.} \end{aligned} \tag{6.3.15}$$

It is curious that we need two algebras in this context, whereas there is only one in the case of noncommutative differentials (i).

In Example 6.2.4 we saw that $\Omega^k_A = 0$ for all $k > n$ for $A = \mathbf{C}[X_1, \dots, X_n]$. Likewise, for an n-dimensional manifold, the cotangent bundle has $\wedge^k T^*M = 0$ for all $k > n$. However, there is no particular reason to suppose that the noncommutative differentials should have $\Omega^k A = 0$ for large k.

Exercise 6.3.5 (Integrable operators) There are various concrete models for $\Omega^1 A$ for specific A. Let $J = \cup_{j=0}^k [a_j, b_j]$ be a union of real intervals and $A = C(J, M_m)$ the space of continuous functions $f\colon J \to M_m(\mathbf{C})$. Let $H = L^2(J; dx, \mathbf{C}^m)$ be the space of square integrable functions $\psi\colon J \to \mathbf{C}^m$. Let $M\colon H \to H$ be $M\psi(x) = x\psi(x)$. For continuous $K\colon J \times J \to M_m$, there is an integral operator $K\colon H \to H$

$$K\psi(x) = \int_J K(x,y)\psi(y)\,dy \qquad (\psi \in H).$$

(i) Consider

$$K(x,y) = \sum_{j=0}^{k} \frac{f_j(x)g_j(y)}{x-y}$$

where $f_j, g_j \in A$ and

$$\sum_{j=0}^{k} f_j(x)g_j(x) = 0 \qquad (x \in J).$$

Show that $[M, K]$ arises from an element of $\Omega^1 A$.

(ii) Show that the set $\mathcal{A}$ of all operators of the form $\lambda I + K$, with K as in (i) and $\lambda \in \mathbf{C}$, forms an algebra under composition.

(iii) Let $\lambda I + K$ be an invertible element of $\mathcal{A}$ in $\mathcal{L}(H)$. By considering

$$[M, (\lambda I + K)^{-1}] = -(\lambda I + K)^{-1}[M, K](\lambda I + K)^{-1},$$

show that $(\lambda I + K)^{-1} = \lambda^{-1} I + L_\lambda$ also belongs to $\mathcal{A}$, and find an expression for L_λ. See [26, lemma 2.8].

(iv) Let $F(X, Z) = Z^2 - \prod_{j=0}^k (X - a_j)(X - b_j)$, and consider the algebra $\mathbf{C}[X, Z]/(F)$, which has maximal ideal space $\mathbf{X}$, the Riemann surface given by the curve $F = 0$. Discuss how $\mathbf{X}$ is modelled on $(\{\infty\} \cup \mathbf{C}) \setminus \cup_{j=0}^k [a_j, b_j]$.

The main application of $(\Omega A, d)$ is in defining the Cuntz algebra and a special algebra RA, which we consider in Section 6.7. The homology of $(\Omega A, d)$ does not lead to many profound results. In Section 7.2, we consider $(\Omega A, b)$ which gives the Hochschild homology, which is generally more interesting. For commutative algebras, we show that the Hochschild homology matches with the homology of Kähler differentials in a natural way.

6.4 $\Omega^1 A$ as an A-bimodule

The A-bimodule $\Omega^1 A$ has a special role, which we consider in this section.

Definition 6.4.1 (Free module) An A-module F is said to be free if F has a linearly independent system of generators. (This is consistent with our definitions of Section 2.10, except we allow the generating set to be infinite.) An A-module M is said to be projective if there exists a free module F and an A-module M' such that $F = M \oplus M'$.

Suppose that M is a quotient module of a free module, so there is an exact sequence

$$0 \to M' \overset{\iota}{\longrightarrow} F \overset{\pi}{\longrightarrow} M \longrightarrow 0 \tag{6.4.1}$$

of A modules. The sequence of modules is said to split if there exists an A-module map $j\colon M \to F$ such that $\pi \circ j = id$ on M. Then $e = j \circ \pi$ is an idempotent as an A-module map on F, hence $id - e$ is also an idempotent as an A-module map on F, so $F = eF \oplus (id - e)F$ gives $F = M \oplus M'$.

Definition 6.4.2 (Separable) We say that a unital algebra A is separable if the multiplication map m gives a split exact sequence:

$$0 \to \Omega^1 A \overset{\iota}{\longrightarrow} A \otimes A \overset{m}{\longrightarrow} A \to 0. \tag{6.4.2}$$

Separability is the noncommutative analogue of étale. For smooth varieties X and Y over $\mathbf{C}$, a morphism $f\colon Y \to X$ is étale if f induces an isomorphism of the tangent spaces $df\colon T_y Y \to T_{f(y)} X$. The following is useful for verifying examples.

Lemma 6.4.3 *A unital complex algebra is separable if it satisfies any of the following equivalent conditions.*

(i) *There exists a bimodule splitting of* $0 \to \Omega^1 A \to A \otimes A \to A \to 0$.

(ii) *There exists Z in the centre of $A \otimes A$ as an A-bimodule such that* $m(Z) = 1$.

(iii) *There exists $Y \in \Omega^1 A$ such that $da = [a, Y]$.*

Proof $(i) \Leftrightarrow (ii)$ Now (i) is equivalent to the existence of an A-bimodule map $s\colon A \to A \otimes A$ such that $\mu \circ s = id$, so we let $Z = s(1)$ with $m(Z) = \mu \circ s(1) = 1$ and $aZ = as(1) = s(a) = s(1)a = Za$ for all $a \in A$. Conversely, Z determines s by $s(a) = aZ$. Hence (i) and (ii) are equivalent.

$(i) \Leftrightarrow (iii)$ Let $\iota\colon \Omega^1 A \to A \otimes A$ and observe that (i) is equivalent to the existence of $\pi\colon A \otimes A \to \Omega^1 A$ such that $\pi \circ \iota = id$, so we seek $Y \in \Omega^1 A$ such that $a \otimes 1 - 1 \otimes a = a\iota(Y) - \iota(Y)a$. Let $Y = \pi(1 \otimes 1)$, so

$$\pi(a_1 \otimes a_2) = a_1\pi(1 \otimes 1)a_2 = a_1 Y a_2; \tag{6.4.3}$$

then

$$\pi \circ \iota(a_0 da_1) = a_0\pi \circ \iota(da_1) = a_0\pi(a_1 \otimes 1 - 1 \otimes a_1) = a_0(a_1 Y - Y a_1); \tag{6.4.4}$$

so $\pi \circ \iota = id$ if and only if $da = [a, Y]$. Note that Y and Z correspond to the same splitting if and only if $id = \iota \circ \pi + s \circ m$, so $1 \otimes 1 = \iota(Y) + Z$. □

Example 6.4.4 (Separable algebras) (i) The algebra $A = M_n(\mathbf{C})$ is separable. The multiplication map $m\colon A \otimes A \to A\colon a \otimes b \mapsto ab$ splits as an A-bimodule map. In terms of the standard matrix units e_{jk}, the element

$$Z = \frac{1}{n}\sum_{j,k=1}^{n} e_{jk} \otimes e_{k,j} \tag{6.4.5}$$

has $m(Z) = \sum_{j=1}^{n} e_{jj} = I \in M_n$, and $aZ = Za$, so let $s(a) = aZ$. The retraction of ι has the form $A \otimes A \to \Omega^1 A\colon a_1 \otimes a_2 \to a_1 Y a_2$ for some $Y \in \Omega^1 A$. We have $\iota(Y) = 1 \otimes 1 - Z$ and

$$\sum_j a_j \otimes b_j \mapsto \sum_j a_j \otimes b_j - \Big(\sum_j a_j b_j\Big) Z. \tag{6.4.6}$$

In this case, the interesting cyclic theory is contained in $A/[A, A]$.

(ii) Let G be a finite group, with identity element $1 \in G$, and consider the group algebra $A = \mathbf{C}[G] = \{\sum_{g\in G} a_g g\colon a_g \in \mathbf{C}\}$ as in Proposition 2.9.19. Then

$$Z = \frac{1}{\sharp G}\sum_{g\in G} g^{-1} \otimes g \tag{6.4.7}$$

is the separability element such that $m(Z) = 1$, the identity of A, and Z is in the centre of $A \otimes A$ as an A-bimodule. To see this, we consider

$$\sum_g g^{-1} \otimes gh = \sum_g h(gh)^{-1} \otimes gh$$

to deduce that $hZ = Zh$. As in Proposition 2.9.19, all finite-dimensional $\mathbf{C}[G]$-modules are projective by Maschke's theorem. Alternatively, one can use Proposition 1.3.3 to show that $\mathbf{C}[G]$ is isomorphic to a finite direct sum of finite matrix algebras $M_n(\mathbf{C})$, and then use (i) to show that $\mathbf{C}[G]$ is separable.

(iii) See Example 7.2.2 for more examples and discussion of Hochschild homology.

Definition 6.4.5 (Ext(P,Q)) Let A be a unital algebra over $\mathbf{k}$, and consider short exact sequences of finitely generated A-bimodules. For P and Q finitely generated A-bimodules, we consider

$$\mathcal{E}(P,Q) = \Big\{ 0 \longrightarrow Q \overset{\alpha}{\longrightarrow} E \overset{\beta}{\longrightarrow} P \longrightarrow 0 \Big\}$$

the set of all extensions of P by Q. Certainly, we can choose $E = P \oplus Q$ to give the split exact sequence in $\mathcal{E}(P,Q)$. We introduce an equivalence relation, in the style of Yoneda [33, p. 652],

$$\Big(0 \longrightarrow Q \overset{\alpha}{\longrightarrow} E \overset{\beta}{\longrightarrow} P \longrightarrow 0\Big) \sim \Big(0 \longrightarrow Q \overset{\gamma}{\longrightarrow} F \overset{\delta}{\longrightarrow} P \longrightarrow 0\Big),$$

if and only if there exists an A-bimodule map ϕ such that the diagram

$$\begin{array}{ccccccccc} 0 & \longrightarrow & Q & \overset{\alpha}{\longrightarrow} & E & \overset{\beta}{\longrightarrow} & P & \longrightarrow & 0 \\ & & =\downarrow & & \phi\downarrow & & \downarrow= & & \\ 0 & \longrightarrow & Q & \overset{\gamma}{\longrightarrow} & F & \overset{\delta}{\longrightarrow} & P & \longrightarrow & 0 \end{array} \tag{6.4.8}$$

commutes; then we define $\mathrm{Ext}^1_{A\otimes A}(P,Q)$ as the space of equivalence classes. Then in the category of finitely generated A bimodules:

$$P \text{ is projective } \Leftrightarrow \mathrm{Hom}_{A\otimes A}(P,\cdot) \text{ is exact } \Leftrightarrow \mathrm{Ext}^1_{A\otimes A}(P,\cdot) = 0.$$

Proposition 6.4.6 *A unital algebra A is separable if A satisfies any of the following equivalent properties:*

(i) *$H^n(A,M) = 0$ for all A-bimodules M and all $n \geq 1$, so A has cohomological dimension ≤ 0 with respect to Hochschild cohomology;*

(ii) *A is a projective A-bimodule;*

(iii) *every derivation $D\colon A \to M$ is inner.*

Proof Compare (i) with (iii) via Section 2.6. The equivalence of (ii) and (iii) has a concrete version. All derivations are inner if and only if the universal derivation $d\colon A \to \Omega^1 A$ is inner. Given a derivation $D\colon A \to M$, there is a universal A-bimodule homomorphism $u\colon \Omega^1 A \to M$. Consider the diagram in which $da = [a,Y]$ with $Y \in \Omega^1 A$, and then $Da = [a, u(Y)]$:

$$\begin{array}{ccc} A & \overset{d}{\longrightarrow} & \Omega^1 A \\ & {\scriptstyle D}\searrow & \downarrow{\scriptstyle u} \\ & & M. \end{array} \tag{6.4.9}$$

Also the linear map $A \to A \oplus M\colon a \mapsto (a, D(a))$ is a homomorphism if

$$(a, D(a))(b, D(b)) = (ab, aD(b) + D(a)b) \tag{6.4.10}$$

or equivalently $D(ab) = aD(b) + D(a)b$. □

Definition 6.4.7 (Quasi-free) We say that A is quasi-free if A satisfies any of the following equivalent properties:

(i) For any algebra R with nilpotent ideal I, any homomorphism $R/I \to A$ with linear splitting $\ell\colon A \to R/I$ lifts to a homomorphism $A \to R$.

(ii) $H^n(A, M) = 0$ for all bimodules M over A and $n \geq 2$, so A has cohomological dimension ≤ 1 with respect to Hochschild cohomology.

(iii) $\Omega^1 A$ is a projective bimodule over A, so

$$0 \longrightarrow \Omega^2 A \longrightarrow \Omega^1 A \otimes A \longrightarrow \Omega^1 A \to 0 \tag{6.4.11}$$

splits as an exact sequence of A-bimodules.

Lemma 6.4.8 *The following conditions refer to a unital algebra A:*

(i) *A is separable;*

(ii) *A is projective as an A-bimodule;*

(iii) *$\Omega^1 A$ is projective as an A-bimodule;*

(iv) *A is quasi-free.*

Then (i) $\Leftrightarrow$ (ii) $\Rightarrow$ (iii) $\Leftrightarrow$ (iv).

Proof The properties are addressed by considering the sequence

$$0 \longrightarrow \Omega^1 A \longrightarrow A \otimes A \longrightarrow A \to 0 \tag{6.4.12}$$

and how it splits. □

Example 6.4.9 (Quasi-free algebras) (1) The free tensor algebra $A = T(V) = \mathbf{C} \oplus V \oplus V^{\otimes 2} \oplus \dots$ is quasi-free. Here there is a canonical lifting homomorphism of $A \to R/I$ given by $\ell\colon A \to RA$ which is the unique homomorphism such that $\ell(v) = v$ for all $v \in V$. In this case, the interesting cyclic theory is contained in $\Omega^1 A/[A, \Omega^1 A]$, as we discuss in Chapter 12.

(2) The algebra T_n of upper triangular $n \times n$ complex matrices is quasi-free. In particular

$$T_3 = \left\{ \begin{bmatrix} a_{11} & a_{12} & a_{13} \\ 0 & a_{22} & a_{23} \\ 0 & 0 & a_{33} \end{bmatrix} : a_{ij} \in \mathbf{C} \right\}$$

is quasi-free. The diagonal matrices in T_3 give a separable subalgeba $\mathbf{C}^3$ spanned by three orthogonal idempotents.

(3) The algebra $M_n(\mathbf{C})$ is separable by Example 6.4.4, hence quasi-free. If A is quasi-free, then $M_n(A)$ is also quasi-free by [22].

(4) The class of quasi-free algebras is not closed under tensor products, notwithstanding (3). Exercise 6.4.13 leads towards a suitable example.

(5) The group algebra $\mathbf{C}[\mathbf{Z}]$ is quasi-free, since $\mathbf{Z}$ is the free group on one generator and we can produce a lifting homomorphism in Definition 6.4.7(i). The compact group $SU(2) = \{U \in M_2(\mathbf{C}) : UU^* = I, \det U = 1\}$ has for each $n \in \mathbf{N}$ a unique irreducible unitary representation on $\mathbf{C}^n$. It follows that the group C^*-algebra generated by $SU(2)$ is isomorphic to $\oplus_{n=1}^{\infty} M_n(\mathbf{C})$. The inductive limit of quasi-free algebras is also quasi-free. See [22, proposition 5.3].

Let A be an algebra and M an A-bimodule. Then we can form the algebra

$$T_A M = A \oplus M \oplus (M \otimes_A M) \oplus (M \otimes_A M \otimes M) \oplus \dots, \qquad (6.4.13)$$

which combines the tensor algebra of Section 1.4 with the square zero extension of Section 1.8.

Proposition 6.4.10 (i) *For any algebra R, the set* $\mathrm{Hom}_{alg}(T_A M, R)$ *is equivalent to the set of pairs (u, v) where $u : A \to R$ is an algebra homomorphism and $v : M \to R$ is an A-bimodule map related to u.*

(ii) *Let A be a quasi-free algebra, and M a projective A-bimodule. Then $T_A M$ is also quasi-free.*

(iii) *Let A be a separable algebra, and M any finitely generated A-bimodule. Then $T_A M$ is quasi-free.*

Proof (i) Given a homomorphism $\rho : T_A M \to R$, we consider the restriction $\rho | A \oplus M$ so $\rho(a, m) = u(a) + v(m)$ where $u : A \to R$ and $v : M \to R$ are linear maps such that $\rho((a, m) \cdot (b, n)) = \rho(a, m)\rho(b, n)$, so

$$(u(ab) + v(an + mb)) + \rho(n \otimes m) = (u(a) + v(m))(u(b) + v(n)),$$

which gives $u(ab) = u(a)u(b)$, $v(an) = u(a)v(n)$ and $v(mb) = v(m)u(b)$.

Conversely, given such a pair (u, v), we define $\rho(a, m) = u(a) + v(m)$ for $a \in A$ and $m \in M$, and extend to higher order tensors by using $\rho(ma \otimes n) = v(m)u(a)v(n)$, etc.

(ii) We introduce a square-zero extension

$$0 \longrightarrow I \longrightarrow R \longrightarrow T_A M \longrightarrow 0. \qquad (6.4.14)$$

Considering the first summand of (6.4.12), we obtain a natural homomorphism $A \to T_A M$, and since A is quasi-free, there is a lifting homomorphism $A \to T_A M$. Likewise, from the second summand on (6.4.12), there is a natural A-bimodule inclusion map $M \to T_A M$, which lifts to an A-bimodule map $M \to R$ since R is projective. Given these maps, we have an A-bimodule map $A \oplus M \to R$, and by the universal property of tensor algebras as expressed in Proposition 1.1.4, there is a lifting homomorphism $T_A M \to R$. Hence $T_A M$ is quasi-free.

(iii) As in the Artin–Wedderburn theorem, M is projective and A is quasi-free, so we can apply (ii). □

Remarks 6.4.11 (i) Suppose that A is quasi-free. Then for any A-bimodule M, there is an exact sequence

$$0 \longrightarrow \Omega^1 A \otimes_A M \longrightarrow A \otimes M \longrightarrow M \longrightarrow 0, \tag{6.4.15}$$

where $\Omega^1 A \otimes M$ and $A \otimes M$ are projective. It follows that M has a finite projective resolution of length ≤ 1, hence the projective dimension of M is ≤ 1.

(ii) In Lemma 8.5.2 we show that A is quasi-free if and only if there exists a connection mapping $\nabla_r : \Omega^1 A \to \Omega^2 A$ commuting with left multiplication.

(iii) Note that when A is a unital commutative algebra, we can form the noncommutative differentials ΩA as in Theorem 6.3.1 and the Kähler commutative differentials $\Omega_A = I/I^2$ as in Lemma 6.2.1. There is a canonical homomorphism $\mu : \Omega A \to \Omega_A$ of differential graded algebras. Whereas the space of Kähler differentials Ω_A satisfies relations such as $\alpha\omega = (-1)^{|\omega|}\omega\alpha$, the space of noncommutative differentials ΩA generally does not. See Section 7.7 for further discussion of how ΩA and Ω_A are related.

(iv) In Section 7.2, we describe the Hochschild homology of ΩA.

In Proposition 6.1.2, we introduced a compact complex algebraic curve $\mathbf{X}$ (Riemann surface) and the algebra $\mathbf{C}[\mathbf{X}]$ of coordinate functions on $\mathbf{X}$. The following Exercise 6.4.12 and Example 7.2.4 show that $\mathbf{C}[\mathbf{X}]$ is the prototype of a quasi-free commutative algebra. However, when $\mathbf{Y}$ is a smooth complex algebraic variety of dimension $d \geq 2$, the algebra $\mathbf{C}[\mathbf{Y}]$ can be smooth but not quasi-free.

Exercise 6.4.12 (Graded algebras) Suppose that A is a unital, commutative and finitely generated algebra over $\mathbf{C}$, which is also quasi-free. Then

$$0 \longrightarrow (F_1, \dots, F_m) \longrightarrow \mathbf{C}[x_1, \dots, x_n] \longrightarrow A \longrightarrow 0. \tag{6.4.16}$$

(i) Let I be a maximal ideal in A. Show that $\mathbf{k} = A/I$ is a field.

(ii) In the graded algebra

$$g_I(A) = \oplus_{j=1}^{\infty} I^{j-1}/I^j \tag{6.4.17}$$

prove by induction that there is an isomorphism

$$(I/I^2) \otimes_{\mathbf{k}} (I^{j-1}/I^j) \to I^j/I^{j+1}, \tag{6.4.18}$$

so that I^j/I^{j+1} is isomorphic to $(I/I^2)^{\otimes j}$ over $\mathbf{k}$.

(iii) Let $\mathbf{X}$ be the space of maximal ideals in A, and suppose that I/I^2 has dimension one or zero. Deduce that the connected components of $\mathbf{X}$ are points or non-singular affine curves.

See [8, p. 124] and [22, p. 286].

Exercise 6.4.13 (Graded algebras and rank) (After [8, ex. 11.1]) Let $F \in \mathbf{C}[X_1, X_2, X_3]$ be an irreducible complex polynomial, and let $A = \mathbf{C}[X_1, X_2, X_3]/(F)$, so $\mathbf{X}$ is the algebraic variety $F = 0$. Let I be the maximal ideal in A that corresponds to $(t_1, t_2, t_3) \in \mathbf{X}$ and consider

$$DF(t_1,t_2) = \begin{bmatrix} \frac{\partial F}{\partial X_1}(t_1,t_2) \\ \frac{\partial F}{\partial X_2}(t_1,t_2) \end{bmatrix}. \qquad (6.4.19)$$

(i) Using the Taylor expansion of F show that $F \in I^2$, if and only if the rank of $DF(t_1,t_2)$ equals zero.

(ii) Deduce that if the rank of $DF(t_1,t_2)$ is one, then the dimension of

$$I/I^2 = \big(X_1 - t_1, X_2 - t_2\big)/\big((X_1 - t_1, X_2 - t_2)^2 + (F)\big) \qquad (6.4.20)$$

is equal to 2. Here dimension can be interpreted as the number of generators of I.

(iii) Let $g_I(A) = \oplus_{n=0}^{\infty} I^n/I^{n+1}$, where $I^0 = A$. Deduce that if the rank of $DF(t_1,t_2)$ is one, then $g_I(A)$ is isomorphic to $\mathbf{k}[Y_1, Y_2]$ for indeterminates Y_1, Y_2. (See theorem 11.22 of [8].)

Exercise 6.4.14 (Separable algebras are finite–dimensional) In the notation of Lemma 6.4.3, let $Z = \sum_{j=1}^{n} x_j \otimes y_j$, and let $V = \mathrm{span}\{x_j : j = 1, \dots, n\}$.

(i) Show that V is a left ideal of A.

(ii) For $a \in A$, let $\lambda_a : V \to V$ be the map $\sum_{j=1}^{n} \alpha_j x_j \mapsto \sum_{j=1}^{n} \alpha a x_j$. Show that $a \mapsto \lambda_a$ gives an injective homomorphism $A \to \mathrm{Hom}_{\mathbf{C}}(V)$.

(iii) Deduce that A is finite–dimensional.

Exercise 6.4.15 Show that ϕ in (6.4.8) is bijective, hence verify that $\sim$ is an equivalence relation.

6.5 The Cuntz Algebra with Fedosov's Product

In this section we introduce an alternative view of ΩA, by introducing a special product on A that extends to ΩA. This Fedosov product is $\star$ closely related to the $\star$ operation we introduced in Section 1.9, and we will interpret its properties in quantum mechanics in Section 9.3. The Cuntz algebra $QA = A \star A$ is characterized by the universal property that there are canonical homomorphisms

$\iota\colon A \to QA$ and $\tilde{\iota}\colon A \to QA$ such that, given homomorphisms $u, u'\colon A \to B$ to another algebra B, there exists a unique homomorphism $w\colon QA \to B$ such that $w \circ \iota = u$ and $w \circ \tilde{\iota} = u'$. In particular, there is an automorphism of QA of order two $\omega\colon z \mapsto \tilde{z}$ such that $\omega \circ \iota = \tilde{\iota}$ and $\omega \circ \tilde{\iota} = \iota$. Thus QA is a superalgebra.

We split $\iota(a)$ into odd and even parts, defining

$$a^+ = (1/2)(\iota(a) + \tilde{\iota}(a)) \quad \text{and} \quad a^- = (1/2)(\iota(a) - \tilde{\iota}(a)) \tag{6.5.1}$$

so that $\iota(a) = a^+ + a^-$ and $\tilde{\iota}(a) = a^+ - a^-$; also

$$(a_0a_1)^+ = a_0^+a_1^+ + a_0^-a_1^- \quad \text{and} \quad (a_0a_1)^- = a_0^-a_1^+ + a_0^+a_1^-. \tag{6.5.2}$$

Proposition 6.5.1 *There is a vector space isomorphism $\Psi\colon \Omega A \to QA$ given by*

$$\Psi(a_0da_1 \dots da_n) = a_0a_1^- \dots a_n^- \tag{6.5.3}$$

such that

$$\Psi(\omega \star \eta) = \Psi(\omega)\Psi(\eta) \tag{6.5.4}$$

where $\star$ is the Fedosov product

$$\omega \star \eta = \omega\eta - (-1)^{deg(\omega)} d\omega d\eta. \tag{6.5.5}$$

Proof (i) First we prove that the whole algebra is given by

$$V = \text{span}\{a_0a_1^- \dots a_n^- : a_j \in A; n = 0, 1, \dots\}. \tag{6.5.6}$$

Clearly V contains $\iota(A)$ and $\tilde{\iota}(A)$. Also $a^+V \subseteq V$ for all $a \in A$ since

$$a^+(a_0^+a_1^- \dots a_n^-) = (aa_0)^+a_1^- \dots a_n^- - 1^+a^-a_0^-a_1^- \dots a_n^-; \tag{6.5.7}$$

likewise $a^-V \subseteq V$ for all $a \in A$ since by (6.5.2)

$$a^-(a_0^+a_1^- \dots a_n^-) = 1^+(aa_0)^-a_1^- \dots a_n^- - a^+a_0^-a_1^- \dots a_n^-. \tag{6.5.8}$$

We deduce that V is a left ideal containing $1 = 1^+$, hence $V = QA$.

(ii) We endow ΩA with the structure of a left QA-module. Also ΩA has a $\mathbf{Z}/(2)$ grading given by the direct sum $\Omega A = \Omega^{ev}A \oplus \Omega^{odd}A$, so $\text{End}_{\mathbf{k}}(\Omega)$ is a superalgebra. Note that $(1 + d)(1 - d) = 1 = (1 - d)(1 + d)$ so we write $(1 + d)^{-1} = 1 - d$ and introduce a homomorphism $A \to \text{End}_{\mathbf{k}}(\Omega A)$ by $a \mapsto (1 + d)a(1 - d)$, where a stands for left multiplication by a. Then $[d, a] = da$, namely left multiplication by da, so

$$(1 + d)a(1 - d) = a + da - (da)d. \tag{6.5.9}$$

This homomorphism extends uniquely to a homomorphism of superalgebras $QA \to \mathrm{End}_{\mathbf{k}}(\Omega A)$ such that

$$\iota(a) \mapsto a + da - (da)d \quad \text{and} \quad \tilde{\iota}(a) = a - da - (da)d \tag{6.5.10}$$

so

$$a^+ \mapsto a - (da)d \quad \text{and} \quad a^- \mapsto da. \tag{6.5.11}$$

Thus ΩA becomes a left QA module under $a^+\eta = a\eta - dad\eta$ and $a^-\eta = (da)\eta$. To check that $(\Psi(\omega))\eta = \omega \star \eta$ has the stated formula, we consider $\omega = a_0da_1 \dots da_n$ and compute

$$\begin{aligned}\Psi(\omega)\eta &= a_0^+ a_1^- \dots a_n^- \eta \\ &= (a_0 - (da_0)d)(da_1 \dots da_n\eta) \\ &= a_0da_1 \dots da_n\eta - (-1)^n da_0da_1 \dots da_n d\eta \\ &= \omega\eta - (-1)^n d\omega d\eta \\ &= \omega \star \eta.\end{aligned} \tag{6.5.12}$$

Choosing $\eta = 1$, we have $\Psi(\omega)1 = \omega$, so Ψ is injective and hence is an isomorphism with inverse $\Psi^{-1}: \eta \mapsto \eta 1$. Finally,

$$\Psi^{-1}(\Psi(\omega)\Psi(\eta)) = \Psi(\omega)\Psi(\eta)1 = (\Psi(\omega))\eta = \omega \star \eta. \tag{6.5.13}$$

□

Remarks 6.5.2 (i) Let ΩA be split into even and odd graded components

$$\Omega^{ev} A = \bigoplus_{n=0}^{\infty} \Omega^{2n} A, \tag{6.5.14}$$

$$\Omega^{odd} A = \bigoplus_{n=0}^{\infty} \Omega^{2n+1} A, \tag{6.5.15}$$

so

$$\Omega A = \Omega^{ev} A \oplus \Omega^{odd} A. \tag{6.5.16}$$

The $\mathbf{Z}/(2)$ grading on QA corresponds to the decomposition $\Omega A = \Omega^{ev} A \oplus \Omega^{odd} A$. Note that the degree of $\omega\eta$ has the same parity as the degree of $\omega * \eta$.

(ii) Let $J = \mathrm{Ker}\{A \star A \to A\}$ for the folding map given by $\iota(a) \mapsto a$ and $\tilde{\iota}(a) \mapsto a$, so that J is the ideal generated by a^-. Then J corresponds to $\oplus_{m=1}^{\infty} \Omega^m A$, and more generally, J^n corresponds to $\oplus_{m=n}^{\infty} \Omega^m A$.

(iii) The graded sum of left QA modules

$$g_J(QA) = \oplus_{n=0}^{\infty}\big(J^n/J^{n+1}\big) \tag{6.5.17}$$

is canonically isomorphic to ΩA.

On the graded algebra QA, a supertrace τ has the form $\tau(a_+ + a_-) = \tau_+(a_+) - \tau_-(a_-)$, where $\tau_+ : \Omega^{ev}A \to \mathbf{C}$ and $\tau_- : \Omega^{odd}A \to \mathbf{C}$ are traces.

6.6 Cyclic Cochains on the Cuntz Algebra

Let τ be a supertrace of QA, and define cochains

$$\varphi_n(a_0, \dots, a_n) = \tau(a_0^+ a_1^- \dots a_n^-). \tag{6.6.1}$$

Proposition 6.6.1 (i) *φ_n is a normalized trace in the sense that, if $a_i = 1$ for some $i \geq 1$, then $\varphi_n(a_0, a_1, \dots, a_n) = 0$.*

(ii) *$\varphi_n(1, a_1, \dots, a_n)$ is a cyclic cochain, so*

$$\varphi_n(1, a_n, a_1, \dots, a_{n-1}) = (-1)^{n-1}\varphi_n(1, a_1, \dots, a_n). \tag{6.6.2}$$

(iii) *$b\varphi_n(a_0, a_1, \dots, a_{n+1}) = 2\varphi_{n+2}(1, a_0, \dots, a_{n+1})$.*

Proof (i) This follows from the fact that $1^- = 0$.

(iii) We have

$$\begin{aligned} &b\varphi_n(a_0, a_1, \dots, a_{n+1}) \\ &\quad= \sum_{j=0}^{n}(-1)^j\varphi_n(a_0, a_1, \dots, a_j a_{j+1}, \dots, a_{n+1}) + (-1)^{n+1} \\ &\qquad\times b\varphi_n(a_{n+1}a_0, a_1, \dots, a_n) \\ &\quad= \tau\big((a_0a_1)^+ a_2^- \dots a_{n+1}^-\big) + \sum_{j=1}^{\infty}(-1)^j\tau\big((a_0^+ a_1^- \dots (a_j a_{j+1})^- \dots a_{n+1}^-\big) \\ &\qquad+ (-1)^{n+1}\tau\big((a_{n+1}a_0)^+ a_1^- \dots a_n^-\big) \end{aligned} \tag{6.6.3}$$

in which $(a_0a_1)^+ = a_0^+ a_1^+ + a_0^- a_1^-$ and $(a_ja_{j+1})^- = a_j^+ a_{j+1}^- + a_j^- a_{j+1}^+$, so by (ii) we have

$$\begin{aligned} b\varphi_n(a_0, a_1, \dots, a_{n+1}) &= \tau\big(a_0^+ a_1^+ a_2^- \dots a_{n+1}^-\big) + \tau\big(a_0^- a_1^- a_2^- \dots a_{n+1}^-\big) \\ &\quad+ \sum_{j=1}^{\infty}(-1)^j\Big(\tau\big((a_0^+ a_1^- \dots a_j^+ a_{j+1}^- \cdots a_{n+1}^-\big) \end{aligned}$$

$$
\begin{aligned}
&+\tau\big((a_0^+a_1^-\dots a_j^-a_{j+1}^+\dots a_{n+1}^-)\big)\\
&+(-1)^{n+1}\tau\big(a_{n+1}^+a_0^+a_2\dots a_n^-\big)\\
&+(-1)^{n+1}\tau\big((a_{n+1}^-a_0^-a_2^-\dots a_n^-\big)\\
&=2\tau(a_0^-\dots a_{n+1}^-).
\end{aligned}
\tag{6.6.4}
$$

□

This example suggests a general definition.

Definition 6.6.2 (B-operator) On normalized cochains $\varphi(a_0,\dots,a_n)$, let

$$
B\varphi(a_1,\dots,a_n)=\sum_{i=1}^{n}(-1)^{i(n-1)}\varphi(1,a_{i-1},\dots,a_n,a_1,\dots,a_i). \tag{6.6.5}
$$

6.7 Tensor Algebra with the Fedosov Product

For a unital algebra A, let $T(A)=\oplus_{n=0}^{\infty}A^{\otimes n}$ and let RA be the quotient of $T(A)$ by the ideal generated by $1_{T(A)}-1_A$. There is a short exact sequence

$$
0\longrightarrow \mathbf{k}\overset{\iota}{\longrightarrow}A\overset{\pi}{\longrightarrow}\bar{A}\longrightarrow 0, \tag{6.7.1}
$$

and we choose a splitting $j\colon \bar{A}\to A$, namely a linear map so that $e=\pi\circ j$ has $e^2=e$ and $\iota(\mathbf{k})=\mathrm{Ker}(\pi)$. Then RA is isomorphic to $T(\bar{A})$. We can also characterize RA by the universal property that

$$
\mathrm{Hom}(RA,R)=\{\rho\colon A\to R\colon \rho(1_A)=1_R;\rho\quad\text{linear}\} \tag{6.7.2}
$$

so that the algebra homomorphisms on RA are given by $\mathbf{k}$-linear unital maps. See Section 4.9(iii) for discussion. There is a canonical linear map $A\to RA$ such that $\rho(1_A)=1_R$, given by embedding A as $A^{\otimes 1}$ in $T(A)$ and then quotienting to RA. There is an obvious homomorphism $\pi\colon RA\to A$, so $\pi\circ\phi=id_A$. Thus we obtain an extension of algebras

$$
0\longrightarrow IA\longrightarrow RA\longrightarrow A\longrightarrow 0 \tag{6.7.3}
$$

where $IA=\mathrm{Ker}\{\pi\colon RA\to A\}$. This is the universal extension of A. For the canonical map ρ, let

$$
\omega(a_1,a_2)=-\rho(a_1a_2)+\rho(a_1)\rho(a_2), \tag{6.7.4}
$$

which is loosely analogous to the canonical 2-form ω_2 of Section 1.7.

Proposition 6.7.1 *Let $\Omega^{ev}A$ have the Fedosov product. Then*

$$\Phi\colon \Omega^{ev}A \to RA\colon a_0da_1da_2\dots da_{2n} \mapsto \rho(a_0)\omega(a_1,a_2)\dots\omega(a_{2n-1},a_{2n}) \tag{6.7.5}$$

gives an isomorphism of algebras.

Proof The map is well defined on $A\otimes(\bar{A})^{\otimes 2n}$ and extends to $\oplus_{n=0}^{\infty}\Omega^{2n}A$. In degree zero, $a_1,a_2\in A$ give $a_1\star a_2 = a_1a_2 - da_1da_2$, which is mapped to $\rho(a_1a_2)-\omega(a_1,a_2)=\rho(a_1)\rho(a_2)$. Using this motivation, we use the universal property of RA to define a left RA module structure on $\Omega^{ev}A$. The linear map $A\to \mathrm{End}(\Omega^{ev}A)\colon a\mapsto a-(da)d$ induces a homomorphism $RA\to \mathrm{End}(\Omega^{ev}A)$ by the universal property of RA. This is the unique left RA module structure such that $\rho(a)\eta = a\eta - dad\eta$ for $a\in A$ and $\eta\in\Omega^{ev}A$.

We check that Φ is a homomorphism of RA modules by computing

$$\begin{aligned}
\rho(a)\Phi(a_0da_1da_2\dots da_{2n}) &= \rho(a)\rho(a_0)\omega(a_1,a_2)\dots\omega(a_{2n-1},a_{2n})\\
&= \big(\rho(aa_0)-\omega(a,a_0)\big)\omega(a_1,a_2)\dots\omega(a_{2n-1},a_{2n})\\
&= \Phi(aa_0da_1\dots da_{2n}) - \Phi(dada_0\dots da_{2n})\\
&= \Phi((a-(da)d)a_0da_1\dots da_{2n})\\
&= \Phi(\rho(a)a_0da_1\dots da_{2n}).
\end{aligned}\tag{6.7.6}$$

Now RA is generated by the elements $\rho(a)$, so $x\Phi(\eta)=\Phi(x\eta)$ for all $x\in RA$. Hence $\{\Phi(\eta)\colon \eta\in RA\}$ is a left ideal in RA which contains $1=\Phi(1)$; hence Φ is surjective. To check that Φ is also injective, we consider the map $RA\to\Omega^{ev}A\colon x\mapsto x1$. Now

$$\begin{aligned}
\omega(a_1,a_2)\eta &= (\rho(a_1a_2)-\rho(a_1)\rho(a_2))\eta\\
&= a_1a_2\eta - d(a_1a_2)\eta - (a_1-(da_2)d)(a_2\eta - da_2d\eta)\\
&= a_1a_2\eta - (da_1)a_2d\eta - a_1da_2d\eta - a_1a_2\eta + a_1da_2d\eta\\
&\quad (da_1)da_2\eta + (da_1)a_2d\eta\\
&= da_1da_2\eta
\end{aligned}\tag{6.7.7}$$

and hence

$$\begin{aligned}
\Phi(a_0da_1\dots da_{2n})\eta &= \rho(a_0)\omega(a_1,a_2)\dots\omega(a_{2n-1},a_{2n})\eta\\
&= (a_0-(da_0)d)(da_1da_2\dots da_{2n-1}da_{2n})\\
&= a_0da_1\dots da_{2n}\eta - da_0da_1\dots da_{2n}d\eta
\end{aligned}\tag{6.7.8}$$

so $\Phi(\xi)\eta = \xi \star \eta$ and $\Phi(\xi)1 = \xi \star 1 = \xi$. Hence Φ is a homomorphism of algebras with

$$\Phi(\xi * \eta) = \Phi(\Phi(\xi)\eta) = \Phi(\xi)\Phi(\eta). \tag{6.7.9}$$

□

Corollary 6.7.2 *With the usual product for differential forms, $\Omega^{ev}A$ is isomorphic to*

$$g_{IA}(RA) = \oplus_{n=0}^{\infty}(IA)^n/(IA)^{n+1}. \tag{6.7.10}$$

Proof The algebra isomorphism of Proposition 6.7.1 restricts to an isomorphism on the decreasing IA-adic filtration

$$RA \supseteq IA \supseteq (IA)^2 \supseteq \cdots \tag{6.7.11}$$

so that $\Phi\colon \oplus_{m=n}^{\infty} \Omega^{2m}A \to (IA)^n$. We write $\rho(A)^2$ for $\mathrm{span}\{\rho(a_1)\rho(a_2)\colon a_1, a_2\}$ and we observe that $\rho(A) \subseteq \rho(A)^2$ since $1 = \rho(1) \in \rho(A)$. We can split the inclusion maps in the canonical increasing filtration

$$\mathbf{k} \subseteq \rho(A) \subseteq \rho(A)^2 \subseteq \cdots \tag{6.7.12}$$

and write

$$\rho(A)^{2n+1} \oplus (IA)^{n+1} \cong RA \tag{6.7.13}$$

and then take intersections to obtain

$$\left((IA)^n \cap \rho(A)^{2n+1}\right) \oplus (IA)^{n+1} = (IA)^n, \tag{6.7.14}$$

so the differentials of degree $2n$ are mapped to linear combinations of

$$\rho(a_0)\omega(a_1, a_2)\ldots\omega(a_{2n-1}, a_{2n}) \tag{6.7.15}$$

so

$$\Omega^{2n}A = (IA)^n \cap \rho(A)^{2n+1}. \tag{6.7.16}$$

□

Corollary 6.7.3 *Any square zero algebra extension of A has a lifting homomorphism, if and only if there is a lifting homomorphism $A \to RA/(IA)^2$ where $RA/(IA)^2 = A \oplus \Omega^2 A$ under the Fedosov product.*

Proof Consider an algebra extension

$$0 \longrightarrow I \hookrightarrow R \underset{\pi}{\overset{\rho_0}{\leftrightarrows}} A \longrightarrow 0, \tag{6.7.17}$$

and choose a linear lifting map $\rho_0\colon A \to R$ such that $\rho_0(1) = 1$ and $\pi \circ \rho_0 = id$. Then let $\omega(a_1,a_2) = \rho_0(a_1a_2) - \rho_0(a_1)\rho_0(a_2)$ and introduce $\rho_*\colon RA \to R$ by

$$\rho_*(a_0,\dots,a_{2n}) = \rho_0(a_0)\omega(a_1,a_2)\dots\omega(a_{2n-1},a_{2n}). \tag{6.7.18}$$

With the Fedosov product, we have $da_1da_2 = a_1a_2 - a_1 \star a_2$, so da_1da_2 is the curvature of the linear map $\rho\colon A \to RA$ from the proof of Proposition 6.7.1. In the case $I^2 = 0$, we have

$$A = RA/(IA)^2/(IA)/(IA)^2$$

and we can identify the quotients as $RA/(IA)^2 = A \oplus \Omega^2 A$ and $IA/IA^2 = \Omega^2 A$. In this case, ρ_* kills $(IA)^2$.

$$\begin{array}{ccccccccc}
0 & \longrightarrow & IA/(IA)^2 & \longrightarrow & RA/(IA)^2 & \longrightarrow & A & \longrightarrow & 0 \\
 & & \downarrow = & & \downarrow = & & \downarrow = & & \\
0 & \longrightarrow & \Omega^2 A & \longrightarrow & A \oplus \Omega^2 A & \longrightarrow & A & \longrightarrow & 0 \\
 & & \downarrow & & \downarrow \rho_* & & \downarrow = & & \\
0 & \longrightarrow & M & \longrightarrow & R & \longrightarrow & A & \longrightarrow & 0
\end{array} \tag{6.7.19}$$

□

Proposition 6.7.4 *Let A be quasi-free. Then any square zero algebra extension of A has a lifting homomorphism.*

We prove this in Section 8.7, after we have introduced an alternative formulation of the notion of quasi-free in terms of connections.

Definition 6.7.5 (Cochain) A cochain of degree n is a $\mathbf{k}$-linear map $\Phi\colon A^{\otimes n} \to V$. A cochain $f(a_0,\dots,a_n)$ is normalized if f vanishes whenever $a_i = 1$ for for some $i = 1,\dots,n$. A normalized even cochain is a $\mathbf{k}$-linear map $f\colon A \otimes (\bar{A})^{\otimes n} \to V$ where n is even.

Whereas we can take $V = A^*$ and identify $C^n(A,A^*)$ with $C^{n+1}(A,\mathbf{k})$, we regard the degree as the number of arguments. When the cochain is normalized, we count the number of variables in $\bar{A}$ from $A \otimes (\bar{A})^{\otimes n}$.

Corollary 6.7.6 *A linear functional τ on RA is equivalent to a sequence of normalized even cochains τ_{2n} via*

$$\tau_{2n}(a_0,\dots,a_{2n}) = \tau(\rho(a_0)\omega(a_1,a_2)\dots\omega(a_{2n-1},a_{2n})). \tag{6.7.20}$$

This result suggests examples such as Example 1.7.5 and Proposition 4.6.1.

6.8 Completions

Definition 6.8.1 (Inverse limit) Let $(A_j)_{j=0}^{\infty}$ be a sequence of abelian groups for which there are connecting homomorphisms $\varphi_{kj}\colon A_k \to A_j$ for all $k \geq j$ such that $\varphi_{\ell j} \circ \varphi_{k\ell} = \varphi_{kj}$ for all $k \geq \ell \geq j$. Then (A_j) is said to be an inverse system, with inverse limit

$$\lim_{\leftarrow j} A_j = \Big\{ (a_j) \in \prod_j A_j : \varphi_{kj}(a_k) = a_j, \forall k \geq j \Big\}. \tag{6.8.1}$$

We write

$$\dots \overset{\varphi_{n,n-1}}{\longrightarrow} A_{n-1} \overset{\varphi_{n-1,n-2}}{\longrightarrow} \dots \overset{\varphi_{2,1}}{\longrightarrow} A_1 \overset{\varphi_{1,0}}{\longrightarrow} A_0. \tag{6.8.2}$$

Evidently, the inverse limit is determined by the maps $d^A(a_n) = (a_n - \varphi_{n+1,n}(a_{n+1}))$, so $\operatorname{Ker} d^A = \lim_{\leftarrow j} A_j$. Then we introduce $\lim_{\leftarrow}^{(1)} A_j = \operatorname{Coker} d^A$, as in [8, p. 104].

Let R be an algebra with ideal I, so

$$0 \subseteq \dots \subseteq I^n \subseteq I^{n-1} \subseteq \dots \subseteq I^2 \subseteq I \subseteq R \tag{6.8.3}$$

gives rise to an inverse system

$$\longrightarrow R/I^n \longrightarrow R/I^{n-1} \longrightarrow \dots \longrightarrow R/I^2 \longrightarrow R/I \longrightarrow 0, \tag{6.8.4}$$

and we define the completion to be

$$\hat{R} = \lim_{\leftarrow} R/I^n. \tag{6.8.5}$$

Similarly, we define

$$\widehat{I^k} = \lim_{\leftarrow} I^k/I^n, \tag{6.8.6}$$

and we note that this is different from $(\hat{I})^k$, so the operation of forming completions does not generally commute with the operation of forming powers.

The idea is that R/I^n gives the nth approximation to R, and one uses this in Taylor approximation to functions, as in the following example. See [8, ch. 10] for further information on filtrations and completions.

Example 6.8.2 (Completions) (i) Let $R = \mathbf{C}[x]$ be the algebra of complex polynomials and $I = (x)$ be the maximal ideal $I = \operatorname{Ker}\{R \to \mathbf{C}\colon f(x) \mapsto f(0)\}$. Hence R/I^n is the algebra of complex polynomials of degree $\leq n-1$, with multiplication mod (x^n). Now $\hat{R} = \lim_{\leftarrow} R/I^n = \mathbf{C}[[x]]$ is the algebra of formal power series.

(ii) Now let $A = C^\infty(\mathbf{R};\mathbf{C})$ and $I = \{f \in A\colon f(0) = 0\}$. Borel showed that the map

$$\theta\colon f \mapsto \sum_{j=0}^{\infty} \frac{f^{(j)}(0)}{j!} x^j \tag{6.8.7}$$

is surjective $A \to \mathbf{C}[[x]]$, so all formal power series arise as Taylor series of smooth functions. Then there is a homomorphism $A \mapsto A/I^n$:

$$f \mapsto f(0) + f'(0)x + \cdots + \frac{f^{(n-1)}(0)x^{n-1}}{(n-1)!} \qquad (\mathrm{mod}\ I^n).$$

However, θ is not injective, since $f(x) = e^{-1/x^2}$ produces the zero element of $\hat{A}$. See also [8, ex. 11, p. 115].

Example 6.8.3 (Universal lifting) Let A be quasi-free. Then there is a lifting homomorphism $A \to \widehat{RA}$. This is the universal case which leads to Proposition 6.7.4. We prove this in Section 8.7.

7
Hodge Decomposition and the Karoubi Operator

In our approach to cyclic theory, the graded complexes of cochains and differential forms are also regarded as modules over commutative algebras that are generated by fundamental operators S, B and κ. Hence we can use the general decomposition theory for modules over principal ideal domains or local rings to analyse the graded complexes. In this chapter, we introduce a monodromy operator on the noncommutative differential forms called the Karoubi operator κ. We use this to define other basic operators such as Connes's B operator, which produces the cyclic bicomplex, and Connes's periodicity operator S. The Karoubi operator is also associated with a Green's function, and a type of Laplace operator also produces a decomposition of ΩA which is analogous to the Hodge decomposition of vector bundles over a compact Riemannian manifold. The chapter begins by exhibiting the operators in the classical case of a compact Riemann surface. Then we proceed to introduce the analogous concepts in cyclic theory. We obtain complexes and homology groups:

- $(\Omega A, d)$, which is the main topic of Chapter 6, and led us to the Fedosov product on $\Omega^{ev} A$;
- The Hochschild homology of A given by $HH(A) = H(\Omega A, b)$, which we compute using the Hodge decomposition in this chapter;
- The cyclic homology of A given by $HC(A) = H(\Omega A, b + B)$, which we introduce in this chapter using a double complex.

So we begin with a brief discussion of the geometrical setting, as in [107].

7.1 Hodge Decomposition on a Compact Riemann Surface

The basic ingredients of Hodge theory are:

- a Laplace operator Δ which is self-adjoint and non-negative in Hilbert space H;
- the linear projection P from H onto $\mathrm{Ker}(H)$;
- a Green's operator G, the inverse of $\Delta\colon H \to (I-P)H$, as in Theorem 5.3.2, for instance;
- derivations d and d^*, defined with a special $*$-operation, such that $\Delta = d^*d + dd^*$;
- a monodromy operator κ.

First we consider the Hodge decomposition in the classical case in which $\mathbf{X}$ is a one-dimensional complex manifold, namely a compact Riemann surface $\mathbf{X}$, as in [96, 75]. Suppose that $f(X,Y) \in \mathbf{C}[X,Y]$ is an irreducible complex polynomial, so $\mathbf{X} = \{(x,y) \in \mathbf{C}^2 \colon f(x,y) = 0\} \cup \{\infty\}$ is a compact Riemann surface, and $A = \mathbf{C}[\mathbf{X}] = C[X,Y]/(f(X,Y))$ is the coordinate ring on $\mathbf{X}$. Then A is the prototype of a quasi-free commutative algebra. The field of fractions of A is $\mathbf{C}(\mathbf{X})$, the field of functions of rational character on $\mathbf{X}$. Each non-constant $g \in \mathbf{C}(\mathbf{X})$ gives a cover $g\colon \mathbf{X} \to S^2$ of the projective line, namely the Riemann sphere $S^2 = \mathbf{C} \cup \{\infty\}$. Unless $\mathbf{X}$ is itself conformally equivalent to S^2, the cover will have ramifications, which are described in local coordinates by $y(p) = x(p)^{1/r}$ for some integer $r > 1$. See [75].

By Hopf's Lemma 5.3.1 or Liouville's theorem, the only holomorphic functions on $\mathbf{X}$ are constants, and the only exact holomorphic differential is zero. However, there are non-constant meromorphic functions, and non-constant holomorphic differentials. Let $Z\colon \mathbf{X} \to S^2$ be a non-constant meromorphic function, and suppose that the poles of Z are $p_1, \dots, p_r$, so $Z(p_k) = \infty$. Then $G(z,w) = -2\log|Z(z) - Z(w)|$ is defined on $\mathbf{X} \setminus \{p_1, \dots, p_r\}$ and gives a meromorphic differential

$$\frac{\partial}{\partial z} G(z,w)dz = \frac{-Z'(z)dz}{Z(z) - Z(w)} = \frac{dZ}{Z(w) - Z(z)}, \tag{7.1.1}$$

which has a simple pole at $z = w$ and finitely many other points.

Consider a coordinate chart $\varphi\colon \mathbf{D} \to \mathbf{X}$. We introduce the partial derivatives

$$\frac{\partial}{\partial z} = \frac{1}{2}\left(\frac{\partial}{\partial x} - i\frac{\partial}{\partial y}\right), \qquad \frac{\partial}{\partial \bar{z}} = \frac{1}{2}\left(\frac{\partial}{\partial x} + i\frac{\partial}{\partial y}\right), \tag{7.1.2}$$

and the differentials

$$dz = dx + idy, \quad d\bar{z} = dx - idy,$$

with which we can define ∂ and $\bar\partial$ by $\partial f = \frac{\partial f}{\partial z}dz$ and $\bar\partial f = \frac{\partial f}{\partial \bar z}d\bar z$ so that the usual de Rham differential is $d = \partial + \bar\partial$. The Hodge $*$ operation is the real-linear operation on differential 1-forms

$$\omega = f dz + g d\bar z \mapsto *\omega = i\bar f d\bar z - i\bar g dz, \tag{7.1.3}$$

so $**\omega = -\omega$. We recognize holomorphic functions on open subsets of $\mathbf{X}$ by the condition $\bar\partial f = 0$, which reduces to the Cauchy–Riemann equations in x and y. The following exercise shows how one can solve the inhomogeneous version of the Cauchy–Riemann equations.

Exercise 7.1.1 (Cauchy kernel on a disc) Let $g\colon \mathbf{C} \to \mathbf{C}$ be a compactly supported C^∞ function, and let

$$f(z) = -\frac{1}{\pi}\iint_{\mathbf{C}} \frac{g(\xi + i\eta)}{\xi + i\eta - z}\, d\xi d\eta \qquad (z \in \mathbf{C}).$$

By changing to polar coordinates centred at z, show that $\bar\partial f = g(z)d\bar z$.

The following result shows that one can compute the de Rham cohomology by considering the holomorphic and anti-holomorphic differentials. This splitting of the space is one of the basic ideas underlying Hodge theory.

Proposition 7.1.2 *Let $\Omega^1_{\mathbf{X}}$ be the space of holomorphic differentials on a compact Riemann surface $\mathbf{X}$ of genus g. There is an isomorphism of complex vector spaces*

$$\Omega^1_{\mathbf{X}} \oplus \overline{\Omega^1_{\mathbf{X}}} \to H^1_{DR}(\mathbf{X};\mathbf{C})\colon \omega_1 \oplus \overline{\omega_2} \mapsto \omega_1 + \overline{\omega_2} \tag{7.1.4}$$

so that both spaces have dimension $2g$.

Proof We sketch the main ideas of the proof. First, suppose that $\omega_1, \omega_2 \in \Omega^1_{\mathbf{X}}$ satisfy $\omega_1 + \bar\omega_2 = df$, so $\omega_1 + \bar\omega_2$ is the exact differential $df = \frac{\partial f}{\partial z}dz + \frac{\partial f}{\partial \bar z}d\bar z$ of some $f \in C^\infty(\mathbf{X};\mathbf{C})$.

Now $\omega_1 = f_1 dz$ and $\omega_2 = f_2 dz$ with holomorphic f_1 and f_2, so $\omega_1 \wedge \omega_2 = 0$ and $d\omega_2 = \partial f_2 dz + \bar\partial f_2 d\bar z = 0$. Then

$$\omega_2 \wedge \bar\omega_2 = \omega_2 \wedge (\omega_1 + \bar\omega_2) = \omega_2 \wedge df = d(\omega_2 f). \tag{7.1.5}$$

Hence by Stokes's theorem

$$\iint_{\mathbf{X}} \omega_2 \wedge \bar\omega_2 = \iint_{\mathbf{X}} d(\omega_2 f) = 0. \tag{7.1.6}$$

However,

$$\iint_{\mathbf{X}} \omega_2 \wedge \bar\omega_2 = \iint_{\mathbf{X}} |f_2(z)|^2 dz \wedge d\bar z = -2i \iint_{\mathbf{X}} |f_2(z)|^2 dx \wedge dy, \tag{7.1.7}$$

so $f_2(z) = 0$, and likewise $f_1 = 0$. Hence $\omega_1 = \omega_2 = 0$, so the map in (7.1.4) is injective.

Next, by a special case of the Riemann–Roch theorem, one computes $\dim_{\mathbf{C}} \Omega^1_{\mathbf{X}} = g$. (This is relatively easy to prove in the case of hyperelliptic Riemann surfaces $Z^2 = p(X)$ where $p(X)$ is an irreducible polynomial of degree $2g + 1$, where one can exhibit a basis of $\Omega^1_{\mathbf{X}}$ explicitly as $\{dX/Z, XdX/Z, \dots, X^{g-1}/Z\}$ as in (6.1.13).) The general case is given in [43, p. 109]. Let $\omega_1, \dots, \omega_g$ be a basis of $\Omega^1_{\mathbf{X}}$.

By definition of genus, the rank of $H_1(\mathbf{X}; \mathbf{Z})$ as a $\mathbf{Z}$-module equals $2g$, and there is a natural homology basis $\{\sigma_1, \dots, \sigma_{2g}\}$, as described in texts on Riemann surfaces such as [36]. Using this basis, one uses the pairing between the differential forms in $H^1_{DR}(\mathbf{X}; \mathbf{C})$ and the homology group $H_1(\mathbf{X}; \mathbf{Z})$ from Proposition 6.2.6 to construct the $g \times 2g$ period matrix

$$\Pi = \left[\int_{\sigma_j} \omega_k \right]_{j=1,\dots,2g;\, k=1,\dots,g} = [I_g \quad Q], \tag{7.1.8}$$

where Q is a $g \times g$ complex matrix which is symmetric and $\Im Q$ is positive definite, where $\Im Q = (Q - Q^*)/(2i)$. Hence the rows of Π are independent over $\mathbf{R}$, and we deduce that $\mathrm{Hom}_{\mathbf{C}}(H_1(\mathbf{X}; \mathbf{Z}); \mathbf{C}) = \underline{\mathbf{C}^{2g}}$, which shows that $\dim H^1_{DR}(\mathbf{X}; \mathbf{C}) = 2g$. Hence the image of $\Omega^1_{\mathbf{X}} \oplus \overline{\Omega^1_{\mathbf{X}}}$ in $H^1_{DR}(\mathbf{X}; \mathbf{C})$ has dimension $2g$, so the spaces are equal. See p. 107 of [43]. □

Corollary 7.1.3 *There is an injective group homomorphism $H^1(\mathbf{X}; \mathbf{Z}) \to (\Omega^1_{\mathbf{X}})^*$ so that the image is a lattice Λ.*

Proof We have a group isomorphism $\mathbf{Z}^{2g} \cong H^1(\mathbf{X}; \mathbf{Z})$ via $(n_j)_{j=1}^{2g} \mapsto \sum_{j=1}^{2g} n_j \sigma_j$, where $\gamma = \sum_{j=1}^{2g} n_j \sigma_j$ gives rise to a linear function on the holomorphic differentials via

$$\int_\gamma : \omega \mapsto \sum_{j=1}^{2g} n_j \int_{\sigma_j} \omega \qquad (\omega \in \Omega^1(\mathbf{X}));$$

the map $\gamma \mapsto \int_\gamma$ is obviously a group homomorphism $H^1(\mathbf{X}; \mathbf{Z}) \to \Omega^1(\mathbf{X})^*$ and injective due to the form of the period matrix Π. The image of $(\int_\gamma \omega_k)_{k=1}^g$ for the basis $(\omega_1, \dots, \omega_g)$ and $\gamma \in H^1(\mathbf{X}; \mathbf{Z})$ is a lattice Λ relative to the basis $\{\omega_1, \dots, \omega_g\}$. □

The other idea involved in this chapter is monodromy. Let $A(z)$ be a meromorphic $n \times n$ matrix function on $\mathbf{X}$, with possible simple poles at $p \in \mathbf{X}$, and we consider the differential equation

$$\frac{d\Psi}{dz} + A(z)\Psi(z) = 0. \tag{7.1.9}$$

On an open neighbourhood U, we look for locally meromorphic solutions $\Psi\colon U \subset \mathbf{X} \to M_n(\mathbf{C})$,

$$\{\Psi \in M_n(\mathbf{K}(U))\colon \Psi'(z) + A(z)\Psi(z) = 0\}, \tag{7.1.10}$$

which give sections of a finite-dimensional complex vector bundle over $\mathbf{X}$. As z describes a simple closed curve γ around p, the solution changes $\Psi(z) \mapsto \Psi(z)M_\gamma$, where $M_\gamma \in M_n(\mathbf{C})$ is the monodromy matrix associated with γ. In some cases $M_\gamma^N = 1$ for some integer $N > 0$, so the monodromy matrix has finite order, though this is not always the case. The Birkhoff–Grothendieck classification of vector bundles over the Riemann sphere is described in [85] and in section 6.2 of [99].

In this chapter, we seek to generalize these results to the noncommutative setting. Some fresh subtleties emerge.

7.2 The *b* Operator and Hochschild Homology

Let A be a unital associative algebra over $\mathbf{C}$. We introduce the exact sequence

$$0 \longrightarrow \mathbf{C} \longrightarrow A \longrightarrow \bar{A} \longrightarrow 0 \tag{7.2.1}$$

via the unital embedding. The algebra ΩA of noncommutative differential forms is characterized in Theorem 6.3.2, as we briefly recall. There exists a unique differential graded algebra $\Omega A = \oplus_{n=-\infty}^{\infty} \Omega^n A$ with $\Omega^0 A = A$ and differential $d\colon \Omega^n A \to \Omega^{n+1} A$ with the following properties. Given any differential graded algebra R and a homomorphism $u\colon A \to R$, there is a unique homomorphism of differential graded algebras $\Omega A \to R$ extending u. The map $A \otimes (\bar{A})^{\otimes n} \to \Omega^n A\colon (a_0, \dots, a_n) \mapsto a_0 da_1 \dots da_n$ is an isomorphism for all $n = 1, 2, \dots$. We think of $A \otimes (\bar{A})^{\otimes n} = \Omega^n A$ as normalized n-chains. There is an iterative construction $\Omega^n A \otimes \bar{A} \to \Omega^{n+1} A\colon \omega \otimes a \mapsto \omega da$ which enables us to define the complex $(\Omega A, b)$, which has the interesting homology.

Definition 7.2.1 (i) *(b-operator)* Let $b\colon \Omega^{n+1} A \to \Omega^n A\colon b(\omega da) = (-1)^n(\omega a - a\omega)$ or equivalently

$$\begin{aligned} b(a_0, \dots, a_n) &= \sum_{i=1}^{n-1} (-i)^i (a_0, \dots, a_i a_{i+1}, \dots, a_n) \\ &\quad + (-1)^n (a_n a_0, \dots, a_{n-1}), \end{aligned} \tag{7.2.2}$$

in which the final term is sometimes referred to as the cross-over term. For $n < 0$, let $\Omega^n A = 0$ with $b\colon \Omega^n A \to \{0\}$ for $n \leq 0$.

(ii) *(Hochschild homology)* Then $b^2 = 0$, so we have a complex

$$0 \leftarrow A \leftarrow \Omega^1 A \leftarrow \Omega^2 A \leftarrow \cdots \tag{7.2.3}$$

with differential b of degree (-1). Then the Hochschild homology $HH_n(A)$ is the homology $H_n(\Omega A, b)$ of this complex.

Example 7.2.2 (Hochschild homology) (i) We have $b(a_0 da_1) = [a_0, a_1]$, so $b\Omega^1 A = [A, A]$ and

$$HH_0(A) = A/[A, A] = A_\natural.$$

In particular, $HH_0(A) = A$ for A any commutative algebra.

(ii) Suppose that A is a direct sum of matrix algebras $M_n(\mathbf{C})$. Then A is separable and

$$HH_j(A) = \begin{cases} A/[A, A], & \text{for } j = 0; \\ 0, & \text{for } j > 0. \end{cases} \tag{7.2.4}$$

Here $\dim(A/[A, A])$ is the dimension of the centre of A, namely the number of summands in the direct sum.

(iii) We consider the polynomial algebra $A = \mathbf{C}[x]$ and identify $A \otimes A$ with $\mathbf{C}[x, y]$. Then $M\colon A \to A\colon Mf(x) = xf(x)$ gives $\delta\colon A \otimes A \to A \otimes A$; $\delta(f)(x, y) = [f, M] = (x - y)f(x, y)$, and we have an exact sequence

$$0 \longrightarrow \mathbf{C}[x, y] \xrightarrow{\delta} \mathbf{C}[x, y] \xrightarrow{m} \mathbf{C}[x] \longrightarrow 0. \tag{7.2.5}$$

From this sequence, A has projective dimension one, since $\Omega^1 A$ is projective and $H_i(\Omega^1 A, M) = 0$ for all $i > 0$. Hence one can compute

$$HH_j(A) = \begin{cases} A, & \text{for } j = 0, 1; \\ 0 & \text{else}. \end{cases} \tag{7.2.6}$$

(iv) Let A be either the complex polynomials $\mathbf{C}[x]$, or the Laurent series with only finitely many terms $\mathbf{C}[x][x^{-1}]$. Then we can identify $\Omega^1 A$ with $\delta A \otimes A$, or $\Omega^1 A = (x - y)A \otimes A$ as an ideal in $A \otimes A$; so $\Omega^1_A = (x - y)/((x - y)^2) \cong A$. These results follow from the observation that

$$g(x) \in A \Rightarrow \frac{g(x) - g(y)}{x - y} \in A \otimes A,$$

and the characterization of $\Omega^1 A$ via the formula

$$\sum_j f_j(x)g_j(y) = \sum_j f_j(x)g_j(x) - (x-y) \times \sum_j f_j(x)\frac{g_j(x)-g_j(y)}{x-y} \qquad (f_j, g_j \in A).$$

This may be compared with the difference quotients in Example 6.1.4. To deal with the formal power series $A = \mathbf{C}[[x]]$, Hartshorn [48] considers x-adic completions.

Proposition 7.2.3 *Let A be a unital and commutative complex algebra. Then there is an injective map*

$$\Omega^n_A \to HH_n(A) \qquad (n = 0, \dots). \tag{7.2.7}$$

Proof Let A be a unital and commutative algebra over $\mathbf{C}$ and M an A-bimodule. Then the anti-symmetrization operator $\varepsilon_n : M \otimes \wedge^n A \to M \otimes A^{\otimes n}$ is

$$\varepsilon_n : a_0 \otimes (a_1 \wedge \dots \wedge a_n) \mapsto \sum_{\sigma \in S_n} \operatorname{sign}(\sigma) a_0 \otimes (a_{\sigma(1)}, a_{\sigma(2)}, \dots, a_{\sigma(n)}) \tag{7.2.8}$$

for all $a_0 \in M$ and $a_j \in A$ for $j = 1, \dots, n$. Let $\hat{a}_j$ indicate that the term a_j has been omitted. There is also a map $\delta_n : M \otimes \wedge^n A \to M \otimes \wedge^{n-1} A$

$$\delta_n : a_0 \otimes (a_1 \wedge \dots \wedge a_n) \mapsto \sum_{j=1}^n (-1)^j [a_0, a_j] \otimes (a_1 \wedge \dots \wedge \hat{a}_j \wedge \dots \wedge a_n). \tag{7.2.9}$$

Then as in [68, 1.3.5] the diagram commutes

$$\begin{array}{ccccc} & M \otimes \wedge^n A & \overset{\varepsilon_n}{\longrightarrow} & M \otimes A^{\otimes n} & \\ \delta_n & \downarrow & & \downarrow & b. \\ & M \otimes \wedge^{n-1} A & \underset{\varepsilon_{n-1}}{\longrightarrow} & M \otimes A^{\otimes n-1} & \end{array} \tag{7.2.10}$$

We observe that when M is a symmetric module, so that $[a_0, a_j] = 0$ for all $a_0 \in M$ and $a_j \in A$, then $\delta_n = 0$. In particular, this happens when we take $A = M$, so we can pass to the quotient space $\bar{A} = A/\mathbf{C}1$ and obtain the commuting diagram

$$\begin{array}{ccccc} & A \otimes \wedge^n \bar{A} & \overset{\varepsilon_n}{\longrightarrow} & A \otimes \bar{A}^{\otimes n} & \\ 0 & \downarrow & & \downarrow & b. \\ & A \otimes \wedge^{n-1} \bar{A} & \underset{\varepsilon_{n-1}}{\longrightarrow} & A \otimes \bar{A}^{\otimes n-1} & \end{array} \tag{7.2.11}$$

This gives the maps on the spaces of Kähler and noncommutative differentials

$$\begin{array}{ccccc} & \Omega^n_A & \overset{\varepsilon_n}{\longrightarrow} & \Omega^n A & \\ 0 & \downarrow & & \downarrow & b. \\ & \Omega^{n-1}_A & \underset{\varepsilon_{n-1}}{\longrightarrow} & \Omega^{n-1} A & \end{array} \tag{7.2.12}$$

We also have $\mu_n : A \otimes \bar{A}^n \to A \otimes \wedge^n A \mu_n : a_0 \otimes (\bar{a}_1, \dots, \bar{a}_n) \to a_0 da_1 \dots da_n$, so that $\mu_n \circ \varepsilon_n = n!\,(Id)$, and the map ε_n is injective. We can identify these with maps $\mu_n : \Omega^n A \to \Omega^{n-1} A$ and $\varepsilon_n : \Omega^n_A \to \Omega^n A$. Hence we obtain an injective map $\Omega^n_A \to H_n(\Omega A)$, where $HH_n(A) = H_n(\Omega A)$ for the b differential. □

Example 7.2.4 (Smooth, but not quasi-free) (i) For $A = \mathbf{C}[x_1, \dots, x_n]$, the map from Proposition 7.2.3 is an isomorphism, so $\Omega^k_A = HH_k(A)$ for all k. In this case, we have a free A-module of rank n given by $\Omega^1_A = \sum A dx_j$ as in Example 6.2.4, so

$$\Omega_A = \bigwedge \Omega^1_A = A \otimes \wedge \mathbf{C}^n$$

and one can consider the Koszul complex for this A-bimodule.

(ii) If A is quasi-free, then $HH_i(A) = 0$ for all $i \geq 2$.

(iii) Let $A = \mathbf{C}[V]$ be the coordinate ring on a smooth affine complex variety V of dimension $n \geq 2$. Then by the Hochstadt–Konstant–Rosenberg Theorem [68, p. 103]

$$HH_n(A) = \Omega^n_A, \tag{7.2.13}$$

so $HH_n(A) \neq 0$ where $n \geq 2$. Hence A is smooth but not quasi-free. A non-singular manifold of dimension $n \geq 2$ is singular in the noncommutative setting.

7.3 The Karoubi Operator

The Karoubi operator is a monodromy operator on the noncommutative differential forms. A cochain $f \in (A^{\otimes n+1})^*$ is said to be normalized if $f(a_0, \dots, a_n)$ vanishes whenever $a_i = 1$ for some $i = 1, \dots, n$; equivalently, a normalized cochain is a $\mathbf{C}$-linear functional on $A \otimes (\bar{A})^{\otimes n}$. Such cocycles are not cyclically invariant, and the operator λ is only defined on unnormalized cochains. The Karoubi operator κ on ΩA is defined on normalized cochains and is a substitute for the operator λ of Section 2.2. We introduce the operators first on chains, and then dualize to cochains. We sometimes abbreviate $\Omega^n A$ by Ω^n.

Later in this section, we use the Karoubi operator to create a Hodge decomposition of ΩA. We can make ΩA into a $\mathbf{C}[X]$ module via $X \mapsto \kappa$, and use the decomposition theory for modules over principal ideal domains. The Karoubi operator reappears in Section 11.6, when we discuss homotopy and deformation of Fredholm modules.

Definition 7.3.1 (Karoubi operator) The Karoubi operator $\kappa: \Omega A \to \Omega A$ is defined to be the operation given by Theorem 6.3.1(ii) for the linear map such that $\kappa(a) = a$ for $a \in \Omega^0 A$, or more explicitly

$$\kappa(\omega da) = (-1)^n da\omega \qquad (\omega \in \Omega^n A). \tag{7.3.1}$$

By Theorem 6.3.1, there is a unique operator with this property. In terms of chains, we have the more explicit formula

$$\kappa(a_0, \dots, a_n) = (-1)^{n-1}(1, a_n a_0, \dots, a_{n-1}) + (-1)^n (a_n, a_0, \dots, a_{n-1}) \tag{7.3.2}$$

since

$$\begin{aligned}\kappa(a_0 da_1 \dots da_n) &= (-1)^{n-1} da_n (a_0 da_1 \dots da_{n-1})\\ &= (-1)^{n-1} d(a_n a_0) da_1 \dots da_{n-1} + (-1)^n a_n da_0 \dots da_{n-1}.\end{aligned} \tag{7.3.3}$$

Proposition 7.3.2 *The Karoubi operator $\kappa: \Omega A \to \Omega A$ satisfies the following properties:*

(1) $bd + db = 1 - \kappa$;
(2) *κ commutes with b and d;*
(3) *κ is an automorphism;*
(4) *on $\Omega^n A$, κ satisfies:*

  (i) $\kappa^n = 1 + b\kappa^{-1} d$;
  (ii) $\kappa^{n+1} = 1 - db$;
  (iii) $(\kappa^n - 1)(\kappa^{n+1} - 1) = 0$.

Proof (1) In degree $n + 1$, we have

$$\begin{aligned}(bd + db)(\omega a) &= b(d\omega da) + d\big((-1)^n(\omega a - a\omega)\big)\\ &= (-1)^{n+1}(d\omega\, a - ad\omega)\\ &\quad + (-1)^n d\omega\, a - da\omega\big) + (-1)^n(\omega da - ad\omega\big)\\ &= \omega da - (-1)^n da\,\omega\\ &= \omega da - \kappa(\omega da).\end{aligned} \tag{7.3.4}$$

(2) By (1), we have

$$b(1-\kappa) = b^2d + bdb = bdb,$$
$$(1-\kappa)b = bdb + db^2 = bdb, \tag{7.3.5}$$

so $1-\kappa$ commutes with b, and likewise with d.

(3) On the subspace $d\Omega^{n-1} \subset \Omega^n$, a typical element is in the span of $da_0 \dots da_{n-1}$, namely the image of $(1, a_0, \dots, a_{n-1}) \in 1 \otimes (\bar{A})^{\otimes n}$ under the isomorphism $(\bar{A})^{\otimes n} \to d\Omega^{n-1}$. On $(\bar{A})^{\otimes n}$, the operator λ of cyclic permutation with sign has $\kappa = \lambda$, hence $\kappa^n = \lambda^n = 1$.

Also we have the exact sequence

$$\begin{array}{ccccccccc} 0 & \longrightarrow & d\Omega^{n-1} & \longrightarrow & \Omega^n & \longrightarrow & d\Omega^n & \longrightarrow & 0 \\ & & \uparrow & & \uparrow & & \uparrow & & \\ 0 & \longrightarrow & \bar{A}^{\otimes n} & \longrightarrow & A \otimes (\bar{A})^{\otimes n} & \longrightarrow & (\bar{A})^{\otimes n+1} & \longrightarrow & 0 \end{array} \tag{7.3.6}$$

so that $\Omega^n/d\Omega^{n-1} \longrightarrow (\bar{A})^{\otimes n+1}$ is an isomorphism, and κ on $\Omega^n/d\Omega^{n-1}$ corresponds to λ on $(\bar{A})^{\otimes n+1}$. Hence κ is an automorphism on $\Omega^n/d\Omega^{n-1}$ of order $n+1$, so

$$(\kappa^n - 1)(\kappa^{n+1} - 1)\Omega^n \subseteq (\kappa^n - 1)d\Omega^{n-1} = 0,$$

hence $(\kappa^n - 1)(\kappa^{n+1} - 1) = 0$. From this we obtain

$$\kappa(\kappa^n + \kappa^{n-1} - \kappa^{2n}) = 1, \tag{7.3.7}$$

so κ is an automorphism on Ω^n. This proves (3), and 4(iii).

(4) (i) We repeatedly use the definition of κ to obtain

$$\begin{aligned} \kappa^n(a_0da_1 \dots da_n) &= (-1)^{n-1}\kappa\big(da_n\, a_0da_1 \dots da_{n-1}\big) \\ &= \dots \\ &= (-1)^{n(n-1)}da_1 \dots da_n\, a_0 \\ &= a_0da_1 \dots da_n + (-1)^n b(da_1 \dots da_n da_0) \\ &= a_0da_1 \dots da_n + b\kappa^{-1}(da_0 \dots da_n) \\ &= a_0da_1 \dots da_n + b\kappa^{-1}d\big(a_0da_1 \dots da_n\big). \end{aligned} \tag{7.3.8}$$

(4) (ii) We multiply $\kappa^n = 1 + b\kappa^{-1}d$ by κ to deduce $\kappa^{n+1} = \kappa + bd = 1 - bd$. □

7.4 Connes's B-operator

In this section, we also introduce Connes's B-operator from the Karoubi operator κ. The operators b and B on Ω determine the cyclic homology of A, as discussed in Section 7.7. The definition of B in Connes's theory provides a stronger motivation in terms of cobordism of cycles; see [19]. As in Section 2.9(iv), $\mathbf{C}[B]$ is a local ring by Proposition 7.4.2(iv).

Definition 7.4.1 (B-operator) Connes's B-operator is $B : \Omega \longrightarrow \Omega$ of degree one such that $B = \sum_{j=0}^{n} \kappa^j d$ on $\Omega^n A$.

Proposition 7.4.2 *The B operator satisfies:*
(i) $\kappa^{n(n+1)} = 1 + bB = 1 - bB$ *on* Ω^n;
(ii) $Bd = dB = 0$;
(iii) $\kappa B = B\kappa = B$;
(iv) $B^2 = 0$.

Proof (i) By geometric series, and using $b\kappa = \kappa b$, we have

$$\begin{aligned}\kappa^{n(n+1)} - 1, &= \sum_{j=0}^{n} (\kappa^n)^j (\kappa^n - 1)\\ &= \sum_{j=0}^{n} (\kappa^n)^j b\kappa^{-1} d\\ &= b \sum_{j=0}^{n} \kappa^{nj+n} d,\end{aligned} \tag{7.4.1}$$

where $\kappa^n = \kappa^{-1}$ on $d\Omega^n$, so $\kappa^{n(n+1)} - 1 = bd$.

Likewise

$$\begin{aligned}\kappa^{n(n+1)} - 1 &= \sum_{j=0}^{n-1} \left(\kappa^{n+1}\right)^j (\kappa^{n+1} - 1)\\ &= -\sum_{j=0}^{n-1} \left(\kappa^{n+1}\right)^j db,\end{aligned} \tag{7.4.2}$$

where $\kappa^{n+1} = \kappa$ on $d\Omega^{n-1}$, so $\kappa^{n(n-1)} - 1 = -Bb$. □

By Propositions 7.3.2 and 7.4.2, the Karoubi operator satisfies $\kappa^{n(n+1)} = 1 - Bd$ where $(Bd)^2 = 0$; so $\kappa^{n(n+1)} - 1$ is nilpotent and κ is quasi-unipotent. In general, κ does not have finite order on $\Omega = \oplus_{n=0}^{\infty} \Omega^n A$.

Example 7.4.3 (Algebra with one idempotent) The complex algebra generated by a non-trivial idempotent has interesting Hochschild homology. Let $e \neq 0, 1$ satisfy $e^2 = e$ and let $A = \mathbf{C} + \mathbf{C}e$, or equivalently the commutative algebra $A = \mathbf{C}[X]/(X^2 - X)$. Now $e^2 = e$ gives $(de)e + e(de)e = de$, so

$$e(de) = (de)(1-e), \qquad (de)e = (1-e)(de), \qquad e(de)^2 = (de)^2 e, \tag{7.4.3}$$

which give us the basic relations for multiplying on the right by elements of A, and also showing that $(de)^2 = (de)(de)$ satisfies $e(de)^2 = (de)^2 e$. Let b be the operator on ΩA given by

$$b(\omega da) = (-1)^{|\omega|}(\omega a - a\omega). \tag{7.4.4}$$

We deduce that

$$\begin{aligned} b((de)^{2n}) &= b((de)^{2n-1}(de) \\ &= -(de)^{2n-1}e + e(de)^{2n-1} \\ &= (2e-1)(de)^{2n-1} \end{aligned} \tag{7.4.5}$$

and likewise

$$\begin{aligned} b((2e-1)(de)^{2n}) &= b((2e-1)(de)^{2n-1}de) \\ &= -[(2e-1)(de)^{2n-1}, e] \\ &= -(2e-1)(de)^{2n-1}e + e(2e-1)(de)^{2n-1} \\ &= -(2e-1)(1-e)(de)^{2n-1} + e(de)^{2n-1} \\ &= (de)^{2n-1}. \end{aligned} \tag{7.4.6}$$

We deduce that $b\colon \Omega^{2n}A \to \Omega^{2n-1}A$ is an isomorphism for $n = 1, 2, \dots$. Hence the Hochschild homology is

$$HH_n(A) = H_n(\Omega A, b) = \begin{cases} A, & \text{for } n = 0; \\ 0, & \text{for } n = 1, 2, \dots. \end{cases} \tag{7.4.7}$$

The expression $b + B$ is the sum of an operator b of degree -1 and an operator B of degree $+1$. Now we take $e \in A = \Omega^0 A$ and look for a cycle $\zeta \in \Omega^{ev} A$ for $b + B$ to represent de; the candidate is

$$\zeta = c_0 e + \sum_{n=1}^{\infty} c_n (2e-1)(de)^{2n}, \tag{7.4.8}$$

where the constants c_n are to be chosen. Now

$$B\big((2e-1)(de)^{2n}\big) = \sum_{j=0}^{2n} \kappa^j d\big((2e-1)(de)^{2n}\big)$$
$$= 2(2n+1)(de)^{2n+1}. \tag{7.4.9}$$

Hence we have

$$(B+b)\zeta = c_0 be + c_0 Be + \sum_{n=1}^{\infty} c_n (de)^{2n-1} + \sum_{n=1}^{\infty} 2(2n+1)c_n (de)^{2n+1}, \tag{7.4.10}$$

so the coefficient of $(de)^{2n+1}$ vanishes when

$$c_{n+1} + 2(2n+1)c_n = 0; \tag{7.4.11}$$

the required solution is given by

$$c_n = \frac{(-1)^{n-1}(2n)!}{2(n!)} c_1, \tag{7.4.12}$$

so

$$(b+B)\zeta = c_0 be + c +_0 Be - de = de. \tag{7.4.13}$$

7.5 The Hodge Decomposition

Let R be a principal ideal domain and M a left R-module. Then a submodule Z of M is cyclic if $Z = Rm = (m)$ for some $m \in M$. Then we define the order of Z to be z where $(z) = \{x \in R : xm = 0\}$ is the generator of the module that annihilates m, so that z is determined up to a unit factor in R. (This definition is suggested by the case in which M is a finite abelian group, and $R = \mathbf{Z}$.) The decomposition theorem asserts that if M is finitely generated, then M may be written as a direct sum $M = Z_1 \oplus \cdots \oplus Z_t$, where Z_j is a non-zero cyclic submodule of M with order z_j, where z_j divides z_{j+1} for $j = 1, \dots, t-1$. See [47, p. 124].

Lemma 7.5.1 *There is a direct sum decomposition*

$$\Omega^n A = \mathrm{Ker}\big((1-\kappa)^2\big) \oplus \bigoplus_{\zeta} \mathrm{Ker}\big(\zeta 1 - \kappa\big), \tag{7.5.1}$$

where the final sum is taken over all complex roots of unity $\zeta \neq 1$ such that $\zeta^n = 1$ or $\zeta^{n+1} = 1$.

Proof We can regard $\Omega^n A$ as a module over the principal ideal domain $\mathbf{C}[z]$ via $(p(z), \omega) \mapsto p(\kappa)\omega$. Suppose for the moment that A is finite dimensional, $\Omega^n A$ is also finite dimensional, so by Proposition 7.3.2(4), $\kappa \colon \Omega^n \to \Omega^n$ has a minimal polynomial which divides $(X^n - 1)(X^{n+1} - 1)$, which we can factorize as

$$(X-1)^2 \prod_{j=1}^{n} (X - e^{2\pi i j/n}) \prod_{\ell=1}^{n+1} (X - e^{2\pi i j/(n+1)}). \tag{7.5.2}$$

Suppose for notational simplicity that this is indeed the minimal polynomial. Then the Jordan decomposition consists of a generalized eigenspace $\mathrm{Ker}((1-\kappa)^2)$ for the double root $X = 1$, and eigenspaces $\mathrm{Ker}(\zeta - \kappa)$ for the simple roots $\zeta = \zeta_n^j = e^{2\pi i j/n}$ for $j = 1, \dots, n-1$ and $\zeta = \zeta_{n+1}^j = e^{2\pi i j/(n+1)}$ for $j = 1, \dots, n$.

In the case where A is infinite dimensional, a similar result holds. Let $\mathbf{C}[\kappa]$ be the unital algebra of linear operators on $\Omega^n A$ generated by κ. Indeed, the proof in [47, p. 174] applies. Then there is a surjective homomorphism $\mathbf{C}[X] \to \mathbf{C}[\kappa]$, where the kernel is the ideal generated by some factor of $(X^n - 1)(X^{n+1} - 1)$. Let $\zeta_j (j = 1, \dots, 2n-1)$ be a complete list of the roots of $(X^n - 1)(X^{n+1} - 1) = 0$, excluding $X = 1$. Then as in Lagrange interpolation, let

$$f_j(X) = \frac{(X-1)^2}{(\zeta_j - 1)^2} \prod_{k=1}^{2n-1} \frac{X - \zeta_k}{\zeta_j - \zeta_k}, \tag{7.5.3}$$

so that $f_j(\zeta_j) = 1$ and $f_j(\zeta_k) = 0$ for $j \neq k$, hence

$$f_j(\kappa) f_k(\kappa) = 0 \qquad (j \neq k); \tag{7.5.4}$$

$$f_j(\kappa)^2 = f_j(\kappa). \tag{7.5.5}$$

Now let $P = 1 - \sum_{j=1}^{2n-1} f_j(\kappa)$, so that $P^2 = P$ on Ω^n, and we can introduce $P^\perp = 1 - P$ as the orthogonal idempotent. Note that $g(\kappa) f_k(\kappa) = g(\zeta_j) f_j(\kappa)$ for all $g \in \mathbf{C}[X]$, so $\mathrm{Im}(f_j(\kappa)) = \mathrm{Ker}\big(\zeta_j - \kappa\big)$. (Possibly this subspace is zero.) □

Theorem 7.5.2 (Hodge decomposition) (i) *There is a decomposition*

$$\Omega = P\Omega \oplus P^\perp \Omega, \tag{7.5.6}$$

which is stable under b and d and such that $P^\perp \Omega$ is contractible with respect to b and d.

(ii) *There is a further decomposition*

$$P^\perp \Omega = b P^\perp \Omega \oplus d P^\perp \Omega, \tag{7.5.7}$$

where $b \colon dP^\perp \Omega \to bP^\perp \Omega$ and $d \colon bP^\perp \Omega \to dP^\perp \Omega$ are isomorphisms.

Proof (i) By the Lemma, we can introduce orthogonal idempotents on Ω^n such that $P_n\colon \Omega \to \mathrm{Ker}\big((1-\kappa)^2\big)$ and

$$P_n^\perp\colon \Omega^n \to \bigoplus_\zeta \mathrm{Ker}\big(\zeta 1 - \kappa\big), \tag{7.5.8}$$

where the sum is taken over all nth and $(n+1)$st roots of unity other than 1 itself. Then we form $P = \oplus_{n=0}^\infty P_n$ and $P^\perp = \oplus_{n=0}^\infty P_n^\perp$, so that P and $P^\perp$ are orthogonal idempotents as operators on Ω, such that $P\Omega = \mathrm{Ker}\big((1-\kappa)^2\big)$,

$$P^\perp\Omega = \bigoplus_\zeta \mathrm{Ker}\big(\zeta 1 - \kappa\big), \tag{7.5.9}$$

where the sum is taken over all roots of unity such that $\zeta \neq 1$.

The operators b and d commute with κ, hence b maps Ker $(\zeta - \kappa)$ to itself and d maps Ker $(\zeta - \kappa)$ to itself. On $\mathrm{Ker}(\zeta - \kappa)$, the identity $bd + db = 1 - \kappa$ reduces to

$$d\left(\frac{1}{1-\zeta}b\right) + \left(\frac{1}{1-\zeta}b\right) = 1, \tag{7.5.10}$$

which gives a contraction of the differential d. Hence $P^\perp\Omega$ is contractible.

(ii) We take the Green's operator to be 0 on $P\Omega$ and $G\omega = (1-\zeta)^{-1}\omega$ for $\omega \in \mathrm{Ker}(\zeta - \kappa)$, to $P^\perp = G(1-\kappa) = G(db + db)$; hence

$$P^\perp = b(Gd) + d(Gb) \tag{7.5.11}$$

is a sum of orthogonal idempotents. Indeed, $(bGd)(dGb) = 0$ since $d^2 = 0$ and likewise $(dGb)(dGb) = 0$ since $b^2 = 0$; also

$$(bGd)(bGd) = bG(1 - \kappa - bd)Gd = bGd - bGbdGd = bGd \tag{7.5.12}$$

and likewise

$$(dGb)(dGb) = dG(1 - \kappa - db)Gb = dGb. \tag{7.5.13}$$

Also $\mathrm{Im}(b) = \mathrm{Ker}(d)$ and $\mathrm{Im}(d) = \mathrm{Ker}(b)$, so $b\colon dP^\perp\Omega \to bP^\perp\Omega$ is surjective and hence an isomorphism; so bGd projects onto $bP^\perp\Omega$. Likewise, dGb projects onto $dP^\perp\Omega$. □

Complements 7.5.3 (i) In the Hodge decomposition, the summand $P\Omega = \mathrm{Ker}\big((1-\kappa)^2\big)$ is called the space of harmonic forms, while $P^\perp\Omega = \mathrm{Im}\big((1-\kappa)^2\big)$ is decomposed so that

$$\Omega = P\Omega \oplus (1-\kappa)d\Omega \oplus b(1-\kappa)\Omega. \tag{7.5.14}$$

(ii) On Ω^n, we have

$$\Omega^n = P\Omega^n + (1-\lambda)(\bar{A})^{\otimes n} + b(1-\lambda)(\bar{A})^{\otimes n+1}.$$

(iii) We have $P^\perp = (Gd)b + b(Gd)$ where on Ω^n

$$Gd = \frac{1}{n+1}\sum_{j=0}^{n}\Big(\frac{n}{2} - j\Big)\kappa^j d. \tag{7.5.15}$$

To see this, it suffices to consider $\omega \in \mathrm{Ker}(\zeta - \Omega)$, where $\zeta \neq 1$ has $\zeta^{n+1} = 1$. Then

$$\frac{1}{n+1}\sum_{j=0}^{n}\Big(\frac{n}{2} + 1 - (j+1)\Big)\zeta^j = \frac{1}{1-\zeta} \tag{7.5.16}$$

by elementary summation, so

$$d\frac{1}{n+1}\sum_{j=0}^{n}\Big(\frac{n}{2} - j\Big)\kappa^j\omega = \frac{1}{1-\zeta}d\omega = (1-\kappa)^{-1}d\omega. \tag{7.5.17}$$

Remarks 7.5.4 (i) The operator κ has eigenvalues that are roots of unity, contained in $\{\zeta_n^j : j = 0, \dots, n-1\} \cup \{\zeta_{n+1}^j ; j = 0, \dots, n\}$, where $\zeta_n^n = \zeta_{n+1}^{n+1} = 1$; also $(\lambda - \kappa)^{-1}$ has a double pole at $\lambda = 1$, which is the pole of highest order. These properties are familiar from the theory of positive matrices, as we now recall from [93] theorem 2.7.

(ii) Suppose that $K = [\kappa_{jk}]$ is a $N \times N$ real matrix with $\kappa_{jk} \geq 0$ for all j, k, and spectral radius $r = \sup\{|\lambda| : \lambda \in \sigma(K)\}$. Without loss of generality we suppose $r = 1$ and consider the peripheral spectrum $\sigma_{per}(K) = \{\lambda \in \sigma(K) : |\lambda| = 1\}$. Then

(1) for all $\zeta \in \sigma_{per}(K)$, the group $\{\zeta^j : j = 0, 1, \dots, N\}$ is contained in $\sigma_{per}(K)$;

(2) $1 \in \sigma_{per}(K)$, and $(\lambda I - K)^{-1}$ has a pole at $\lambda = 1$ which has the largest order amongst poles in $\sigma_{per}(K)$.

We remark that operators with this property arise in the theory of Markov chains, and are described by a classical theorem of Frobenius.

Exercise 7.5.5 (Matrix proof of Karoubi identities) One can summarize the calculations of this section in some matrix identities, which are concise but not intuitive. Note that $A = \bar{A} \oplus \mathbf{C}$, and

$$\Omega^n A = A \otimes \bar{A}^{\otimes n} = \bar{A}^{\otimes(n+1)} \oplus \bar{A}^{\oplus n}. \tag{7.5.18}$$

(i) Introduce the matrices

$$\tilde{d} = \begin{bmatrix} 0 & 0 \\ 1 & 0 \end{bmatrix}, \qquad \tilde{b} = \begin{bmatrix} b & I - \lambda \\ 0 & -b' \end{bmatrix}, \tag{7.5.19}$$

and compute $\tilde{\kappa} = I - (\tilde{d}\tilde{b} + \tilde{b}\tilde{d})$. This gives

$$\tilde{\kappa} = \begin{bmatrix} \lambda & 0 \\ b' - b & \lambda \end{bmatrix}. \tag{7.5.20}$$

Show that $\tilde{b}^2 = 0 = \tilde{d}^2 = 0$, and

$$\tilde{\kappa}^j \tilde{d} = \begin{bmatrix} 0 & 0 \\ \lambda^j & 0 \end{bmatrix}. \tag{7.5.21}$$

(ii) Now compute

$$\tilde{G}\tilde{d} = \begin{bmatrix} 0 & 0 \\ G_\lambda & 0 \end{bmatrix} \tag{7.5.22}$$

by the obvious extension of (7.5.15), and likewise find $\tilde{B} = \sum_{j=0}^{n} \tilde{\kappa}^j \tilde{d}$.

(iii) Show that the matrices

$$\widetilde{P} = \begin{bmatrix} P_\lambda & 0 \\ b'G_\lambda - G_\lambda b & P_\lambda \end{bmatrix}, \ \widetilde{(bGd)} = \begin{bmatrix} P_\lambda^\perp & 0 \\ -b'G_\lambda & 0 \end{bmatrix}, \ \widetilde{(Gdb)} = \begin{bmatrix} 0 & 0 \\ G_\lambda b & P_\lambda^\perp \end{bmatrix} \tag{7.5.23}$$

are annihilating idempotents which sum up to the identity matrix.

Remarks 7.5.6 (Abstract versus concrete Hodge decomposition) We compare Theorem 7.5.2 with the Hodge decomposition for manifolds. Let M be a compact C^∞ manifold, and let $\Omega(M) = \Gamma(M, \wedge T^*M)$ be the space of differential forms on M. With a suitable inner product on T^*M, the de Rham differential $d\colon \Omega(M) \to \Omega(M)$ has an adjoint d^* such that

$$\Delta = d^*d + dd^*, \tag{7.5.24}$$

$$[\Delta, d] = 0 = [\Delta, d^*], \tag{7.5.25}$$

expressing the Hodge decomposition of the Laplace operator. Let $P\colon \Omega M \to \mathrm{Ker}(\Delta)$ be the projection onto the harmonic forms. We have

$$\langle \Delta f, f \rangle = \langle df, df \rangle + \langle d^* f, d^* f \rangle,$$

where $\Delta \geq 0$, so that $\mathrm{Ker}(\Delta) = \mathrm{Ker}(d) \cap \mathrm{Ker}(d^*)$. Also, $\mathrm{Ker}(\Delta)$ in $T^{*n}M$ is a finite-dimensional vector space.

The Hodge decomposition $\Omega(M) = P\Omega(M) \oplus P^{\perp}\Omega(M)$ has a further splitting

$$(\mathrm{Im}(P))^{\perp} = d\Omega(M) \oplus d^{*}\Omega(M) \tag{7.5.26}$$

such that $d^{*}\colon d\Omega(M) \to d^{*}\Omega(M)$ is invertible with inverse Gd, and $d\colon d^{*}\Omega(M) \to d\Omega(M)$ is invertible with inverse Gd^{*}.

7.6 Harmonic Forms

By analogy with the Hodge theory over a smooth connected compact Riemannian manifold M without boundary, we call $P\Omega$ the space of harmonic forms. The following result is analogous to the identity $\mathrm{Ker}(\Delta) = \mathrm{Ker}(d) \cap \mathrm{Ker}(d^{*})$ from the Riemannian case. Nevertheless, there is a substantial difference between $P\Omega$ from Theorem 7.5.2 and the one-dimensional space of harmonic functions on M, as described in Hopf's Lemma. In the abstract Hodge decomposition, the eigenspace corresponding to eigenvalue 1 has algebraic multiplicity greater than 1. In the noncommutative case, d need not be zero on $P\Omega$, which is the analogue of $\mathrm{Ker}\Delta$.

Proposition 7.6.1 *The differential ω is harmonic if and only if $d\omega$ and $db\omega$ are fixed under κ.*

Proof ($\Rightarrow$) To say that ω is harmonic means that ω belongs to the generalized eigenspace for κ that corresponds to the generalized eigenvalue $\zeta = 1$. Hence $d\omega$ and $bd\omega$ also belong to the 1-generalized eigenspace for κ. Now κ is of finite order on $d\Omega^{n}$ for all n, so $d\omega$ and $bd\omega$ are invariant for κ.

($\Leftarrow$) One applies powers of κ to the identity

$$\omega = P\omega + bGd\omega + Gdb\omega. \tag{7.6.1}$$

□

In this section we use the Hodge decomposition to compute some homology groups. For the decomposition $\Omega = P\Omega \oplus P^{\perp}\Omega$, we have differentials b, d and B, and the following result takes these one at a time.

Proposition 7.6.2 (i) *$H(P^{\perp}\Omega, b) = 0 = H(P^{\perp}\Omega, d)$ for the b and d differentials.*

(ii) *$H(P\Omega, d) = H(P\Omega, B) = \mathbf{C}[0]$, so the only homology is a copy of $\mathbf{C}$ in degree zero.*

Proof (i) First, we have $H(P^{\perp}\Omega) = 0$ for the b and d differentials since $P^{\perp}\Omega$ is contractible with respect to b and d by Theorem 7.5.2.

(ii) Hence the homology resides in the harmonic forms. There we have the Hochschild homology

$$H(P\Omega, b) = H(\Omega, b) = HH(A) \tag{7.6.2}$$

and

$$H(P\Omega, d) = H(\Omega, d) = \mathbf{C}[0]. \tag{7.6.3}$$

□

This states that the homology of (Ω, d) is nearly exact; the only non-zero component is in degree zero. In the next section, we will tidy up this statement by passing to the reduced differential forms.

From the identity

$$Pd = \frac{1}{n+1}\sum_{j=0}^{n} \kappa^j d = \frac{1}{n+1} B \tag{7.6.4}$$

we have $B = \nu P d$ with $\nu(\omega) = |\omega|\omega$, the grading operator. Hence $B = 0$ on $P^{\perp}\Omega$, so $H(P^{\perp}\Omega, B) = P^{\perp}\Omega$. Meanwhile, the homology of the harmonic forms is $H(P\Omega, B) = H(P\Omega, d)$ since $B = d$ up to non-zero scalar factors, so $H(P\Omega, B) = \mathbf{C}[0]$. Hence B is nearly exact on $P\Omega$.

In the next section, we consider the homology of complexes where differentials are considered jointly. In particular, we will consider $H(\Omega A, b + B)$.

7.7 Mixed Complexes in the Homology Setting

Definition 7.7.1 (Mixed complex) A mixed complex is a graded vector space M equipped with operators b of degree -1 and B of degree $+1$ that satisfy

$$b^2 = B^2 = bB + Bb = 0. \tag{7.7.1}$$

Usually $M = \oplus_{n=-\infty}^{\infty} M_n$, where $M_n = 0$ for $n < 0$. We have the point of view that (M, b, B) is primarily a complex (M, b) with the extra structure afforded by B. The ordinary homology is

$$H(M) = H(M, b). \tag{7.7.2}$$

Example 7.7.2 (Hochschild homology) The differential graded algebra $(\Omega A, b, B)$ is a mixed complex. Here the Hochschild homology is the homology of the columns

$$HH(A) = H(\Omega A, b). \tag{7.7.3}$$

In the double complex, we use p for the horizontal coordinate and q for the vertical coordinate in a Cartesian array with vertices $K_{p,q} = M_{q-p}$, so that the cross diagonals are constant. The entry at $(0,0)$ is M_0, and the indices increase as we go upwards or leftwards. The vertical differential is b downwards, and the horizontal differential is B leftwards. The following is the portion of the array in the first quadrant $q \geq p \geq 0$.

$$\begin{array}{ccccccc} \dots & \overset{B}{\longleftarrow} & M_2 & \overset{B}{\longleftarrow} & M_1 & \overset{B}{\longleftarrow} & M_0 \\ & & b\downarrow & & b\downarrow & & \\ \dots & \overset{B}{\longleftarrow} & M_1 & \overset{B}{\longleftarrow} & M_0 & & \\ & & b\downarrow & & & & \\ \dots & \overset{B}{\longleftarrow} & M_0 & & & & \end{array} \tag{7.7.4}$$

We can slide the entire array down the cross diagonal $p = q$, without changing the array. This gives an obvious automorphism S of degree $(-1, -1)$. This is a version of Connes's S operator, where $\mathbf{C}[S]$ is a principal ideal domain. Later, we will obtain a formula for S on chains.

The portion of the array in the first quadrant gives the total complex $(\mathcal{B}_{\geq}(M), b + B)$, where the component in degree n is

$$\mathcal{B}_{\geq}(M)_n = M_n \oplus M_{n-2} \oplus M_{n-4} \oplus \dots, \tag{7.7.5}$$

which is the finite sum of subspaces along the nth leading diagonal in the first quadrant. Note that we have a double complex in the sense of Section 2.4 since

$$(b + B)^2 = b^2 + bB + Bb + B^2 = 0. \tag{7.7.6}$$

Definition 7.7.3 (Cyclic homology) (i) The cyclic homology of M is defined to be

$$HC(M) = H(\mathcal{B}_{\geq}(M), b + B). \tag{7.7.7}$$

(ii) The cyclic homology of A is $HC(A) = H(\Omega A, b + B)$; see Example 7.7.5. We have an exact sequence of complexes induced by the shift in the periodic complex

$$0 \longrightarrow (M, b) \longrightarrow \mathcal{B}_{\geq}(M) \longrightarrow \mathcal{B}_{\geq}(M) \longrightarrow 0.$$

The map is $x_n + x_{n-2} + \cdots \mapsto x_{n-2} + x_{n-4} + \cdots$.

Definition 7.7.4 (Connes's exact sequence) The corresponding long exact sequence for this short exact sequence (as in Lemma 2.1.2) is

$$H_n(M) \xrightarrow{I} HC_n(M) \xrightarrow{S} HC_{n-2}(M) \xrightarrow{B} H_{n-1}(M) \xrightarrow{I} HC_{n-1}(M). \tag{7.7.8}$$

Example 7.7.5 (Cyclic homology) (i) The cyclic homology of A is $HC_n(A) = H_n(\Omega A, b + B)$. The normalized (b, B) complex is the mixed complex

$$\begin{array}{ccccccc} \dots & \overset{B}{\longleftarrow} & \Omega^2 A & \overset{B}{\longleftarrow} & \Omega^1 & \overset{B}{\longleftarrow} & \Omega^0 A \\ & & b \downarrow & & b \downarrow & & \\ \dots & \overset{B}{\longleftarrow} & \Omega^1 A & \overset{B}{\longleftarrow} & \Omega^0 A & & \\ & & b \downarrow & & & & \\ \dots & \overset{B}{\longleftarrow} & \Omega^0 A & & & & \end{array} \tag{7.7.9}$$

especially when written in terms of $\Omega^n A = A \otimes (\bar{A})^{\otimes n}$.

(ii) A module M over a commutative algebra A' is said to be flat if tensoring by M, as in $N \mapsto N \otimes_{A'} M$, which transforms all exact sequences of A'-modules into exact sequences; see [8, p. 29]. Suppose that M is a separable and flat algebra over a field $\mathbf{k}$. Then Kassel [109, Proposition 3.9] has shown that

$$HC_n(A \otimes M) = HC_n(A) \otimes M/[M, M] \qquad (n = 0, 1, 2 \dots)$$

for all $\mathbf{k}$-algebras A. This applies to the M in Example 6.4.4, such as $M = M_n(\mathbf{C})$.

Let $HH(A) = H(\Omega A, b)$. Then Connes's exact sequence for cyclic and Hochschild homology is

$$\dots \longrightarrow HH_n(A) \xrightarrow{I} HC_n(A) \xrightarrow{S} HC_{n-2}(A) \xrightarrow{B} HH_{n-1}(A) \longrightarrow \cdots. \tag{7.7.10}$$

Corollary 7.7.6 *Let A be quasi-free. Then*

(i) $HH_j(A) = 0$ for $j = 2, 3, 4, \dots$;

(ii) $HC_3(A) = HC_1(A)$;

(iii) $HC_2(A) = \mathrm{Ker}(B : HC_0(A) \to HH_1(A))$.

Proof (i) This is immediate from the definitions as in Definition 6.4.7.

(ii) Given (i), the exact sequence (7.7.10) reduces to the exact sequence

$$0 \overset{I}{\longrightarrow} HC_3(A) \overset{S}{\longrightarrow} HC_1(A) \overset{B}{\longrightarrow} 0 \overset{I}{\longrightarrow} \dots$$
$$\dots \overset{I}{\longrightarrow} HC_2(A) \overset{S}{\longrightarrow} HC_0(A) \overset{B}{\longrightarrow} HH_1(A) \overset{I}{\longrightarrow} \dots . \quad (7.7.11)$$

□

Remark 7.7.7 The definition of B is indirect and takes account of the asymmetry of the terms in $A \otimes (\bar{A})^{\otimes n}$. Suppose that $\mathcal{A}$ is a non-unital complex algebra and let $A = \mathcal{A}^+ = \mathcal{A} + \mathbf{C}1$ be its augmentation to a unital algebra. We recall λ be the cyclic permutation operator with sign $\lambda(a_0, \dots, a_n) = (-1)^n (a_n, a_0, \dots, a_{n-1})$ and $N = \sum_{j=0}^{n-1} \lambda^j$, the cyclic symmetrization operator, so that

$$\mathrm{Ker}(N) = \mathrm{Im}(1-\lambda), \qquad \mathrm{Im}(N) = \mathrm{Ker}(1-\lambda).$$

Next we introduce

$$s(a_0, \dots, a_n) = (1, a_0, \dots, a_n),$$

which satisfies

$$b's + sb' = id. \quad (7.7.12)$$

Then we can define $B = (1-\lambda)sN \colon A^{\otimes n} \to A^{\otimes(n+1)}$. Tsygan considered the following complex and showed it is a double complex.

$$\begin{array}{ccccccc}
\mathcal{A}_\lambda^{\otimes 3} & \overset{N}{\longleftarrow} & \mathcal{A}^{\otimes 3} & \overset{1-\lambda}{\longleftarrow} & \mathcal{A}^{\otimes 3} & \overset{N}{\longleftarrow} & \mathcal{A}^{\otimes 3} \\
\downarrow b & & \downarrow b & & \downarrow -b' & & \downarrow b \\
\mathcal{A}_\lambda^{\otimes 2} & \overset{N}{\longleftarrow} & \mathcal{A}^{\otimes 2} & \overset{1-\lambda}{\longleftarrow} & \mathcal{A}^{\otimes 2} & \overset{N}{\longleftarrow} & \mathcal{A}^{\otimes 2} \\
\downarrow b & & \downarrow b & & \downarrow -b' & & \downarrow b \\
\mathcal{A} & \overset{N}{\longleftarrow} & \mathcal{A} & \overset{1-\lambda}{\longleftarrow} & \mathcal{A} & \overset{N}{\longleftarrow} & \mathcal{A}
\end{array} \quad (7.7.13)$$

The column on the left gives the cyclic complex $(\mathcal{A}_\lambda, b)$. The squares anticommute, and the complex is periodic via shifts to the right. The rows are exact, so $\mathrm{Ker}(N) = \mathrm{Im}(1-\lambda)$ and $\mathrm{Im}(N) = \mathrm{Ker}(1-\lambda)$. There is a Connes's exact sequence

$$HH_n(\mathcal{A}) \overset{I}{\longrightarrow} HC_n(\mathcal{A}) \overset{S}{\longrightarrow} HC_{n-2}(\mathcal{A}) \overset{B}{\longrightarrow} HH_{n-1}(\mathcal{A}). \quad (7.7.14)$$

The notion of quasi-free algebras was intended as a generalization of smooth algebras in the commutative setting, so we consider how the homology theories are related, as in section 13 of [23]. We momentarily ignore the assumption that A is commutative, and form the noncommutative differential 1-forms

$$0 \longrightarrow \Omega^1 A \longrightarrow A \otimes A \longrightarrow A \longrightarrow 0, \tag{7.7.15}$$

and then form ΩA as in Theorem 6.3.1(i). By Theorem 6.3.1(ii), the identity homomorphism $A \to \Omega_A$ extends to a homomorphism of differential graded algebras $\Omega A \to \Omega_A$, so there are maps $\Omega^{ev} A \to \Omega_A^{ev}$ and $\Omega^{odd} A \to \Omega_A^{odd}$. This gives a diagram

$$\begin{array}{ccccc} & \widehat{\Omega}_A^{ev} & \overset{b+B}{\longrightarrow} & \widehat{\Omega}_A^{odd} & \\ \mu & \downarrow & & \downarrow & \mu, \\ & \Omega_A^{ev} & \underset{d}{\longrightarrow} & \Omega_A^{odd} & \end{array} \tag{7.7.16}$$

where

$$\mu(a_0 da_1 \dots da_n) = \frac{1}{n!} a_0 da_1 \dots da_n \tag{7.7.17}$$

and

$$B(a_0 da_1 \dots da_n) = \sum_{i=0}^{n} (-1)^{in} da_i \dots da_n da_0 \dots da_{i-1}. \tag{7.7.18}$$

The operators d, b, κ, B, P and d on ΩA descend to d, $b = 0$, $\kappa = 1$, $B = Nd$ and $P = 1$ on Ω_A.

We can introduce $\widehat{\Omega}^{ev} A = \prod_{n=0}^{\infty} \Omega^{2n} A$ and $\widehat{\Omega}^{odd} A = \prod_{n=0}^{\infty} \Omega^{2n+1} A$, which of course combine to give $\widehat{\Omega} A = \widehat{\Omega}^{ev} A \oplus \widehat{\Omega}^{odd} A$; then we have a complex

$$\widehat{\Omega}^{ev} A \underset{\underset{b+B}{\longrightarrow}}{\overset{b+B}{\longleftarrow}} \widehat{\Omega}^{odd} A, \tag{7.7.19}$$

since b has degree -1, B has degree 1 and $(b+B)^2 = 0$. We define the periodic cyclic homology of A to be the homology of this complex, denoted $HP_i(A)$ for $i = 0, 1$. Periodic cyclic homology is a generalization of de Rham cohomology. Recalling the terminology of the Zariski–Grothendieck theorem, we have:

Theorem 7.7.8 *Let A be smooth and commutative, with maximal ideal space* **X**.

Then

$$HH_i(A) = H_i(\Omega A, b) = \Omega_A^i. \tag{7.7.20}$$

There is a quasi-isomorphism μ in the diagram

$$\begin{array}{ccc} \widehat{\Omega}^{ev} A & \overset{b+B}{\underset{b+B}{\leftrightarrows}} & \widehat{\Omega}^{odd} A \\ \mu \downarrow & & \downarrow \mu \\ \Omega_A^{ev} & \overset{d}{\underset{d}{\leftrightarrows}} & \Omega_A^{odd} \end{array}$$

so that there is isomorphism of the $\mathbf{Z}/(2)$ graded homology

$$HP_i(A) = H^i(\Omega_A) = H^i(\mathbf{X}) \qquad (i = 0, 1).$$

Proof For a smooth and commutative algebra, we can invoke the Hochschild–Konstant–Rosenberg theorem sections 3.4.4 and 5.1.12 of [68] to show that μ gives an isomorphism of complexes $(\Omega A, b, B) \rightarrow (\Omega_A, 0, d)$. For further discussion, see section 13 of [23], and Chapter 12. □

7.8 Homology of the Reduced Differential Forms

In this section we introduce the space $\bar{\Omega}$ of reduced differential forms, so that we can replace statements about near exactness of homology for Ω as in Proposition 7.6.2 by statements of exactness of homology for $\bar{\Omega}$.

Recall that $\Omega\mathbf{C} = \mathbf{C}[0]$, so we can introduce the reduced space of differential forms $\bar{\Omega} = \bar{\Omega}A$ by the exact sequence

$$0 \longrightarrow \Omega\mathbf{C} \longrightarrow \Omega A \longrightarrow \bar{\Omega} A \longrightarrow 0.$$

Hence

$$\bar{\Omega}^n A = \begin{cases} \Omega^n A, & \text{for } n \neq 0; \\ \bar{A}, & \text{for } n = 0. \end{cases} \tag{7.8.1}$$

The operators b, B, κ, P and G all descend from Ω to $\bar{\Omega}$. Hence the reduced differential forms have a Hodge decomposition

$$\bar{\Omega} = P\bar{\Omega} \oplus P^{\perp}\bar{\Omega} \tag{7.8.2}$$

in which $P\bar{\Omega} = P\Omega/\mathbf{C}[0]$ and $P^{\perp}\bar{\Omega} = P^{\perp}\Omega$. Hence B is exact on $P\bar{\Omega}$ and $B = 0$ on $P^{\perp}\bar{\Omega}$.

Connes's Lemma 7.8.1 *For B on $\bar{\Omega}$, the inclusion* $\mathrm{Im}(B) \longrightarrow \mathrm{Ker}(B)$ *induces an isomorphism in the homology with respect to b.*

Proof The Hodge decomposition $\bar{\Omega} = P\bar{\Omega} \oplus P^{\perp}\bar{\Omega}$ has $B = 0$ on $P^{\perp}\bar{\Omega}$ and B exact on $P\bar{\Omega}$, so

$$\mathrm{Ker}(B) = \mathrm{Im}(B) \oplus P^{\perp}\bar{\Omega}, \tag{7.8.3}$$

so $P^{\perp}\bar{\Omega} = \mathrm{Ker}(B)/\mathrm{Im}(B)$. Also, $P^{\perp}\bar{\Omega}$ is acyclic with respect to b since $b(Gd) + (Gd)b = 1$ gives a homotopy to 0. Hence

$$H(\mathrm{Ker}(B), b) = H(\mathrm{Im}(B), b). \tag{7.8.4}$$

□

Definition 7.8.2 (Reduced homology) The reduced Hochschild homology is

$$\bar{H}H(A) = H(\bar{\Omega}A, b).$$

The reduced cyclic homology is $\bar{H}C(A) = H(\bar{\Omega}A/\mathrm{Ker}(B), b)$.

Proposition 7.8.3 *The following is a long exact sequence:*

$$\longrightarrow \bar{H}C_1(A) \longrightarrow \bar{H}H_2(A) \longrightarrow \bar{H}C_2(A) \overset{S}{\longrightarrow} HC_0(A) \overset{B}{\longrightarrow} \bar{H}C_1(A) \longrightarrow 0. \tag{7.8.5}$$

Proof This is the long exact sequence in the sense of Lemma 2.1.2 that corresponds to the short exact sequence

$$0 \longrightarrow \mathrm{Ker}(B) \longrightarrow \bar{\Omega} \longrightarrow \bar{\Omega}/\mathrm{Ker}(B) \longrightarrow 0. \tag{7.8.6}$$

By the Lemma, the inclusion $\mathrm{Ker}(B) \to \mathrm{Im}(B)$ induces an isomorphism in homology, so

$$B : \bar{\Omega}/\mathrm{Ker}(B) \longrightarrow \mathrm{Im}(B) \longrightarrow \mathrm{Ker}(B) \tag{7.8.7}$$

induces an isomorphism in homology that shifts the degree by one. □

7.9 Cyclic Cohomology

Consider the normalized cochains on a unital algebra A as in Section 2.6; we write as in Section 6.3

$$C^n(A) = \left(\Omega^n A\right)^* = \left(A \otimes \bar{A}^{\otimes n}\right)^*$$

for the space of complex linear functionals

$$f(a_0 da_1 \dots da_n) = f(a_0, a_1, \dots, a_n), \tag{7.9.1}$$

such that $f = 0$ whenever $a_j = 1$ for some $j = 1, \dots, n$. Also, let $C^n(A) = 0$ for $n < 0$. We define b as in Section 1.2 and B via

$$Bf(a_0, \dots, a_n) = f\big(B(a_0 da_1 da_2 \dots da_n)\big), \tag{7.9.2}$$

where B is defined on $\Omega^n A$ as in Section 7.4. Then Proposition 7.4.2 gives

$$b^2 = 0, \quad B^2 = 0, \quad bB + Bb = 0. \tag{7.9.3}$$

Hence there is a double complex of cochains

$$\begin{array}{ccccccccc}
\dots & \xrightarrow{B} & C^3(A) & \xrightarrow{B} & C^2(A) & \xrightarrow{B} & C^1(A) & \xrightarrow{B} & C^0(A) \\
 & & b\uparrow & & b\uparrow & & b\uparrow & & \\
\dots & \xrightarrow{B} & C^2(A) & \xrightarrow{B} & C^1(A) & \xrightarrow{B} & C^0(A) & & \\
 & & b\uparrow & & b\uparrow & & & & \\
\dots & \xrightarrow{B} & C^1(A) & \xrightarrow{B} & C^0(A) & & & &
\end{array} \tag{7.9.4}$$

in which arrows run left to right along rows and upwards on columns. Then we let $K^n = \prod_p C^{n-2p}(A)$ be the total complex of cochains with arbitrary support, and note that $b + B \colon K^n \to K^{n+1}$ satisfies $(b + B)^2 = 0$. The North East corner gives $K_{\geq 0}$.

Definition 7.9.1 (Cyclic cohomology) The cyclic cohomology of A is defined by the homology of the double complex

$$HC^i(A) = H^i(K_{\geq 0}). \tag{7.9.5}$$

Then $HC^i(A) = HC_i(A)^*$, the dual of the cyclic homology.

7.10 Traces on RA and Cyclic Cocycles on A

In this section we return to the algebra RA as in Section 6.7, which is isomorphic to $\Omega^{ev} A$. The idea is to consider chains and traces on a universal object. Let A be a non-unital algebra over a field $\mathbf{k}$ of characteristic zero. We write $C^n(A) = (A^{\otimes n})^*$ for the space of linear functionals $f \colon A^{\otimes n} \to \mathbf{k}$, or equivalently the space of multilinear maps $f \colon A^{\times n} \to \mathbf{k}$. The main task is to obtain an explicit description of the traces on RA.

We have the following operations on cochains: first b' is given by

$$b' f(a_1, \dots, a_{n+1}) = \sum_{i=1}^{n} (-1)^{i-1} f(a_1, \dots, a_i a_{i+1}, \dots, a_n); \tag{7.10.1}$$

and with the cross-over term added, we have b, as in

$$bf(a_1,\dots,a_{n+1}) = b'f(a_1,\dots,a_{n+1}) + (-1)^n f(a_{n+1}a_1,a_2,\dots,a_n); \tag{7.10.2}$$

then the cyclic permutation with sign

$$\lambda f(a_1,\dots,a_n) = (-1)^{n-1} f(a_n,a_1,\dots,a_{n-1}) \tag{7.10.3}$$

and finally the symmetrization operator

$$Nf(a_1,\dots,a_n) = \sum_{i=0}^{n-1} \lambda^i f(a_1,\dots,a_n). \tag{7.10.4}$$

These satisfy the identities

$$b^2 = (b')^2 = 0; \tag{7.10.5}$$

$$(\lambda - 1)N = N(\lambda - 1) = 0; \tag{7.10.6}$$

$$b'(1-\lambda) = (1-\lambda)b; \tag{7.10.7}$$

$$bN = Nb'; \tag{7.10.8}$$

so the following gives a double complex. The reader may pause to compare this with Tsygan's complex (7.7.13).

$$\begin{array}{ccccccc}
\uparrow -b & & \uparrow b' & & \uparrow -b & & \uparrow b' \\
C^3(A) & \overset{N}{\longrightarrow} & C^3(A) & \overset{\lambda-1}{\longrightarrow} & C^3(A) & \overset{N}{\longrightarrow} & C^3(A) \overset{\lambda-1}{\longrightarrow} \\
\uparrow -b & & \uparrow b' & & \uparrow -b & & \uparrow b' \\
C^2(A) & \overset{N}{\longrightarrow} & C^2(A) & \overset{\lambda-1}{\longrightarrow} & C^2(A) & \overset{N}{\longrightarrow} & C^2(A) \overset{\lambda-1}{\longrightarrow} \\
\uparrow -b & & \uparrow b' & & \uparrow -b & & \uparrow b' \\
C^1(A) & \overset{N}{\longrightarrow} & C^1(A) & \overset{\lambda-1}{\longrightarrow} & C^1(A) & \overset{N}{\longrightarrow} & C^1(A) \overset{\lambda-1}{\longrightarrow}
\end{array} \tag{7.10.9}$$

The rows are exact, so the interesting homology lies in the vertical differentials b and b'. Let $R = RA$ and $I = IA$ be the algebras from Section 6.7, where I is the kernel of the map $R \to A$, so

$$0 \longrightarrow I \longrightarrow R \longrightarrow A \longrightarrow 0$$

is an exact sequence of algebra maps. Then we have a canonical inclusion $\rho\colon A \to R$ and we consider cochains $\omega = \delta\rho + \rho^2$, and

$$\omega^n(a_1,\dots,a_{2n}) = \omega(a_1,a_2)\omega(a_3,a_4)\dots\omega(a_{2n-1},a_{2n}); \tag{7.10.10}$$

$$\rho\omega^n(a_0,\dots,a_{2n}) = \rho(a_0)\omega(a_1,a_2)\omega(a_3,a_4)\dots\omega(a_{2n-1},a_{2n}). \tag{7.10.11}$$

Now R is isomorphic as a $\mathbf{k}$ vector space to $\oplus_{p=1}^{\infty} A^{\otimes p}$. A linear functional f on $\oplus_{p=1}^{\infty} A^{\otimes p}$ is a formal expression $f = \sum_{p=1}^{\infty} f_p$, where $f_p : A^{\otimes p} \to \mathbf{k}$ is a linear functional; that is, $f_p \in C^p(A)$. For $p = 2n$ even the components of $A^{\otimes 2n}$ are given by sums of ω^n, whereas for $p = 2n+1$ odd, the components of $A^{\otimes(2n+1)}$ are given by sums of $\rho\omega^n$. Let $\tau : R \to \mathbf{C}$ be a linear functional, and let

$$f_{2n}(a_1, a_2, \dots, a_{2n-1}, a_{2n}) = \tau(\omega(a_1, a_2) \dots \omega(a_{2n-1}, a_{2n})); \tag{7.10.12}$$

$$f_{2n+1}(a_0, a_1, \dots, a_{2n-1}, a_{2n}) = \tau(\rho(a_0)\omega(a_1, a_2) \dots \omega(a_{2n-1}, a_{2n})). \tag{7.10.13}$$

We now compare the properties of τ restricted to commutator subspaces with the sum of cochains $f = \sum_{p=1}^{\infty} f_p$.

Theorem 7.10.1 (i) *The functional τ is a trace on R, if and only if*

(1) $b' f_{2n} = (1 - \lambda) f_{2n+1}$, *for all $n \geq 1$;*
(2) $b f_{2n+1} = (n+1)^{-1} N f_{2n+2}$, *for all $n \geq 1$;*
(3) $\lambda^2 f_{2n} = f_{2n}$, *for all $n \geq 1$.*

(ii) *Let τ be a linear functional on I^m for some $m \geq 1$. Then τ vanishes on $[I^m, R]$, if and only if (1) and (2) hold for all $n \geq m$, and (3) holds for all $n > m$.*

(iii) *Let τ be a linear functional on I^m for some $m \geq 1$. Then τ vanishes on $[I^{m-1}, I]$, if and only if (1) and (2) hold for all $n \geq m$, and (3) holds for all $n \geq 1$.*

Proof (i) Assume that τ is a trace. Then by Bianchi's identity Proposition 2.8.2(i), we have

$$\delta\omega^n + \rho\omega^n - \omega^n\rho = 0, \tag{7.10.14}$$

so

$$\big(b'(\omega^n) - (1 - \lambda)(\rho\omega^n)\big)(a_1, \dots, a_{2n+1}) = -[\omega^n(a_1, \dots, a_{2n}), \rho(a_{2n+1})]. \tag{7.10.15}$$

We also have

$$\big(b(\rho\omega^n) - (1 + \lambda)\omega^{n+1}\big)(a_0, \dots, a_{2n+1}) = [\rho(a_0)\omega^n(a_1, \dots, a_{2n}), \rho(a_{2n+1})]. \tag{7.10.16}$$

Taking the trace, the terms in the commutators vanish and we deduce

$$b'\tau(\omega^n) = (1 - \lambda)\tau(\rho\omega^n),$$

as in (i), so

$$b\tau(\rho\omega^n) = (1 + \lambda)\tau(\omega^{n+1}).$$

From the formula

$$f_{2n}(a_1, a_2, \dots, a_{2n-1}, a_{2n}) = \tau(\omega(a_1, a_2) \dots \omega(a_{2n-1}, a_{2n})); \tag{7.10.17}$$

since τ is a trace, we deduce that $\lambda^2 f_{2n} = f_{2n}$ as in (3), and we obtain (2) from

$$N f_{2n} = \sum_{j=0}^{2n} \lambda^j f_{2n} = n(1+\lambda) f_{2n}. \tag{7.10.18}$$

Conversely, we assume that (1), (2) and (3) hold and aim to prove that τ is a trace. Given (1), we have

$$\tau\big[\omega^n(a_1, \dots, a_{2n}), \rho(a_{2n+1})\big] = 0, \tag{7.10.19}$$

$$\tau\big[\rho(a_0)\omega^n(a_1, \dots, a_{2n}), \rho(a_{2n+1})\big] = 0; \tag{7.10.20}$$

so that

$$\tau[r, \rho(a)] = 0 \qquad (a \in A, r \in R). \tag{7.10.21}$$

But the elements of the form $\rho(a)$ generate R, so $\tau|[R, R] = 0$ and τ is a trace.

(ii) The proof of (ii) is similar to (i).

(iii) There is a subtle point in this case regarding $n = m$. We assume that $f_{2n} = \tau\omega^n$ is λ^2-invariant and $\tau|[R, I^m] = 0$, and aim to prove that $\tau[I, I^{m-1}] = 0$. The proof uses the identity

$$\begin{aligned} b(1+\lambda)\omega^n(a_1, \dots, a_{2n}) &= \big[\rho(a_0)\omega^{n-1}(a_1, \dots, a_{2n-2}), \omega(a_{2n-1}, a_{2n})\big] \\ &\quad + \big[\omega^n(a_0, \dots, a_{2n-1}), \rho(a_{2n})\big] \\ &\quad + \big[\omega^n(a_{2n}, a_0, \dots, a_{2n-2}), \rho(a_{2n-1})\big]. \end{aligned} \tag{7.10.22}$$

Then

$$\frac{1}{m} N f_{2m} = \frac{1}{m} bN f_{2m+1} = \frac{1}{m} Nb' f_{2m} = \frac{1}{m} f_{2m+1} = 0, \tag{7.10.23}$$

so $b(1+\lambda) f_{2n} = 0$. □

Definition 7.10.2 (Cyclic cohomology) Consider the space of cyclic cochains

$$C^n_\lambda(A) = \{f \in C^{n+1}(A) : \lambda f = f\}; \tag{7.10.24}$$

note the change of index. Then $(C_\lambda(A), b)$ is a complex, and the associated homology $HC(A)$ is the cyclic cohomology of A.

The following result describes the cyclic cocycles of odd order.

Corollary 7.10.3 *For the cochains f_p as in Theorem 7.10.1, the map*

$$\sum_{p=2m}^{\infty} f_p \mapsto Nf_{2m} \in C_\lambda^{2m-1}(A)$$

gives an exact sequence

$$(R/[R,R])^* \to (I^m/[I,I^{m-1}])^* \to HC^{2m-1}(A) \to 0. \qquad (7.10.25)$$

Proof Given a trace τ on R, there is a cochain $f = \sum_{p=1}^{\infty} f_p$ where the $f_{2n} = \tau(\omega^n)/n!$ and $f_{2n+1} = \tau(\rho\omega^n)$. We consider in particular f_{2m} and Nf_{2m} which satisfies $(1-\lambda)Nf_{2m} = 0$ and $Nf_{2m} = bf_{2m-1}$, so $bNf_{2m} = b^2 f_{2m-1} = 0$; hence $Nf_{2m} \in C_\lambda^{2m-1}(A)$ is a coboundary.

Now suppose that $\tau\colon I^m/[I,I^{m-1}] \to \mathbf{k}$ is linear, and introduce as in Theorem 7.10.1(iii) a cochain $f = \sum_{p=2m}^{\infty} f_p$. Then Nf_{2m} again satisfies $(1-\lambda)Nf_{2m} = 0$, so $Nf_{2m} \in C_\lambda^{2m-1}(A)$; also, $Nf_{2m} = mbf_{2m+1}$, so $bNf_{2m} = mb^2 f_{2m+1} = 0$, and Nf_{2m} is a cocycle. □

8

Connections

The fundamental concept introduced in this chapter is the Chern character of a connection over a manifold. The chapter reviews the notion of curvature of a connection on a vector bundle over a compact manifold, then introduces Chern classes. As an illustration, we compute the first Chern class for complex line bundles over a compact Riemann surface. A basic result of Chern and Weil describes how characteristic classes vary when the connection changes smoothly. The basic results in this chapter are closely related to results in [28]. In Sections 8.1, 8.2 and 8.3, we consider classical cases where A is a unital and commutative algebra over $\mathbf{C}$, and interpret the classical Chern–Weil theory in terms of K_0 and cyclic theory. In Sections 8.4, 8.5, 8.6 and 8.7, we consider algebras that are generally noncommutative, and introduce definitions that are analogous to the commutative case. The formulas will be extended in subsequent chapters to cyclic cocycles over noncommutative algebras. Interesting cyclic classes, especially those related to K-theory, are generally Chern character forms or variants of these when suitably interpreted; see [88].

8.1 Connections and Curvature on Manifolds

Definition 8.1.1 (Connection on manifold) Consider scalar functions on a smooth manifold M with tangent space $T(M)$. Let $C = C^\infty(M,\mathbf{R})$ and V be the fibre of $T(M)$, so V is a left A-module. Then an affine connection on M is a map $V \to V$ denoted $X \mapsto \nabla_X$ such that

(1) $Y \mapsto \nabla_Y$ is linear, so $\nabla_{aY+bZ}X = a\nabla_Y X + b\nabla_Z X$ for all $a,b \in \mathbf{R}$;

(2) $X \mapsto \nabla_Y X$ is a derivation in the sense that $\nabla_Y(fX) = (D_Y f)X + f\nabla_Y X$, so

$$\nabla(fX) = (df)X + f\nabla X.$$

The space of affine connections is a left $C^\infty(M)$ module, in the sense that one can define $h\nabla_Y$ by the usual product $h\nabla_Y X$ on V. The idea is that the tangent vector X is mapped to the operator ∇_X, which enables us to differentiate along the direction of X.

There is a special connection associated with the Riemannian metric structure, which also satisfies:

(3) $\nabla_X Y - \nabla_Y X = [X, Y]$, so the connection is torsion free; and

(4) $D_Z\langle X, Y\rangle = \langle \nabla_Z X, Y\rangle + \langle X, \nabla_Z Y\rangle$, where $g(X, Y) = \langle X, Y\rangle$.

Theorem 8.1.2 *There exists a unique connection ∇ that satisfies (1)–(4).*

Proof This result is known as the fundamental theorem of Riemannian geometry [83, p. 28]. □

The unique connection ∇ is known as the Levi-Civita connection. When one replaces $T(M)$ by other vector bundles over M, there are other possibilities for connections, as we discuss next. The formalism resembles the discussion of Maxwell's equation and the Dirac operator from Sections 5.4 and 5.5.

Definition 8.1.3 (Connection) Let M be a C^∞ connected n-dimensional manifold with cotangent bundle T^*M. Also, let E be a complex vector bundle over M with fibres V linearly isomorphic to $\mathbf{C}^m$; let $\Gamma(M, V)$ be the space of smooth sections, so that locally each section looks like a column of complex-valued C^∞ functions. A connection is a $\mathbf{C}$-linear operator $\nabla\colon \Gamma(M; E) \to \Gamma(M; T^* \otimes E)$ such that

$$\nabla(fs) = f\nabla s + (df)s, \tag{8.1.1}$$

where $f \in \Gamma(M, \mathbf{C})$, $s \in \Gamma(M, E)$ and d is the usual de Rham differential on $\Gamma(M, E)$ from Section 2.1.

Example 8.1.4 (Connections on a trivial bundle) Let $\tilde{V}$ be the trivial bundle over a coordinate patch U of M, given by $M \times V \to M$, so the space of smooth sections is simply $\Gamma(M, \tilde{V}) = C^\infty(M, V)$. Then a connection is $\nabla = d + \theta$ where θ is a matrix 1-form; more explicitly, a connection may be expressed as

$$\nabla = \sum_{j=1}^{n} \left(\frac{\partial}{\partial x_j} + \theta_j \right) dx_j, \tag{8.1.2}$$

where $\frac{\partial}{\partial x_j}$ is differentiation in the jth direction in a coordinate chart and θ_j is an $m \times m$ complex matrix. In calculations below, it is sometimes possible to reduce to this case.

Resuming the general discussion, we specify how ∇ on E can be reduced locally to a trivial bundle. Let G be a Lie subgroup of $GL_m(\mathbf{C})$ and let

$(\varphi_\alpha : U_\alpha \to M)$ be an atlas of coordinate charts for M. On each U_α, there exists an $m \times m$ complex matrix 1-form $\theta^{(\alpha)}$, and we suppose that there exist smooth transition functions $u_{\alpha,\beta} : U_\alpha \cap U_\beta \to G$ such that $u_{\alpha,\beta} = u_{\beta,\alpha}^{-1}$ so that transiting from U_α to U_β is inverse to transiting from U_β to U_α, and the consistency condition $u_{\alpha,\beta} u_{\beta,\gamma} = u_{\alpha,\gamma}$ holds on $U_\alpha \cap U_\beta \cap U_\gamma$. Then we require that

$$\theta^{(\beta)} = u_{\alpha,\beta}\theta^{(\alpha)} u_{\alpha,\beta}^{-1} - (du_{\alpha,\beta}) u_{\alpha,\beta}^{-1},$$

where $du_{\alpha,\beta}$ is a 1-form in the Lie algebra of G. Let $\Omega^p(M,E) = \Gamma(M, \wedge^p T^* \otimes E)$. Then there is a unique sequence of maps $\nabla : \Omega^p(M,E) \to \Omega^{p+1}(M,E)$ such that

(i) ∇ on $\Omega^0(M,E)$ is as in (1);

(ii) $\nabla(\omega \wedge \theta) = (\nabla\omega) \wedge \theta + (-1)^p \omega \wedge \nabla\theta$ for all $\omega \in \Omega^p(M,\mathbf{C})$ and $\theta \in \Omega^q(M,E)$. Note that the final term includes a sign as the ∇ goes through the term ω of degree p.

Whereas $d^2 = 0$, we generally have $\nabla^2 \neq 0$. Let g be the space of endomorphisms of E, interpreted as matrices acting on columns. Then the operation of ∇_F on $\Gamma(M, g)$ is given by $\nabla W = dW + [\theta, W]$, where the commutator is just the matrix product on g, and d is the exterior derivative on matrix entries. Observe that

$$\begin{aligned}
[\nabla, W]f &= \nabla(Wf) - W\nabla f \\
&= (d + \theta)(Wf) - W(df + \theta f) \\
&= (dW + \theta W - W\theta)f.
\end{aligned} \tag{8.1.3}$$

Definition 8.1.5 (Curvature) Then $\nabla = d + \theta\wedge$ is the exterior covariant derivative. There exists $F_\theta \in \Omega^2(M, \mathrm{End}_{\mathbf{C}} E)$ called the curvature of the connection ∇ such that

$$\nabla\nabla s = F_\theta s. \tag{8.1.4}$$

In the case where $F_\theta = 0$, the connection is said to be flat.

Note that flatness of the connection describes the vector bundle lying above M and the θ_j, not just about the Levi-Civita connection on the underlying manifold M. As in Example 5.3.3, in terms of local coordinates, we can write $F_\theta = \sum_{j<k} [F_\theta]_{jk} dx_j dx_k$ where

$$[F_\theta]_{i<j} = \frac{\partial\theta_j}{\partial x_i} - \frac{\partial\theta_i}{\partial x_j} + [\theta_i, \theta_j]. \tag{8.1.5}$$

Given such a matrix $[F_\theta]_{jk}$ we can introduce characteristic classes by introducing the invariant polynomial functions on matrices as in Section 5.6.

In terms of cyclic theory, $b' + ad(\rho)$ corresponds to ∇ and ω corresponds to F_θ.

Proposition 8.1.6 (i) *The curvature is given in terms of the connection matrices* $F = d\theta + \theta \wedge \theta$;

(ii) *Bianchi's identity holds:* $\nabla F = 0$; *that is,* $dF + [\theta, F] = 0$.

Proof (i) We have

$$\begin{aligned} \nabla\nabla s &= (d + \theta\wedge)(ds + \theta \wedge s) = d\theta \wedge s - \theta \wedge ds + \theta \wedge ds + \theta \wedge \theta s \\ &= (d\theta + \theta \wedge \theta)s. \end{aligned} \tag{8.1.6}$$

(ii) We compute $\nabla F = dF + [\theta, F]$. From (i), we have

$$\begin{aligned} dF &= d(d\theta + \theta \wedge \theta) = (d\theta) \wedge \theta - \theta \wedge (d\theta) \\ &= (F - \theta \wedge \theta) \wedge \theta - \theta \wedge (F - \theta \wedge \theta) \\ &= F \wedge \theta - \theta \wedge F = -[\theta, F]. \end{aligned} \tag{8.1.7}$$

□

Exercise 8.1.7 (Gauge groups) Let $\mathbf{K} = \mathbf{C}(x_1, \dots, x_n)$ and for $j = 1, \dots, n$, let A_j be $m \times m$ matrices with entries in $\mathbf{K}$ and $A = \sum_{j=1}^n A_j dx_j$ be a matrix-valued 1-form. Then we define $F = dA + A \wedge A$.

(i) Let G be a subgroup of $GL_m(\mathbf{C})$ with the usual left multiplication of matrices on column vectors in $\mathbf{C}^m$. Then we define

$$\varphi_g : A \mapsto \hat{A} = gAg^{-1} + g(dg^{-1}). \tag{8.1.8}$$

Show that $g \mapsto \varphi_{g^{-1}}$ defines an action of G on the potential field. Also show that

$$d\hat{A} + \hat{A} \wedge \hat{A} = g(dA + A \wedge A)g^{-1}, \tag{8.1.9}$$

so

$$F \mapsto \hat{F} = gFg^{-1}. \tag{8.1.10}$$

We call G the gauge group, A the potential field and F the field intensity or curvature form. The transformations are referred to as gauge transformations.

(ii) Let $G = U(m)$ be the group of unitary complex matrices, which has Lie algebra $u(m) = \{X \in M_m(\mathbf{C}) : A^* = -A\}$, the space of skew-symmetric matrices. Show that for $A \in u(m)$,

$$\text{trace}\,(A^2) = -\text{trace}(AA^*) \leq 0 \tag{8.1.11}$$

with equality if and only if $A = 0$. Next we identify the 2-form with its matrix with respect to some basis. Deduce that $\text{trace}(F^2) = \text{trace}(\hat{F}^2)$, so $\text{trace}(F^2)$ is invariant under the action of the gauge group. Deduce also that $\mathrm{trace}(F^2) = 0$ if and only if the field intensity is zero.

The Bianchi identity for the connection form F generalizes the homogeneous Maxwell equations from Section 5.4, which reduced to $dF = 0$. Usually the gauge group is taken to be a compact Lie group, and in the discussion below, we take $G = U(m)$. In this context, F is known as the field strength and can be expressed in terms of the signed commutator as $F = d\theta + (1/2)[\theta,\theta]$. Also, F transforms according to the adjoint representation under gauge transformations, and we have $[d + \theta\wedge, F^n] = 0$.

Example 8.1.8 (Connections on Riemann surfaces) The notion of a connection is important for complex manifolds, where the complex analytic structure is of primary interest.

(i) Let U be a domain in $\mathbf{C}$, and E the space of holomorphic functions $\Psi\colon U \to \mathbf{C}^n$. Let S be a finite subset of U, and p the monic polynomial with simple zeros at the points of S. Let $A\colon U \to M_n(\mathbf{C})$ be a meromorphic function with at most simple poles, all in S. Then

$$\nabla\colon \Psi \to \Psi'(z)dz + A(z)\Psi(z)dz$$

determines a map $E \to Edz/p(z)$ such that $\nabla(f\Psi) = f'\Psi dz + f\nabla\Psi$. We say that ∇ is a regular singular connection, and the points of S are regular singular points. This reflects the traditional terminology of ordinary differential equations, where

$$\mathrm{Ker}\nabla = \{\Psi\colon \Psi'(z) + A(z)\Psi(z) = 0\}$$

gives the vector space of solutions of an ordinary differential equation with regular singular points, as in Example 5.4.5.

(ii) Let $\mathbf{X}$ be a connected Riemann surface, and S a finite subset of $\mathbf{X}$. A regular singular connection is a complex vector bundle E on $\mathbf{X}$ and $\nabla\colon E \to \Omega^1(\mathbf{X}; S) \otimes E$ where

(1) $\Omega^1(\mathbf{X}; S)$ is the space of meromorphic 1-forms on $\mathbf{X}$ with simple poles in S;

(2) Leibniz's rule $\nabla(f\Psi) = df \otimes \Psi + f\nabla\Psi$ holds on open subsets U of $\mathbf{X}$ for all $\Psi \in E$ and holomorphic $f\colon U \to \mathbf{C}$.

There is a significant question as to how the algebraic and analytic structures relate to one another. In section 6.2 of [99], vector bundles and connections on compact Riemann surfaces are discussed in terms of divisors and the classical Fuchsian differential equations. This leads to the simplest instance of the GAGA principle, that the algebraic and analytic vector bundles give an equivalent description of $\mathbf{X}$.

8.2 The Chern Character

The Chern–Weil complex extends the de Rham complex in the following sense. As in Example 8.1.4, let $\tilde{V}_0$ be the trivial bundle over M with fibres $\tilde{V}_0 \cong \mathbf{C}^m$, given by $\tilde{V}_0 \times M \to M$. Then $\Gamma(M, \tilde{V}_0) = C^\infty(M, \tilde{V}_0)$, and the de Rham differential extends to elements of $C^\infty(M, \tilde{V}_0)$ when they are written as column vectors with entries in $C^\infty(M, \mathbf{C})$. Thus we can extend d to $\Omega^p(M, \tilde{V}_0) = \Gamma(M, \wedge^p T^* \otimes \tilde{V}_0)$ with $\theta = 0$, so $F_\theta = 0$.

In Chern–Weil theory, de Rham cohomology classes are represented by invariant polynomials in the curvature. Here we review the basic result in the current formalism. Let $\tau\colon M_n(\mathbf{C}) \to \mathbf{C}$ be the usual matrix trace, and the Chern polynomials are defined as in Section 5.9.

Proposition 8.2.1 (Weil) *Let E be a complex vector bundle over M with fibres linearly homeomorphic to $\mathbf{C}^m$ and with connection ∇.*

(i) *Then $\tau\big((\nabla^2)^n\big) \in \Omega^{2n}(M, \mathbf{C})$ is closed for the de Rham differential d over M.*

(ii) *The de Rham cohomology class of $\tau\big((\nabla^2)^n\big)$ in*

$$H^{2n}_{DR}(M) = Ker(d\colon \Omega^{2n} \to \Omega^{2n+1})/Im(d\colon \Omega^{2n-1} \to \Omega^{2n}) \qquad (8.2.1)$$

is independent of the choice of connection.

Proof (i) This is a local calculation, so we can take $\nabla = d + \theta$ on $C^\infty(M; T^{*p} \wedge \mathbf{C}^m)$ by $d\tau\alpha = \tau[\nabla, \alpha]$. Indeed, $\tau[\theta, \alpha] = 0$, so we have

$$d\tau\alpha = \tau d\alpha = \tau[d, \alpha] = \tau[d + \theta, \alpha] = \tau[\nabla, \alpha] \qquad (0 \le t \le 1); \qquad (8.2.2)$$

so in particular, $d\tau((\nabla^2)^n) = \tau[\nabla, \nabla^{2n}] = 0$.

(ii) The space of connections with fibre V is convex in the sense that for ∇_0 and ∇_1 satisfying (8.1.1), the sum $\nabla_t = (1-t)\nabla_0 + t\nabla_1$ also satisfies (8.1.1) for all $0 \le t \le 1$. In particular, we can choose ∇_1 for our original connection and ∇_0 for the trivial connection. Then $d\tau(\nabla_1^{2n})$ and $d\tau(\nabla_0^{2n})$ represent the de Rham cohomology classes associated with ∇_1 and the trivial connection, but they are cohomologous since

$$\tau\left(\nabla_1^{2n}\right) - \tau\left(\nabla_0^{2n}\right) = d\left(\int_0^1 n\tau\left(\left(\nabla_t^2\right)^{n-1} \dot{\nabla}_t\right) dt\right), \qquad (8.2.3)$$

where we write $\dot{}$ for $\partial/\partial t$. To see this, we compute

$$\frac{\partial}{\partial t}\tau\left((\nabla_t^2)^n\right) = \tau \sum_{i=1}^{n} (\nabla_1^2)^i \left(\frac{\partial}{\partial t}\nabla_t^2\right) (\nabla_1^2)^{n-i-1}$$

$$= \tau \sum_{i=1}^{n} (\nabla_t^2)^i [\nabla_t, \dot{\nabla}_t] (\nabla_t^2)^{n-i-1}$$

$$= \tau \left[\nabla_t, \sum_{i=1}^{n} (\nabla_t^2)^i \dot{\nabla}_t (\nabla_t^2)^{n-i-1} \right] \tag{8.2.4}$$

since $[\nabla_t, \nabla_t^2] = 0$. By (8.2.2), we obtain

$$\frac{\partial}{\partial t} \tau \Big((\nabla_t^2)^n \Big) = d\tau \left(\sum_{i=1}^{n} (\nabla_t^2)^i \dot{\nabla}_t (\nabla_t^2)^{n-i-1} \right)$$

$$= n d\tau \Big((\nabla_t^2)^{n-1} \dot{\nabla}_t \Big). \tag{8.2.5}$$

Integrating this expression, we recover (8.2.3). □

This calculation is important in cyclic theory as it suggests the formulas which arise in the noncommutative case. In Sections 8.3 and 8.5, we obtain homotopy formulas for deforming cyclic cochains. Also, we can make a basic definition in the classical case of de Rham homology of the commutative algebra $C^\infty(M)$.

Let E be a complex vector bundle over a real smooth manifold M with connection ∇. Then we define $\tau(e^{\nabla^2})$ to be the formal series in ∇^2, as in

$$\tau\big(e^{\nabla^2}\big) = \sum_{n=0}^{\infty} \frac{\tau(\nabla^{2n})}{n!} \tag{8.2.6}$$

with the nth summand in $H_{DR}^{2n}(M)$. The curvature is evaluated at each point in M, then each summand in (8.2.6) is integrated over M with respect to the volume measure. Suppose that the curvature form is F; then the nth summand is

$$\frac{\tau(\nabla^{2n})}{n!} = \frac{1}{n!} \sum F_{j_1, j_2} \wedge F_{j_2, j_3} \wedge \cdots \wedge F_{j_n, j_1} \in H_{DR}^{2n}(M)$$

by the discussion in [63]. The determinant rank of a matrix F is the number of rows in the largest square minor of F that has a non-zero determinant; the determinant rank is equal to the row rank. Suppose E has fibres $\mathbf{C} \otimes T_x^* M$ at $x \in M$, and $F \in \mathrm{End}(E)$; then for n greater than the dimension of M, we have $\tau(\nabla^{2n}) = 0$, so the formal series terminates.

Normalization 8.2.2 (i) As in Example 6.2.4(iii), the tangent bundle to the sphere in $\mathbf{R}^3$ is topologically non-trivial. Take the product vector bundle $\mathbf{P}^1 \times \mathbf{C}^2$ over the complex projective line or Riemann sphere $\mathbf{P}^1 = \mathbf{C} \cup \{\infty\}$, and for each $z \in \mathbf{P}^1$, we assign a line L_z in $\mathbf{C}^2$. This gives the tautological line bundle

L over $\mathbf{P}^1$, often denoted $\theta(-1)$; see [63]. The normalization axiom 4 in [63] requires that the first Chern class $c_1(L)$, when integrated over the fundamental 2-cycle $\mathbf{P}^1$, is equal to -1; equivalently $-c_1(L)$ is the generator of $H^2(\mathbf{P}^1; \mathbf{Z})$. Thus we tidy up the theory by having integer values for the c_1 classes.

(ii) To interpret the constants that we require, we let $L = \{(\zeta, \eta) \in \mathbf{C}^2\}$, so that (ζ, η) are the homogeneous coordinates of a point in $\mathbf{P}^1$, and the inhomogeneous coordinate is $z = \eta/\zeta$. The curvature form is

$$F = -\frac{dz \wedge d\bar{z}}{(1 + |z|^2)^2}.$$

Converting to polar coordinates $z = re^{i\theta}$, we compute

$$\int_{\mathbf{P}^1} F = \int_0^\infty \int_0^{2\pi} \frac{2irdrd\theta}{(1 + r^2)^2} = 2\pi i.$$

In order to be consistent with axiom 4, we introduce a factor of $-1/2\pi i$ and use $c_1(L) = -F/(2\pi i)$ instead of F. This differs in sign from the formulas arising from Cauchy's residue theorem, but bearing this in mind, we make the following definition.

Definition 8.2.3 (Chern character) The Chern character of E with connection ∇ is

$$\mathrm{ch}(E) = \tau\big(e^{-\nabla^2/2\pi i}\big) = \sum_{n=0}^{\infty} \frac{c_n(E)}{n!} \in \prod_{n=0}^{\infty} H^{2n}_{DR}(M), \tag{8.2.7}$$

where the nth Chern class is $c_n(E) = \tau((-\nabla^2/(2\pi i))^n)$, and the total Chern class is

$$c(E) = \det\left(I - \frac{\nabla^2}{2\pi i}\right) = 1 + c_1(E) + \cdots.$$

We also define the direct sum of de Rham cohomology groups with even index and only finitely many non-zero terms,

$$H^{ev}_{DR}(M) = \bigoplus_{n=0}^{\infty} H^{2n}_{DR}(M).$$

Corollary 8.2.4 *The Chern character determines a ring homomorphism*

$$\mathrm{ch}\colon K_0(C^\infty(M)) \to H^{ev}_{DR}(M) \tag{8.2.8}$$

so that vector bundles E_1 and E_2 satisfy

(i) $ch(E_1 \oplus E_2) = ch(E_1) + ch(E_2)$;

(ii) $ch(E_1 \otimes E_2) = ch(E_1)ch(E_2)$.

Proof (i) In Proposition 8.2.1(ii), it was shown that the summands in (8.2.6) are independent of the particular choice of ∇, so $\mathrm{ch}(E)$ depends only upon E. For instance, one can use the trivial connection to compute this. Since E is finite dimensional, the series terminates, so $\mathrm{ch}\,(E) \in H^{\bullet}_{DR}(M, \mathbf{Q})$. In Proposition 2.9.12, we saw that $K_0(A)$ is a unital and commutative ring for any unital and commutative algebra A. In Exercise 5.6.3 we saw that the addition rule (i) holds. See [63] page 33 for further discussion of ch on complex vector bundles. For example, the tangent bundle over the elliptic curve (torus) is trivial as in [7, p. 312] and Example 6.1.4.

(ii) We follow [54]. By a general result in topology known as the splitting principle, we can assume that there are finite direct sums $E_1 = \oplus_j L_1^j$ and $E_2 = \oplus_j L_2^j$ where L_1^j and L_2^j are line bundles. Then the tensor becomes a sum of line bundles

$$E_1 \otimes E_2 = \sum_{j,k} L_1^j \otimes L_2^k$$

and by (i), we have

$$\mathrm{ch}(E_1 \otimes E_2) = \sum_{j,k} \mathrm{ch}\left(L_1^j \otimes L_2^k\right), \tag{8.2.9}$$

where $L_1^j \otimes L_2^k$ is a line bundle and $c_1(L_1^j \otimes L_2^k) = c_1(L_1^j) + c_2(L_2^k)$ by Section 5.9.1. Also $c_n(L_1^j) = 0$ for all $n \geq 2$ since L_1^j is a line bundle. Hence $\mathrm{ch}(L_1^j) = e^{c_1(L_1^j)}$ and we deduce that

$$\begin{aligned}\mathrm{ch}(E_1 \otimes E_2) &= \sum_{j,k} e^{c_1(L_1^j)+c_1(L_2^k)}\\ &= \sum_{j,k} \mathrm{ch}(L_1^j)\mathrm{ch}(L_2^k)\\ &= \sum_{j} \mathrm{ch}(L_1^j)\sum_{k} \mathrm{ch}(L_2^k)\\ &= \mathrm{ch}(E_1)\mathrm{ch}(E_2),\end{aligned} \tag{8.2.10}$$

where we have used (i) at the final step. □

Example 8.2.5 (Characteristic classes) (i) Let

$$\det(1 + s\theta) = \sum_{j=0}^{m} s^j \varphi_j(\theta)$$

for $\theta \in M_m(\mathbf{C})$, so in particular $\varphi_1(\theta) = \mathrm{trace}(\theta)$ and $\varphi_m(\theta) = \det\theta$. Then $\varphi_j(U\theta U^{-1}) = \varphi_j(\theta)$ for all $U \in GL_m(\mathbf{C})$ and $\varphi_j(s\theta) = s^j\varphi_j(\theta)$. Then φ_j is an invariant homogeneous polynomial of degree j in the eigenvalues of θ, and we can apply these φ_j to ∇^2 and F_θ^2. The classes are evaluated at each $x \in M$, then integrated over M with respect to μ.

(ii) Consider the connections $\nabla_\theta = d + \theta$ and $\nabla_{\alpha+\theta} = d + \alpha + \theta$. The 4-forms associated with the curvatures satisfy

$$\tau F_{\theta+\alpha}^2 - \tau F_\theta^2 = d\Big(\tau\big(\alpha \wedge \nabla_\theta \alpha + (2/3)\alpha \wedge \alpha \wedge \alpha\big)\Big), \tag{8.2.11}$$

thus $\tau F_{\theta+\alpha}^2$ and τF_θ^2 represent the same class in $H^4(M)$. This example is of particular importance in Chern–Simons theory; see [28].

To compute the first Chern class in a significant specific case, we look at the case of a one-dimensional complex manifold, namely a compact Riemann surface $\mathbf{X}$, and consider holomorphic vector bundles with fibres isomorphic to $\mathbf{C}$, namely line bundles. This example is important since calculations can be carried out explicitly using tools from classical algebraic geometry; see [7; Lemma 6.6]. This situation is traditionally described in terms of divisors. See [96].

Definition 8.2.6 (Divisors) (i) A divisor on a compact Riemann surface $\mathbf{X}$ is a finitely supported function $D: \mathbf{X} \to \mathbf{Z}$, and the space $\mathcal{D}(\mathbf{X})$ of such divisors forms a group under pointwise addition. With $\varepsilon_x(y) = 1$ for $x = y$ and $\varepsilon_x(y) = 1$ for $x \neq y$, we can write $D = \sum_j n_j \varepsilon_{x_j}$ and define the degree $\delta(D) = \sum_j n_j$. Thus we obtain an exact sequence

$$0 \to \mathcal{D}_0(\mathbf{X}) \to \mathcal{D}(\mathbf{X}) \to \mathbf{Z} \to 0,$$

where $\mathcal{D}_0(\mathbf{X})$ is the subgroup of divisors of degree zero.

(ii) The meromorphic functions on a compact Riemann surface gives a commutative algebra $\mathbf{K}(X)$, and by Liouville's theorem, each non-constant $f \in \mathbf{K}(X)$ determines a function $f: \mathbf{X} \to \mathbf{P}^1$ with a non-empty set of zeros $f^{-1}(0)$ and non-empty set of poles $f^{-1}(\infty)$. For a non-constant meromorphic function $f \in \mathbf{K}(\mathbf{X})$ with zeros of order m_j at $z_j \in \mathbf{X}$ and poles of order n_j at $p_j \in \mathbf{X}$, there is a principal divisor $D(f) = \sum_j m_j \varepsilon_{z_j} - \sum_j n_j \varepsilon_{p_j}$; we interpret this sum to be zero when f is constant. By a theorem of Abel, the set $\mathcal{D}_P(\mathbf{X}) = \{D(f) : f \in \mathbf{K}(\mathbf{X})\}$ of principal divisors forms a subgroup of $\mathcal{D}_0(\mathbf{X})$; see [96, p. 141]. We say that divisors D_1 and D_2 are linearly equivalent if $D_1 - D_2 = D(f)$ for some meromorphic f.

It is helpful to think of ε_z as the unit point charge at $z \in \mathbf{X}$ and $\sum_j n_j \varepsilon_{x_j}$ as a signed measure with total charge $\sum_j n_j$. Indeed, with this interpretation for $f \in \mathbf{K}(\mathbf{X})$, the distributional Laplacian $\Delta = -\mathrm{div}\,\mathrm{grad}$ satisfies

$$D(f) = -\Delta(2\pi)^{-1}\log|f(z)| = \sum_j m_j\varepsilon_{z_j} - \sum_j n_j\varepsilon_{p_j} \tag{8.2.12}$$

by the calculation of Exercise 5.3.4.

(iii) For any divisor $D = \sum_j n_j\varepsilon_{x_j}$ with $n_j \geq 0$ for all j, we write $D \geq 0$. If $(f) \geq 0$, then f is constant. Nevertheless, for any non-empty finite $S \subset \mathbf{X}$, the set of meromorphic functions with poles only in S gives a ring $\Theta_S = \{f \in \mathbf{K}[\mathbf{X}] : (f)|_{\mathbf{X}\setminus S} \geq 0\}$ to which one can apply Proposition 2.9.5.

Example 8.2.7 (The Picard group for a compact Riemann surface) (i) We can consider projective modules over subalgebras of the commutative algebra $\mathbf{K}(\mathbf{X})$, using the methods of Proposition 2.9.5 as in [92, p. 141].

(ii) The quotient group $C\ell(\mathbf{X}) = \mathcal{D}(\mathbf{X})/\mathcal{D}_P(\mathbf{X})$ is the group of equivalence classes of divisors, and is called the divisor class group. One can show that every equivalence class of divisors defines an equivalence class of holomorphic line bundles θ and conversely; see [96, p. 277]. So $C\ell(\mathbf{X})$ may be identified with the multiplicative group of non-zero holomorphic line bundles θ^* on $\mathbf{X}$, which coincides with the Picard group as in Exercise 2.9.17. Theorem 3 page 141 of [96] shows how the various interpretations of $\mathrm{Pic}(\mathbf{X})$ are reconciled, and deduces that $C\ell(\mathbf{X}) \cong H^1(\mathbf{X};\theta^*)$; we call this $\mathrm{Pic}(\mathbf{X})$.

(iii) The quotient group $C\ell^0(\mathbf{X}) = \mathcal{D}_0(\mathbf{X})/\mathcal{D}_P(\mathbf{X}) = \mathrm{Pic}_0(\mathbf{X})$ is the Jacobian of $\mathbf{X}$, which we realize in terms of holomorphic differentials and abelian integrals, as in Exercise 6.1.9. To any continuous $\gamma : [0,1] \to \mathbf{X}$, with $\gamma(0) = y$ and $\gamma(1) = x$, we associate $\varepsilon_x - \varepsilon_y \in \mathcal{D}_0(\mathbf{X})$. Likewise, to any $\gamma = \sum_j \gamma_j$ with continuous $\gamma_j : [a_j, b_j] \to \mathbf{X}$ and $\gamma_j(a_j) = y_j$ and $\gamma_j(b_j) = x_j$, we associate $\sum_j(\varepsilon_{x_j} - \varepsilon_{y_j}) \in \mathcal{D}_0(\mathbf{X})$. Conversely, given $D \in \mathcal{D}_0(\mathbf{X})$, we can write $D = \sum_j(\varepsilon_{x_j} - \varepsilon_{y_j})$ and introduce some continuously differentiable $\gamma_j : [a_j, b_j] \to \mathbf{X}$ such that $\gamma_j(a_j) = y_j$ and $\gamma_j(b_j) = x_j$. Recall from Proposition 7.1.2 that the space of holomorphic differentials $\Omega^1_{\mathbf{X}}$ has a basis $(\omega_k)_{k=1}^g$. Then we define a map

$$\Phi : \mathcal{D}_0(\mathbf{X}) \to \mathbf{C}^g : \sum_j(\varepsilon_{x_j} - \varepsilon_{y_j}) \mapsto \left(\sum_j \int_{\gamma_j} \omega_k\right)_{k=1}^g. \tag{8.2.13}$$

The image depends upon the choice of γ_j, but we proceed to deal with this matter. Let Λ be as in Corollary 7.1.3. Then by Abel's theorem [34, p. 88], D is the divisor of some $f \in \mathbf{K}(\mathbf{X})$ if and only if the degree of D equals zero and $\Phi(D) \in \Lambda$; this identifies the image of $\mathcal{D}_P(\mathbf{X})$ with Λ in $\mathbf{C}^g$. The x_j can be permuted without changing the image, so for fixed y_j, the map Φ is defined on $\mathbf{X}^g/\sim$, where $\sim$ indicates the action by permutations of the symmetric group on g symbols. By considering the derivative of Φ, one can show that the map Φ is surjective. Hence $\mathrm{Pic}_0(\mathbf{X}) \cong \mathbf{C}^g/\mathbf{Z}^{2g}$. See [36] for details.

For $g > 1$, it is important to note that Λ is a special lattice such that $\mathbf{C}^g/\Lambda$ inherits a complex structure; see [76]. In particular, if $\mathbf{X}$ is a hyperelliptic curve of genus g, Mumford constructs a special meromorphic function $\wp$ on the Jacobian $Jac(\mathbf{X})$ and an invariant vector field D_∞ on $Jac(\mathbf{X})$. Then $\wp$ has poles on $\Theta \subset \mathbf{X}$ and derivatives $\wp^{(1)} = D_\infty \wp, \dots, \wp^{(2g-1)} = D_\infty^{(2g-1)}\wp$ such that $\wp, \wp^{(1)}, \dots, \wp^{(2g-1)}$ generate the affine ring of $Jac(\mathbf{X}) \setminus \Theta$. As in theorem 10.3 of [79], this gives an embedding $Jac(\mathbf{X}) \setminus \Theta \to \mathbf{C}^{2g}$.

(iv) We pursue this further by momentarily assuming some results from sheaf theory, as in [36, p. 123]. Now we loosen up and allow C^∞ sections, so we can patch together sections of line bundles on $\mathbf{X}$ by means of C^∞ partitions of unity on $\mathbf{X}$. Let U be an open neighbourhood in $\mathbf{X}$ and $f : U \to \mathbf{C}$ a nowhere vanishing C^∞ function. Then there exists a neighbourhood $V \subseteq U$ and a C^∞ function $g : V \to \mathbf{C}$ such that $f = e^{2\pi i g}$. Hence there is a short exact sequence of groups

$$0 \longrightarrow \mathbf{Z} \overset{\iota}{\longrightarrow} \theta \overset{e^{2\pi i \cdot}}{\longrightarrow} \theta^* \longrightarrow 0. \tag{8.2.14}$$

This gives rise to an exact sequence of cohomology groups

$$H^1(\mathbf{X};\mathbf{Z}) \longrightarrow H^1(\mathbf{X};\theta) \longrightarrow H^1(\mathbf{X};\theta^*) \overset{\delta}{\longrightarrow} H^2(\mathbf{X};\mathbf{Z}) \longrightarrow H^2(\mathbf{X};\theta), \tag{8.2.15}$$

so the connecting map δ gives a group homomorphism $\delta : H^1(\mathbf{X};\theta^*) \to H^2(\mathbf{X};\mathbf{Z})$ which gives a version of the first Chern class $c_1(E) = \delta(E) \in H^2(\mathbf{X};\mathbf{Z})$ for $E \in H^1(\mathbf{X};\theta^*)$ by [63, p. 33]. Then by general results about fine sheaves, we have $H^1(\mathbf{X};\theta) = H^2(\mathbf{X};\theta) = 0$, so in the C^∞ category there is an isomorphism of groups $\delta : H^1(\mathbf{X};\theta^*) \to H^2(\mathbf{X};\mathbf{Z})$.

For a compact Riemann surface $\mathbf{X}$ of genus g, there are four interpretations of the Picard group:

(i) $\mathrm{Pic}(\mathbf{X})$ is the multiplicative group $H^1(\mathbf{X};\theta^*)$ of complex line bundles over $\mathbf{X}$;

(ii) $\mathrm{Pic}(\mathbf{X})$ is the divisor class group, with subgroup $\mathrm{Pic}_0(\mathbf{X})$;

(iii) $\mathrm{Pic}_0(\mathbf{X})$ is isomorphic to the Jacobi variety $\mathbf{C}^g/\Lambda \cong \mathbf{X}^g/\sim$;

(iv) $\mathrm{Pic}(\mathbf{X})$ is the group of first Chern classes $c_1(E)$ in $H^2(\mathbf{X};\mathbf{Z})$.

Example 8.2.8 (Second Chern class) Donaldson and Kronheimer [28] give a detailed discussion of the second Chern class c_2 for a four-dimensional manifold.

8.3 Deforming Flat Connections

Suppose that $V = E \oplus E'$ is a linear direct sum of $\mathbf{C}$-vector spaces with a specified inner product. Usually we take $\dim E = \dim E'$, so $\dim V$ is even. An involution is $\varepsilon \in \mathrm{End}(V)$ such that $\varepsilon^2 = id$. We have embeddings ι and j with corresponding projections ι^* and $j*$, as in the following array:

$$E \underset{\iota}{\overset{\iota^*}{\leftrightarrows}} V \underset{j}{\overset{j^*}{\rightleftarrows}} E', \tag{8.3.1}$$

which may be expressed as a block matrix

$$I = \begin{bmatrix} \iota^* & 0 \\ 0 & j^* \end{bmatrix} \begin{bmatrix} \iota & 0 \\ 0 & j \end{bmatrix} \quad \begin{matrix} E \\ E'. \end{matrix} \tag{8.3.2}$$

We write d for the trivial connection for the bundle $V \times M \to M$ with fibre V, and introduce connections

$$\nabla = \iota^* d\iota, \quad \nabla' = j^* dj, \tag{8.3.3}$$

so we can express ∇^2 and likewise $(\nabla')^2$ as the product of two tensors

$$\nabla^2 = (\iota^* d\iota)(\iota^* d\iota) = \iota^* d(1 - jj^*)d\iota = -(\iota^* dj)(jd\iota^*), \tag{8.3.4}$$

where $\iota^* dj \in \Omega^1(M, \mathrm{Hom}(E', E))$ and $j^* d\iota \in \Omega^1(M, \mathrm{Hom}(E, E'))$.

We change notation and suppose that we have a (flat) connection D where

$$\begin{aligned} D &= \begin{bmatrix} \iota^* D\iota & 0 \\ 0 & j^* Dj \end{bmatrix} + \begin{bmatrix} 0 & \iota^* Dj \\ j^* D\iota & 0 \end{bmatrix} \\ &:= \nabla + \alpha \end{aligned} \tag{8.3.5}$$

and

$$\varepsilon = \begin{bmatrix} \iota^* \iota & 0 \\ 0 & -j^* j \end{bmatrix},$$

so that $\varepsilon^2 = I$, $\varepsilon\nabla = \nabla\varepsilon$ and $\alpha\varepsilon = -\varepsilon\alpha$. Then

$$\begin{aligned} D^2 &= \nabla^2 + \nabla\alpha + \alpha\nabla + \alpha^2 \\ &= \nabla^2 + \alpha^2 + [\nabla, \alpha], \end{aligned} \tag{8.3.6}$$

where $\nabla^2 + \alpha^2$ commutes with ε, whereas $[\nabla, \alpha]$ anti-commutes with ε. If $D^2 = 0$, then $\nabla^2 + \alpha^2 = 0 = [\nabla, \alpha]$. Suppose that $(u_t)_{t \in [0,1]}$ is a continuously differentiable family of unitary operators on V. Then $\varepsilon_t = u_t \varepsilon u_t^*$ is a continuously differentiable family of involutions on V, so the complementary subspaces E and E' change as t varies. We wish to investigate the effect on the cohomology classes.

Proposition 8.3.1 *Let D be a flat connection on a bundle V over m with fibre $\mathbf{C}^m$ and ε an endomorphism of V such that $\varepsilon^2 = id$. Then*

$$D = \nabla + \alpha, \tag{8.3.7}$$

where (i) $\nabla = (1/2)(D + \varepsilon D\varepsilon)$ *is a connection which commutes with* ε*;*

(ii) $\alpha = (1/2)(D - \varepsilon D\varepsilon)$ *is a* 1*-form anti-commuting with* ε*;*

(iii) *the forms* $\mathrm{tr}(\alpha^{2n})$ *and* $\mathrm{tr}(\varepsilon\alpha^{2n})$ *are closed and their de Rham cohomology class does not change as* ε *is changed smoothly.*

Proof By flatness, we have

$$0 = D^2 = \nabla^2 + \nabla\alpha + \alpha\nabla + \alpha^2 = \nabla^2 + \alpha^2 + [\nabla,\alpha], \tag{8.3.8}$$

where $\nabla^2 + \alpha^2$ commutes with ε, whereas $[\nabla,\alpha]$ anti-commutes with ε. Hence $\nabla^2 + \alpha^2 = 0 = [\nabla,\alpha]$, so

$$d\mathrm{tr}\big((\nabla^2)^n\big) = \mathrm{tr}[\nabla,(\nabla^2)^n] = 0, \tag{8.3.9}$$

so $\mathrm{tr}(\nabla^2)^n$ is closed. Hence

$$\mathrm{tr}(\nabla^2)^n = \mathrm{tr}(-\alpha^2)^n = (-1)^n\mathrm{tr}\alpha^{2n} = 0, \tag{8.3.10}$$

since $\mathrm{tr}\,\alpha^{2n} = \mathrm{tr}\,\alpha\alpha^{2n-1} = -\mathrm{tr}\,\alpha^{2n-1}\alpha = -\mathrm{tr}\,\alpha^{2n}$.

Also

$$d\mathrm{tr}(\varepsilon\alpha^{2n}) = \mathrm{tr}\,d(\varepsilon\alpha^{2n}) = \mathrm{tr}([\nabla,\varepsilon\alpha^{2n}]) = 0, \tag{8.3.11}$$

since $[\alpha,\nabla] = 0 = [\nabla,\varepsilon]$.

Now suppose that $\varepsilon = \varepsilon_t$ is a smoothly varying family of involutions on the bundle V with connection D fixed. (We can easily produce these by taking a one-parameter subgroup of $u_t \in U(m)$ and letting $\varepsilon_t = u_t\varepsilon_0 u_t^*$.) Then we have $\varepsilon\dot{\varepsilon} + \dot{\varepsilon}\varepsilon = 0$, so

$$\begin{aligned}\dot{\alpha} &= (-1/2)(\dot{\varepsilon}D\varepsilon + \varepsilon D\dot{\varepsilon})\\ &= (-1/2)(\dot{\varepsilon}(\nabla+\alpha)\varepsilon + \varepsilon(\nabla+\alpha)\dot{\varepsilon})\\ &= (-1/2)(\dot{\varepsilon}\nabla\varepsilon + \varepsilon\nabla\dot{\varepsilon}) + (1/2)(\dot{\varepsilon}\alpha\varepsilon + \varepsilon\alpha\dot{\varepsilon}),\end{aligned} \tag{8.3.12}$$

where the first summand anti-commutes with ε, whereas the final summand commutes with ε. The anti-commuting part is $(-1/2)[\nabla,\varepsilon\dot{\varepsilon}]$.

Now

$$\begin{aligned}\frac{\partial}{\partial t}\mathrm{tr}(\varepsilon\alpha^{2n}) &= tr(\dot{\varepsilon}\alpha^{2n}) + \sum_{i=1}^{n}\mathrm{tr}(\varepsilon\alpha^i\dot{\alpha}\alpha^{2n-1-i})\\ &= \mathrm{tr}(\dot{\varepsilon}\alpha^{2n}) + 2n\mathrm{tr}(\varepsilon\dot{\alpha}\alpha^{2n-1}).\end{aligned} \tag{8.3.13}$$

The first summand here vanishes, while the second depends only upon the part of $\dot{\alpha}$ that anti-commutes with ε. Hence

$$\begin{aligned}\frac{\partial}{\partial t}\mathrm{tr}(\varepsilon\alpha^{2n}) &= (-1/2)2n\mathrm{tr}(\varepsilon[\nabla,\varepsilon\dot{\varepsilon}]\alpha^{2n-1})\\ &= -n\mathrm{tr}([\nabla,\dot{\varepsilon}\alpha^{2n-1}])\\ &= -nd\mathrm{tr}(\dot{\varepsilon}\alpha^{2n-1}),\end{aligned}\tag{8.3.14}$$

since $[\alpha,\nabla]=0=[\nabla,\varepsilon]$, which gives an exact differential form, so

$$\tau(\varepsilon\alpha^{2n})_{t=1}=\tau(\varepsilon\alpha^{2n})_{t=0}-d\int_0^1 n\tau(\dot{\varepsilon}\alpha^{2n-1})\,dt,\tag{8.3.15}$$

and the cocycles $\tau(\varepsilon\alpha^{2n})_{t=1}$ and $\tau(\varepsilon\alpha^{2n})_{t=0}$ are cohomologous. □

Example 8.3.2 Let $A=C^\infty(M)$ and let $e\in M_r(A)$ be an idempotent, so we can introduce a vector bundle over M as the range of e. Then $\mathbf{C}I+\mathbf{C}e$ is an algebra generated by one idempotent as in Example 7.4.3, and we consider

$$\frac{c_n([e])}{n!}=(-1)^{n-1}\frac{(2n)!}{2(n!)}\mathrm{trace}(ede\dots de)\in\Omega_A^{2n},\tag{8.3.16}$$

where the image depends on the class $[e]$ of e in $K_0(A)$. As in Corollary 8.2.4, the sum of these terms gives an element in $H_{DR}^{ev}(M)$. The choice of constants is given by Example 7.4.3. See section 8.3 of [68] for further discussion.

8.4 Universal Differentials

We start by considering universal properties of ΩR, and then use this as a tool for studying connections over manifolds.

Definition 8.4.1 (Derivation) With R an algebra and M an R-bimodule, a derivation $D\colon R\to M$ is a $\mathbf{k}$-linear map such that

$$D(xy)=x(Dy)+(Dx)y.\tag{8.4.1}$$

Example 8.4.2 (Inner derivations) Let $m\in M$ and let D be the inner derivation $D(x)=xm-mx=[x,m]$.

To clarify the various notations, we write

$$Z^1(R,M)=\mathrm{Der}(R,M)=\mathrm{Hom}_{R\otimes R^{op}}(\Omega^1R,M).\tag{8.4.2}$$

We also let $b'\colon R^{\otimes n}\to R^{\otimes n+1}$ be

$$b'(x_1,\dots,x_n)=\sum_{i=1}^n(-1)^{i-1}(x_1,\dots,x_ix_{i+1},\dots,x_n)\tag{8.4.3}$$

and $s\colon R^{\otimes n} \to R^{\otimes n+1}$ be

$$s(x_1, \dots, x_n) = (1, x_1, \dots, x_n). \tag{8.4.4}$$

Proposition 8.4.3 (i) *Then b' is an R-bimodule map such that $b'^2 = 0$;*
(ii) *s is a right R-module map such that $b's + sb' = id$;*
(iii) *the following sequence is exact:*

$$\dots \xrightarrow{b'} R^{\otimes 4} \xrightarrow{b'} R^{\otimes 3} \xrightarrow{b'} R^{\otimes 2} \xrightarrow{b'} R \longrightarrow 0. \tag{8.4.5}$$

Let $\Omega^1 R = \mathrm{Ker}(b'\colon R^{\otimes 2} \to R)$ which may also be expressed as $\mathrm{Im}(b'\colon R^{\otimes 3} \to R^{\otimes 2})$ or as $\mathrm{Coker}(b'\colon R^{\otimes 4} \to R^{\otimes 3})$. In the following result, we use the latter interpretation.

Proposition 8.4.4 (i) *$\Omega^1 R$ is an R-bimodule, and $\partial\colon R \to \Omega^1 R\colon \partial y = (1, y, 1)$ satisfies*

$$f\partial(gh)k = f(\partial g)hk + f(g\partial h)k. \tag{8.4.6}$$

(ii) *The map $Hom_{R\otimes R}(\Omega^1 R, M) \to Der(R, M)\colon u \mapsto u \circ \partial$ is surjective.*

Proof (i) We have

$$b'(x, y, z, w) = (xy, z, w) - (x, yz, w) + (x, y, zw), \tag{8.4.7}$$

so

$$\partial(yz) = (1, yz, 1) = (y, z, 1) + (1, y, z) \quad \mathrm{mod}(b'R^{\otimes 4}), \tag{8.4.8}$$

so

$$\partial(yz) = y(\partial z) + (\partial y)z. \tag{8.4.9}$$

(ii) For any D there is a unique u such that

$$\begin{array}{ccc} R & \xrightarrow{D} & M \\ & \partial \searrow & \uparrow u \\ & & \Omega^1 R \end{array} \tag{8.4.10}$$

We take the first-order differentials in $\Omega^1 R = R^{\otimes 3}/b'R^{\otimes 4}$. The sequence

$$0 \longrightarrow \Omega^1 R \xrightarrow{j} R \otimes R \xrightarrow{b'} R \longrightarrow 0 \tag{8.4.11}$$

is exact, where $j\colon x(\partial y)z \mapsto (xy, z) - (x, yz)$ is the bimodule map which corresponds to the derivation $y \mapsto y \otimes 1 - 1 \otimes y$ and j is induced by $b'\colon R^{\otimes 3} \to R^{\otimes 2}$. □

8.5 Connections on Modules over an Algebra

Having reviewed the concept of a connection in differential geometry in Section 8.1, we proceed to introduce a connection for noncommutative differential forms.

Definition 8.5.1 (Right connection) Let E be an R-bimodule, where R is a unital algebra over $\mathbf{C}$. A right connection is a linear map $\nabla\colon E \to E \otimes_R \Omega^1 R$ that satisfies the Leibniz rule

$$\begin{aligned} \nabla(x\xi) &= x\nabla\xi, \\ \nabla(\xi x) &= (\nabla\xi)x + \xi dx \qquad (\xi \in E, x \in R) \end{aligned} \tag{8.5.1}$$

for the right module operation.

Lemma 8.5.2 *Let R be quasi-free. Then there exists a right connection $\nabla\colon \Omega^1 R \to \Omega^2 R$.*

Proof We have an exact sequence of R-bimodules

$$0 \longrightarrow \Omega^2 R \overset{j}{\longrightarrow} \Omega^1 R \otimes R \overset{\mu}{\longrightarrow} \Omega^1 R \longrightarrow 0, \tag{8.5.2}$$

where $j(\alpha dx) = \alpha x \otimes 1 - \alpha \otimes x$ and $\mu(\alpha \otimes x) = \alpha x$. Now $\Omega^1 R \otimes R \cong R \otimes \bar{R} \otimes R$ is a free R-bimodule, and $\Omega^1 R$ is projective by Definition 6.4.7(iii), so there is a splitting map $s\colon \Omega^1 R \to \Omega^1 R \otimes R$ such that

$$j(\nabla\xi) = \xi \otimes 1 - s(\xi). \tag{8.5.3}$$

Now

$$j\big(\nabla(\xi a) - (\nabla\xi)a\big) = \xi a \otimes 1 - \xi \otimes a - s(\xi a) + s(\xi)a \tag{8.5.4}$$

or equivalently

$$j\big(\nabla(\xi a) - (\nabla\xi)a - \xi da\big) = s(\xi a) - s(\xi)a, \tag{8.5.5}$$

so s is a right R-module map if and only if ∇ is a right connection. □

Example 8.5.3 (Noncommutative tangents) Connections on the A-bimodule $E = \Omega^1 A$ are the noncommutative analogues of connections on the tangent bundle TM to a manifold M. With $A = C^\infty(M)$, and V a vector bundle over M, there is $S_A\Gamma(M, V)$, the symmetric algebra of sections of the bundle V.

Definition 8.5.4 (Cup product) Let L, M, N be A-bimodules and suppose that $L \otimes_A M \to N\colon x \otimes y \mapsto x \cdot y$ is an A-bimodule map. Then the cup product is defined by $C^p(A, L) \otimes C^q(A, M) \to C^{p+q}(A, N)$

$$f(a_1,\dots,a_p)\otimes g(a_{p+1},\dots,a_{p+q})\mapsto f(a_1,\dots,a_p)\cdot g(a_{p+1},\dots,a_{p+q}), \tag{8.5.6}$$

so

$$f\cup g(a_1,\dots,a_{p+q})=f(a_1,\dots,a_p)\cdot g(a_{p+1},\dots,a_{p+q}). \tag{8.5.7}$$

δ is a derivation when

$$\delta(f\cup g)=(\delta f)\cup g+(-1)^{|f|}f\cup\delta g. \tag{8.5.8}$$

Example 8.5.5 (Universal n-cocycle) For $d\in C^1(A,\Omega A)$, the universal derivation, we have $d\cup d\colon (a_1,a_2)\mapsto da_1da_2$. Also $d\cup\dots\cup d$ to n factors is the universal n-cocycle.

Recall from Lemma 6.4.8 that A is quasi-free if and only if Ω^1A is a projective A-bimodule. Also, $\Omega^1A\otimes A=A\otimes\bar{A}\otimes A$ is free and an A-bimodule. The following data exist for A if and only Ω^1A is projective as an A-bimodule, and hence give a further criterion for A to be quasi-free. This completes the proof of Proposition 6.7.4.

Proposition 8.5.6 *The following data are equivalent:*

(i) *a one cochain $\phi\colon\bar{A}\to\Omega^2A$ such that $-\delta\phi=d\cup d$, that is*

$$a_1\phi(a_2)-\phi(a_1a_2)+\phi(a_1)a_2+da_1da_2=0; \tag{8.5.9}$$

(ii) *a lifting homomorphism for $A\to RA/(IA)^2$ (see Corollary 6.7.3);*

(iii) *a bimodule splitting of*

$$0\longrightarrow\Omega^2A\overset{j}{\longrightarrow}\Omega^1A\otimes A\overset{m_r}{\longrightarrow}\Omega^1A\longrightarrow 0,$$

where $j(\omega da)=\omega a\otimes 1-\omega\otimes a$ and $m_r(\omega\otimes a)=\omega a$;

(iv) *a right connection $\nabla_r\colon\Omega^1A\to\Omega^2A$ where $\Omega^2A=\Omega^1A\otimes_A\Omega^1A$.*

Proof *(i)* $\Leftrightarrow$ *(ii)* Recall from Corollary 6.7.3 that $RA/(IA)^2=A\oplus\Omega^2A$ with the Fedosov product. A linear lifting $\phi\colon A\to RA/(IA)^2$ sending $1\mapsto 1$ has the form $a\mapsto a-\phi(a)$, where $\phi\colon A\to\Omega^2A$ is linear. This lifting is a homomorphism for $\star$ when

$$\begin{aligned}a_1a_2-\phi(a_1a_2)&=(a_1-\phi(a_1))\star(a_2-\phi(a_2))\\&=a_1a_2-da_1da_2-\phi(a_1)a_2+d\phi(a_1)da_2-a_1\phi(a_2)\end{aligned} \tag{8.5.10}$$

modulo forms of degree ≥ 4; hence we have a lifting homomorphism if and only if

$$-\phi(a_1a_2)=-da_1da_2-\phi(a_1)a_2-a_1\phi(a_2). \tag{8.5.11}$$

$(i) \Leftrightarrow (iv)$ Let ∇_r be linear and compatible with left multiplication, so $\nabla_r(a_0 da_1) = a_0\phi(a_1)$ where $\phi\colon \bar{A} \to \Omega^2 A$ is linear. Then ∇_r satisfies the Leibniz rule for right multiplication if and only if

$$\nabla_r(\omega a) = \nabla_r(\omega)a + \omega da \tag{8.5.12}$$

if and only if

$$\nabla_r((a_0 da_1)a_2) = \nabla_r(a_0 da_1)a_2 + a_0 da_1 da_2, \tag{8.5.13}$$

that is

$$\nabla_r(a_0 d(a_1a_2)) - \nabla_r(a_0a_1 da_2) = \nabla_r(a_0 da_1)a_2 + a_0 da_1 da_2, \tag{8.5.14}$$

so

$$a_0\phi(a_1a_2) - a_0a_1\phi(a_2) = a_0\phi(a_1)a_2 + a_0 da_1 da_2. \tag{8.5.15}$$

$(i) \Leftrightarrow (iii)$ See Lemma 8.5.2. □

Example 8.5.7 (Tensor algebras) (i) Consider $\nabla\colon \Omega^1 R \to \Omega^1 R \otimes_R \Omega^1 R = \Omega^2 R$ satisfying $\nabla(x\xi) = x\nabla\xi$ and $\nabla(\xi x) = (\nabla\xi)x + \xi dx$. Recall that $R \otimes \bar{R} \otimes R$ is isomorphic to $\Omega^1 R$ via $x \otimes \bar{y} \mapsto xdy$. Then ∇ is determined by $\varphi\colon \bar{R} \to \Omega^2 R$, satisfying $\varphi = \nabla d$ where

$$\nabla(xy) = x\nabla(y) + \varphi(x)y + dxdy. \tag{8.5.16}$$

(ii) Let $R = T(V)$ be the tensor algebra on V, so $R \otimes V \otimes R$ is isomorphic to $\Omega^1 R$ via $x \otimes v \otimes y \mapsto x(dv)y$. Then there is a canonical connection ∇ determined by the requirement that $\nabla dv = 0$ for all $v \in V$. We have an exact sequence

$$0 \longrightarrow \Omega^2 R \longrightarrow \Omega^1 R \otimes R \longrightarrow \Omega^1 R \longrightarrow 0 \tag{8.5.17}$$

and maps $vdy \mapsto v(y \otimes 1 - 1 \otimes y)$ and $(dv)y \otimes 1 \mapsto (dv)y$, with lifting $\ell((dv)y) = dv \otimes y$. Then the requirement $\nabla dv = 0$ gives

$$\nabla(x(dv)y) = x\big(\nabla(dv)y + dvdy\big) = xdvdy, \tag{8.5.18}$$

so there is an extension to $V^{\otimes n}$ for $n > 1$ given by

$$\begin{aligned}\nabla d\big(v_1 \dots v_n\big) &= \nabla \sum_{j=1}^{n} v_1 \dots v_{j-1}(dv_j)v_{j+1} \dots v_n \\ &= \sum_{j=1}^{n-1} v_1 \dots v_{j-1}(dv_j)d(v_{j+1} \dots v_n).\end{aligned} \tag{8.5.19}$$

More generally, suppose that E is a projective R-bimodule. Then there are the following equivalent data:

(i) a right connection $\nabla\colon E \to E \otimes_R \Omega^1 R$;

(ii) a right module splitting of

$$0 \longrightarrow E \otimes_R \Omega^1 R \longrightarrow E \otimes R \longrightarrow E \longrightarrow 0; \tag{8.5.20}$$

(iii) a bimodule section $s\colon E \to E \otimes R$.

Let E be a bimodule over R. A connection on E is a connection on E as a right module such that $\nabla(x\xi) = x\nabla\xi$. These are equivalent to module liftings $\ell\colon E \to E \otimes R$ of the right multiplication $m\colon E \otimes R \to E\colon m(\xi \otimes x) = \xi x$ such that $m \circ \ell = Id$. They exist if and only if E is projective as a bimodule.

Proposition 8.5.8 *Let ∇ be a connection on the bimodule E.*

(i) *Then there is an extension of ∇ to $E \otimes_R \Omega R$ such that*

$$b\nabla + \nabla b = 1 \qquad \text{in degree} \quad j \qquad (j = 1, 2, \ldots). \tag{8.5.21}$$

(ii) *The Hochschild complex is contractible in degree j for $j = 1, 2, \ldots$, and $H_j(R, E) = 0$ for all $j > 0$.*

Proof (i) Given a connection, we can extend ∇ to a degree $(+1)$ operator on $E \otimes \Omega R$ satisfying

$$\nabla(\alpha\omega) = (\nabla\alpha)\omega + (-1)^{|\alpha|}\alpha d\omega \tag{8.5.22}$$

for $\alpha \in E \otimes_R \Omega R$ and $\omega \in \Omega R$.

Let $n > 0$, and consider $E \otimes \Omega^n R$, which is spanned by $\xi dx_1 \ldots dx_n$ with $\xi \in E$ and $x_1, \ldots, x_n$, and on which b is defined by

$$b(\alpha dx) = (-1)^{deg(\alpha)}\big(\alpha x - x\alpha\big). \tag{8.5.23}$$

Then we compute

$$\begin{aligned} \nabla b(\alpha dx) &= (-1)^{|\alpha|}\nabla\big(\alpha x - x\alpha\big) \\ &= (-1)^{|\alpha|}\big((\nabla\alpha)x - (-1)^{|\alpha|}\alpha dx - x\nabla\alpha\big) \end{aligned} \tag{8.5.24}$$

and for comparison

$$\begin{aligned} b\nabla(\alpha dx) &= b\big((\nabla\alpha)dx + \alpha d^2 x\big) \\ &= (-1)^{|\alpha|+1}\big((\nabla\alpha)x - x\nabla\alpha\big), \end{aligned} \tag{8.5.25}$$

so

$$\big(\nabla b + b\nabla\big)(\alpha dx) = \alpha dx. \tag{8.5.26}$$

(ii) This follows from (i). □

Definition 8.5.9 (i) *(Curvature)* One has

$$\nabla^2(\alpha\omega) = (\nabla^2\alpha)\omega,$$

so we define $\nabla^2 : E \to E \otimes_R \Omega^2 R$ to be the curvature.

(ii) *(Chern character classes)* Suppose further that E is finitely generated and projective as a right R-module, so $R^n = E \oplus E'$ for some complementary module, and endomorphisms of E may be regarded as block matrices in $M_n(R)$. Then one can define (Chern character classes)

$$\mathrm{trace}_E\big((\nabla^2)^n\big) \in \Omega^{2n}/[\Omega, \Omega]^{2n}. \tag{8.5.27}$$

In our discussion Normalization 8.2.2 of the Chern character for connections in differential geometry, we introduced a normalizing constant so as to ensure that the Chern classes take integer values. In the noncommutative setting, there is also a natural normalization, chosen so as to fit with results from Kasparov's KK theory. We refer readers to [18] for a statement of the result.

8.6 Derivations and Automorphisms

First, we review the connection between derivations and semigroups of automorphisms of an algebra.

Definition 8.6.1 (Semigroup of automorphisms) Let R be a unital Banach algebra, and $(\rho_t)_{t\geq 0}$ a semigroup such that:

(1) $\rho_t \in \mathcal{L}(R)$ for all $t \geq 0$;

(2) $\rho_t(f) \to f$ in R as $t \to 0+$ for all $f \in R$;

(3) $\rho_t \circ \rho_s = \rho_{s+t}$ for all $s, t \geq 0$;

(4) $\rho_t(fg) = \rho_t(f)\rho_t(g)$ for all $f, g \in R$ and $t \geq 0$. In this case, we say that (ρ_t) is a semigroup of automorphisms.

Proposition 8.6.2 *Suppose that (ρ_t) satisfies (1)–(3).*

(i) *Then there exists a dense linear subspace $\mathcal{D}$ on which the limit exists*

$$-Df = \lim_{t\to 0+} \frac{\rho_t(f) - f}{t} \qquad (f \in \mathcal{D}). \tag{8.6.1}$$

(ii) *Then (ρ_t) also satisfies (4), if and only if $\mathcal{D}$ is a dense subalgebra of R and*

$$D(fg) = D(f)g + fD(g) \qquad (f, g \in \mathcal{D}), \tag{8.6.2}$$

so that D is a derivation, and there is a subalgebra of R given by

$$\ker(D) = \big\{f \in R : \rho_t(f) = f, \forall t > 0\big\}.$$

Proof (i) This is standard Hille–Yoshida theory of semigroups; see [31].

(ii) ($\Rightarrow$) Assume (4) holds. Then for $f, g \in \mathcal{D}$, we have

$$\begin{aligned}\frac{\rho_t(fg) - fg}{t} &= \frac{\rho_t(f) - f}{t}\rho_t(g) + f\frac{\rho_t(g) - g}{t}\\ &\to -(Df)g - fD(g) \qquad (t \to 0+); \end{aligned} \tag{8.6.3}$$

so $fg \in \mathcal{D}$ and

$$D(fg) = (Df)g + f(Dg). \tag{8.6.4}$$

By semigroup theory, $\mathcal{D}$ is dense in R, and $\|f\|_{\mathcal{D}} = \|f\|_R + \|Df\|_R$ gives an algebra norm such that

$$\begin{aligned}\|fg\|_{\mathcal{D}} &= \|fg\|_R + \|(Df)g + f(Dg)\|_R\\ &\leq \|f\|_R\|g\|_R + \|Df\|_R\|g\|_R + \|f\|_R\|Dg\|_R\\ &\leq \|f\|_{\mathcal{D}}\|g\|_{\mathcal{D}}; \end{aligned} \tag{8.6.5}$$

also $\mathcal{D}$ is complete for $\|\cdot\|_{\mathcal{D}}$ since D is closed.

(ii) ($\Leftarrow$) Conversely, suppose that $-D$ generates (ρ_t). Then for $f, g \in \mathcal{D}$ and $0 \leq s \leq t$ we have

$$\begin{aligned}\frac{\partial}{\partial s}\rho_{t-s}\big(\rho_s(f)\rho_s(g)\big) &= -\rho_{t-s}\Big(-D\big(\rho_s(f)\rho_s(g)\big)\Big)\\ &\quad + \rho_{t-s}\big(-D\rho_s(f)\rho_s(g)\big) + \rho_{t-s}\big(\rho_s(f)(-D)\rho_s(g)\big)\\ &= \rho_{t-s}\Big(D\big(\rho_s(f)\rho_s(g)\big)\Big) - D\rho_s(f)\rho_s(g) - \rho_s(f)D\rho_s(g)\Big)\\ &= 0, \end{aligned} \tag{8.6.6}$$

by (8.6.2). Hence $\rho_{t-s}(\rho_s(f)\rho_s(g))$ is constant, and by comparing the values at $s = 0$ with $s = t$, we obtain

$$\rho_t(fg) = \rho_t(f)\rho_t(g). \tag{8.6.7}$$

□

Example 8.6.3 (Automorphisms of the continuous functions) Let X be a compact Hausdorff space. A $*$-homomorphism of $C(X, \mathbf{C})$ is a linear map such that $\alpha(\bar{f}) = \overline{\alpha(f)}$ for all $f \in C(X, \mathbf{C})$. A strongly continuous group of $*$-automorphisms has the form $\rho_t(f) = f \circ \theta_t$, where

(i) $\theta_t : X \to X$ is a homeomorphism of X such that

(ii) $\mathbf{R} \times X \to X$: $(t, x) \mapsto \theta_t(x)$ is continuous, and

(iii) $\theta_t \circ \theta_s(x) = \theta_{t+s}(x)$ for all $s, t \in \mathbf{R}$ and $x \in X$.

This follows from the Banach–Stone theorem.

Exercise 8.6.4 (Automorphisms of the disc algebra) For the disc algebra A of Exercise 3.1.7, let the Poisson semigroup be defined by $\rho_t(f)(z) = f(e^{-t}z)$ for $z \in \mathbf{C}$ with $|z| < 1$.

(i) Verify that $(\rho_t)_{t>0}$ is a strongly continuous semigroup of homomorphisms $\rho_t : A \to A$.

(ii) Show that the generator is $-D$, where $Df(z) = zf'(z)$.

(iii) Show that $\rho_t(f) \to Pf$ as $t \to \infty$, where $Pf(z) = f(0)$.

Exercise 8.6.5 (Connections on tensor algebra) Let A and R be algebras, and M an A-bimodule. Let $\rho_t : T_A M \to R$ be a differentiable one-parameter family of algebra homomorphisms, so that ρ_t corresponds to a pair (u_t, v_t) as in Proposition 6.4.10(i). Use Proposition 8.6.2 to differentiate this pair.

8.7 Lifting and Automorphisms of QA

In this section we consider groups of automorphisms $\rho_t = e^{-tD}$ on the Cuntz algebra. The calculations are simpler to understand if one looks at the Example 5.8.1 first. In that example, we considered the harmonic oscillator semigroup acting on the Hermite polynomials. Here we have a graded algebra $\Omega^{ev}A$ and produce a semigroup of automorphisms on it.

In the following proof, we consider a family of automorphisms e^{-tD} of RA such that the limit $\lim_{t\to\infty} e^{-tD}$ gives an embedded copy of A in RA. So we consider a derivation on RA, and regard $\Omega^{2n}A$ as playing a similar role to He_{2n} in Example 5.8.1. Recall that

$$\mathrm{Hom}_{alg}(RA, R) = \big\{\rho\colon A \to R;\quad \rho \quad \text{linear};\quad \rho(1) = 1\big\}, \tag{8.7.1}$$

where the linear unital map ρ gives rise to a homomorphism ρ_* via

$$\rho_*(a) = \rho(a) \qquad (a \in A); \tag{8.7.2}$$

$$\rho_*(a_0 da_1 da_2 \dots da_{2n}) = \rho(a_0)\omega(a_1,a_2)\dots\omega(a_{2n-1},a_{2n}), \tag{8.7.3}$$

where $\omega(a,b) = \rho(ab) - \rho(a)\rho(b)$ and

$$\omega^n(a_1,\dots,a_{2n}) = \omega(a_1,a_2)\omega(a_3,a_4)\dots\omega(a_{2n-1},a_{2n}). \tag{8.7.4}$$

Now such elements span $(IA)^k$ for $k \le n$, and

$$\begin{aligned}&\mathrm{Hom}_{alg}(RA/(IA)^{n+1}, R)\\ &= \big\{\rho\colon A \to R;\quad \rho \quad \text{linear};\quad \rho(1) = 1;\quad \omega^{n+1} = 0\big\}.\end{aligned} \tag{8.7.5}$$

Theorem 8.7.1 *Let A be quasi-free. Then:*

(i) *there exists a linear map $\phi\colon A \to \Omega^2 A$ such that $-\delta\phi = d \cup d$;*

(ii) *there exists a unique derivation D on RA such that $D(a) = \phi(a)$;*

(iii) *$D\colon RA \to RA$ corresponds to $H + L$ on $\prod_{j=0}^{\infty} \Omega^{2j} A$, where H is the number operator, and L has degree 2, so $L\colon \Omega^{2n} A \to \Omega^{2n+2} A$;*

(iv) *$D = e^{-L} H e^{L}$, and $a \mapsto e^{-L} a$ embeds A as a subalgebra of $\widehat{\Omega}^{ev} A$.*

Proof (i) We consider a linear $\phi\colon A \to \Omega^2 A$ with $D(a) = \phi(a)$, and then consider the condition that D be a derivation for the Fedosov product

$$\begin{aligned} D(da_1da_2) &= D(a_1a_2 - a_1 \star a_2) \\ &= D(a_1a_2) - D(a_1) \star a_2 - a_1 \star D(a_2) \\ &= D(a_1a_2) - (Da_1)a_2 + dDa_1da_2 - a_1(Da_2) + da_1dDa_2 \\ &= \phi(a_1a_2) - a_1\phi(a_2) - a_1\phi(a_2) \end{aligned} \tag{8.7.6}$$

modulo $(IA)^2$; so using the hypothesis that ϕ is quasi-free, we select ϕ so that

$$-a_1\phi(a_2) + \phi(a_1a_2) - \phi(a_1)a_2 = da_1da_2; \tag{8.7.7}$$

then as in Lemma 8.5.2

$$D(da_1da_2) = da_1da_2 + da_1d\phi(a_2) + d\phi(a_1)da_2. \tag{8.7.8}$$

Then we have $d(da_1da_2) = 0$, so we extend D to $\Omega^2 A$ by

$$\begin{aligned} D(a_0da_1da_2) &= D(a_0 \star da_1da_2) \\ &= D(a_0) \star da_1da_2 + a_0 \star D(da_1da_2) \\ &= \phi(a_0)da_1da_2 + a_0da_1da_2, \end{aligned} \tag{8.7.9}$$

and generally D on $\Omega^{2n} A$ satisfies

$$\begin{aligned} &D(a_0da_1da_2\dots da_{2n-1}da_{2n}) \\ &\quad= D(a_0 \star (da_1da_2) \star \dots \star (da_{2n-1}da_{2n})) \\ &\quad= \phi(a_0)da_1da_2\dots da_{2n-1}da_{2n} + a_0da_1da_2\dots da_{2n-1}da_{2n} \\ &\qquad + a_0(d\phi(a_1)da_2 + da_1d\phi(a_2))da_3da_4\dots da_{2n-1}da_{2n} \\ &\qquad + a_0da_1da_2\dots da_{2n-1}da_{2n} \\ &\qquad + \dots \\ &\qquad + a_0da_1\dots da_{2n-2}(d\phi(a_{2n-1})a_{2n} + da_{2n-1}d\phi(a_{2n})) \\ &\qquad + a_0da_1da_2\dots da_{2n-1}da_{2n}. \end{aligned} \tag{8.7.10}$$

(ii) Hence we can write $D = H + L$ where

$$H(a_0da_1da_2\dots da_{2n-1}da_{2n}) = na_0da_1da_2\dots da_{2n-1}da_{2n} \tag{8.7.11}$$

is the number operator and

$$L(a_0da_1da_2\dots da_{2n-1}da_{2n}) = \phi(a_0)da_1\dots da_{2n} + \sum_{j=1}^{2n} a_0da_1\dots da_{j-1}d\phi(a_j)da_{j+1}\dots da_{2n}. \tag{8.7.12}$$

Observe that H acts in $\Omega^{2n}A$ as multiplication by n, while $L\colon \Omega^{2n}A \to \Omega^{2n+2}A$. Note also that L raises the degree by 2, hence $HL\phi_{2n} - LH\phi_{2n} = (n+1)L\phi_{2n} - nL\phi_{2n}$ so

$$L = [H, L].$$

It follows that we can represent D as the lower triangular matrix

$$D \asymp \begin{bmatrix} 0 & 0 & 0 & 0 & \dots \\ 0 & 1 & 0 & 0 & \dots \\ 0 & * & 2 & 0 & \dots \\ \vdots & \vdots & * & 3 & \ddots \\ \vdots & \vdots & \vdots & \ddots & \ddots \end{bmatrix}. \tag{8.7.13}$$

(iii) The idea is to replace D by H, which had matrix $\mathrm{diag}(0, 1, 2, \dots)$. By Duhamel's formula, we have

$$\begin{aligned} e^{-L}He^{L} &= H + e^{-L}[H, e^{L}] \\ &= H + e^{-L}\int_0^1 e^{(1-t)L}[H, L]e^{tL}dt \\ &= H + \int_0^1 e^{-tL}Le^{tL}dt \\ &= H + L \\ &= D. \end{aligned} \tag{8.7.14}$$

Hence $-D$ generates a semigroup e^{-tD} of automorphisms of $RA = \oplus_{n=0}^{\infty}\Omega^{2n}A$. Next we show that e^{-tD} induces a semigroup on $RA/(IA)^{n+1}$ for each n. We have $D(A) \subseteq \Omega^2 A \subseteq IA$, and $(D-1)\Omega^2 A \subseteq (IA)^2$. Hence we have the chain of inclusions

$$\begin{aligned} D(RA) &\subseteq IA; \\ (D-1)D(RA) &\subseteq (D-1)IA \subseteq (IA)^2; \\ &\vdots \qquad\quad \vdots \\ (D-n)\dots(D-1)D(RA) &\subseteq (IA)^{n+1}, \end{aligned} \tag{8.7.15}$$

so

$$(D-n)\dots(D-1)D = 0 \qquad \text{on} \quad RA/(IA)^{n+1}, \tag{8.7.16}$$

hence there is an isomorphism, implemented by $H \mapsto e^{-L}He^{L}$ under which D corresponds to multiplication by j on $\Omega^{2j}A$,

$$RA/(IA)^{n+1} = \mathrm{Ker}((D-n)\dots(D-1)D), \tag{8.7.17}$$

where

$$\bigoplus_{j=0}^{n} \mathrm{Ker}(D-j) \cong \bigoplus_{j=0}^{n} \Omega^{2j}A. \tag{8.7.18}$$

Also,

$$e^{-tD} = e^{-L}e^{-tH}e^{L}, \tag{8.7.19}$$

so that e^{-L} gives an isomorphism between RA and $\Omega^{ev}A$.

The canonical filtration $(RA/(IA)^n)$ has completion $\widehat{RA}$ or equivalently $\widehat{\Omega^{ev}A}$, which we can identify with the product $\prod_{j=0}^{\infty}\Omega^{2j}A$. Now $RA/(IA)^{n+1}$ has a subalgebra isomorphic to A given by $\mathrm{Ker}(D)$, so we have exact sequences of A bimodules

$$0 \longrightarrow A \longrightarrow RA/(IA)^n \longrightarrow \mathrm{Ker}((D-1)D-2)\dots(D-n+1)) \longrightarrow 0 \tag{8.7.20}$$

with completion

$$0 \longrightarrow A \longrightarrow \hat{RA} \longrightarrow \varprojlim \mathrm{Ker}((D-1)(D-2)\dots(D-n+1)) \longrightarrow 0 \tag{8.7.21}$$

by [8, proposition 10.2]. We can regard $A \cong \mathrm{Ker}(D)$ as the range of P where $P(a) = \lim_{t\to\infty} e^{-tD}a$ in the style of Proposition 8.6.2. Also $e^{-tD} = e^{-L}e^{-tH}e^{L}$, so the embedding $A \to \widehat{RA}$ is equivalently given by $A \to \widehat{\Omega^{ev}A}\colon a \mapsto e^{-L}a$. □

We remark in passing that, on the simplex $\Delta = \{(t_1, \dots, t_n)\colon 0 < t_1 < t_2 < \dots < t_n < 1\}$, the expression

$$e^{-t_1L}a_1e^{(t_1-t_2)L}a_2\dots e^{(t_{n-1}-t_n)L}a_ne^{(t_n-1)L} \tag{8.7.22}$$

has all the exponents negative. This may be compared with the expressions (5.7.4) and (11.11.11) which arise in applications of Duhamel's formula.

Corollary 8.7.2 *Let A be quasi-free, let R_0 be an algebra such that $R_0 = R/I$ for some ideal I of R such that $I^2 = 0$. Then any homomorphism $u_0 : A \to R_0$ lifts to a homomorphism $u : A \to \hat{R}$.*

Proof We consider the case of square zero extensions. Now let

$$0 \longrightarrow I \longrightarrow R \longrightarrow A \longrightarrow 0$$

be an extension of A by I such that $I^2 = 0$, and let $\rho : A \to R$ be a linear lifting. Then the universal model for this situation is

$$0 \longrightarrow IA/(IA)^2 \longrightarrow R/(IA)^2 \longrightarrow A \longrightarrow 0 \tag{8.7.23}$$

since $A = RA/IA$ and $IA/(IA)^2$ has square zero in $RA/(IA)^2$. Also we have produced a linear lifting $A \to RA/(IA)^2 : a \mapsto a - \phi(a)$. We have shown that this lifts to a homomorphism $A \to \widehat{RA}$. □

Proposition 8.7.3 *D extends to a derivation on QA.*

Proof We consider the linear maps $A \to QA$ given by

$$a \mapsto \theta(a) = a + da; \quad a \mapsto \theta^\dagger(a) = a - da. \tag{8.7.24}$$

These are analogous to the raising and lowering operators for the quantum harmonic oscillator in Example 5.8.1. A derivation of QA is given by a pair of its retractions to $\theta(A)$ and $\theta^\dagger(A)$, which can be arbitrary derivations $\theta(A) \to QA$ and $\theta^\dagger(A) \to QA$. We choose

$$\begin{aligned} D\theta(a) &= (1/2)da + \phi(a) + d\phi(a), \\ D\theta^\dagger(a) &= (-1/2)da + \phi(a) - d\phi(a), \end{aligned} \tag{8.7.25}$$

so that

$$\begin{aligned} D(a) &= (1/2)D(\theta(a) + \theta^\dagger(a)) = \phi(a), \\ D(da) &= (1/2)D\theta(a) - \theta^\dagger(a)) = (1/2)da + d\phi(a). \end{aligned} \tag{8.7.26}$$

Then $D\theta$ and likewise $D\theta^\dagger$ are derivations for the Fedosov product since

$$\begin{aligned} D\theta(a_1a_2) &= (1/2)d(a_1a_2) + \phi(a_1a_2) + d\phi(a_1a_2) \\ &= (1/2)(da_1)a_2 + (1/2)a_1da_2 + \phi(a_1a_2) + a_1\phi(a_2) + da_1da_2 \\ &\quad + d\phi(a_1)a_2 + \phi(a_1)da_2 + da_1\phi(a_2) + a_1d\phi(a_2) \end{aligned} \tag{8.7.27}$$

while

$$
\begin{aligned}
&D\theta(a_1) \star \theta(a_2) + \theta(a_2) \star D\theta(a_2) \\
&\quad = \big((1/2)da_1 + \phi(a_1) + d\phi(a_1)\big) \star (a_2 + da_2) \\
&\qquad + (a_1 + da_1) \star \big((1/2)da_2 + \phi(a_2) + d\phi(a_2)\big) \\
&\quad = \big((1/2)da_1 + \phi(a_1) + d\phi(a_1)\big)(a_2 + da_2) - d\phi(a_1)da_2 \\
&\qquad + (a_1 + da_1)\big((1/2)da_2 + \phi(a_2) + d\phi(a_2)\big) - da_1 d\phi(a_2), \qquad (8.7.28)
\end{aligned}
$$

so

$$
D\theta(a_1a_2) = D\theta(a_1) \star \theta(a_2) + \theta(a_2) \star D\theta(a_2) \qquad (8.7.29)
$$

and likewise with $D\theta^{\dagger}$. □

Corollary 8.7.4 *There is an isomorphism of filtered algebras*

$$
QA \to \prod_{n=0}^{\infty} \Omega^n A \qquad (8.7.30)
$$

such that D corresponds to multiplication by n on $\Omega^{2n} A$.

9

Cocycles for a Commutative Algebra over a Manifold

The Fedosov product is intended to be a deformation of the standard product on $C^\infty(M)$ in the sense of Section 1.9. There we observed that the Poisson bracket produces the first-order approximation as $\hbar \to 0$, so $f * g = fg + \hbar\{f,g\} + O(\hbar^2)$. In this chapter we recall the concept of Poisson structures over a manifold, as used in classical Hamiltonian mechanics, and the Heisenberg group, which incorporates the canonical anti-commutation relations of quantum mechanics. To interpret quantum mechanics correctly, it is necessary to use unbounded operators on Hilbert space, and interpret position and momentum as generators of unitary groups, and construct a formal Lie algebra from these generators, which can be expressed in terms of pseudo-differential operators. Representations of the Heisenberg group lead to the Weyl calculus for pseudo-differential operators. At the end of this chapter, we state some significant cocycle formulas and commutator formulas in index theory.

9.1 Poisson Structures on a Manifold

We recall the Poisson bracket from Section 1.7. On $T^*(\mathbf{R}^n)$ there are symplectic local coordinates x_j, ξ_j for $j = 1, \dots, n$ and a symplectic form $\sigma = \sum_{j=1}^n d\xi_j \wedge dx_j$. Then the volume form on $T^*(\mathbf{R}^n)$ is given by

$$\frac{1}{n!} d\sigma = d\xi_1 \wedge dx_1 \wedge \cdots \wedge d\xi_n \wedge dx_n. \tag{9.1.1}$$

Given $f_j, g_j \in C^\infty(T^*(\mathbf{R}^n); \mathbf{R})$, we have Poisson brackets

$$\{f_j, g_j\} = \sum_{k=1}^n \left(\frac{\partial f_j}{\partial \xi_k} \frac{\partial g_j}{\partial x_k} - \frac{\partial f_j}{\partial x_k} \frac{\partial g_j}{\partial \xi_k} \right), \tag{9.1.2}$$

such that $\zeta = \sum_{j=1}^{n} df_j \wedge dg_j$ satisfies

$$\zeta^n = \prod_{j=1}^{n} \{f_j, g_j\}\, d\sigma. \tag{9.1.3}$$

The notion of a Poisson structure generalizes the notion of the Poisson bracket.

Definition 9.1.1 (Poisson structure) A Poisson structure on a smooth manifold M is a bilinear relation $C^\infty(M;\mathbf{R}) \times C^\infty(M;\mathbf{R}) \to C^\infty(M;\mathbf{R})$ such that:

(1) $f \mapsto \{f,g\}$ is a derivation, so $\{fh,g\} = f\{h,g\} + h\{f,g\}$;

(2) $C^\infty(M;\mathbf{R})$ is a Lie algebra for $\{f,g\}$, so $\{f,f\} = 0$ and Jacobi's identity holds:

$$\{f,\{g,h\}\} + \{h,\{f,g\}\} + \{g,\{h,f,\}\} = 0. \tag{9.1.4}$$

Proposition 9.1.2 *Let M be a manifold with a Poisson structure, and Ω_M the de Rham space of smooth differential forms on M. Then there is an operator b of degree (-1) on Ω_M such that (Ω_M, b, d) is a mixed complex with*

$$b^2 = 0, \quad [b,d] = 0. \tag{9.1.5}$$

Proof Here $\{f,g\}(x)$ depends only upon $df(x)$ and $dg(x)$, so $\{\cdot,\cdot\}$ is determined by an element of $C^\infty(M, \bigwedge^2 T)$. Let ι be the interior product, then $\iota\omega$ is an operator of degree (-2) on ΩM. We define

$$b = [\iota(\omega), d], \tag{9.1.6}$$

which is of degree (-1) and anti-commutes with d, so $bd + db = 0$. The Jacobi identity is equivalent to $b^2 = 0$. □

In the literature, this is written as $\delta : \Omega_M^n \to \Omega_M^{n-1}$,

$$\begin{aligned} \delta(f_0 df_1 \dots df_n) &= \sum_{i=1}^{n} (-1)^{i+1} \{f_0, f_i\} df_1 \dots \widehat{df_i} \dots df_n \\ &\quad + \sum_{1 \le i < j \le n} (-1)^{i+j} f_0 d\{f_i, f_j\} df_1 \dots \widehat{df_i} \dots \widehat{df_j} \dots df_n, \end{aligned} \tag{9.1.7}$$

where a hat denotes an omitted term. There is a double complex as in (7.7.9), and one can attempt to compute the homology by the methods of Lie algebras.

Example 9.1.3 (i) In [68, 3.3], Loday computes the cyclic homology of the universal enveloping algebra of a complex Lie algebra by means of the Poincaré–Birkhoff–Witt theorem. This applies to the symmetric algebra $S(V)$

on a complex vector space V, and $S(V)$ is isomorphic to a polynomial algebra. As in Exercise 2.1.5, one can compute the homology and cohomology of $\mathbf{C}[x_1, \dots, x_n]$ by formal integration.

(ii) *(Almost symmetric algebra)* Let A be a complex algebra with submodules A_j such that $\dots \subseteq A_j \subseteq A_{j+1} \subseteq \dots$ with $A_i A_j \subseteq A_{i+j}$, $\cup_j A_j = A$, and $\cap_j A_j = \{0\}$. Then we introduce the graded algebra $g(A) = \oplus_j A_{j+1}/A_j$. We say that A is almost symmetric if there exists a complex vector space $V = A_1/A_0$ such that $g(A)$ is isomorphic to the symmetric algebra $S(V)$. Some of these results of (i) extend to the class of almost symmetric algebras. We refer the reader to [68] for these results and proceed to look at some specific examples.

9.2 Weyl Algebras

We introduce the Weyl algebra, which is defined by analogy with the Clifford algebra from Proposition 1.4.3. Recall that for a complex vector space V, we introduced a quadratic form $q \in S^2(V) \to \mathbf{C}$ and imposed the condition $v^2 = q(v)$, then let $C_q = T(V)/(v^2 - q(v); v \in V)$. The representations of the Weyl algebra as operators on Hilbert space are important in quantum mechanics.

Definition 9.2.1 (Weyl algebra) Let V be a finite-dimensional complex vector space and $\omega\colon V \times V \to \mathbf{C}+$ a skew-symmetric form on V, which we identify with $\omega \in (\bigwedge V)^*$. Now let the Weyl algebra be

$$\mathcal{A}_\omega = T(V)/([v_1, v_2] - \omega(v_1, v_2)). \tag{9.2.1}$$

Definition 9.2.2 (Weyl algebra for harmonic oscillator) The following is particularly associated with the quantum harmonic oscillator with Hamiltonian $H = (1/2)p^2 + (1/2)q^2$, as in Example 5.8.1. Let $V = \mathbf{C}p \oplus \mathbf{C}q$ and $\omega(p, q) = -i$. Then let

$$\mathcal{A}_\omega = T(p, q)/(pq - qp + i). \tag{9.2.2}$$

With $p = -i\frac{\partial}{\partial x}$ and $q = x$, regarded as differential operators on $\mathbf{C}[x]$, we have $pq - qp = -i$.

Proposition 9.2.3 *Then A_ω is isomorphic to the complex algebra of differential operators with polynomial coefficients in one variable $\mathbf{C}[x, \frac{\partial}{\partial x}]$.*

Proof The canonical isomorphism is due to Weyl

$$(c_1 q + c_2 p)^n \leftrightarrow \left(c_1 x - i c_2 \frac{\partial}{\partial x}\right)^n \tag{9.2.3}$$

for $c_1, c_2 \in \mathbf{C}$ and $n \in \mathbf{N}$. One uses the universal properties of the Clifford algebra to establish the isomorphism between the algebras. Thus the Weyl algebra is a twisted polynomial algebra. □

The Weyl algebra has a graded structure. Let $A_\omega^j = \mathrm{span}\{p^r q^s : r + s \leq j\}$ so that $A_\omega^j \subseteq A_\omega^{j+1}$, and using the anti-commutation relations one can show that $A_\omega^1 A_\omega^j \subseteq A_\omega^{j+1}$ and $\cup_{j=0}^\infty A_\omega^j = A_\omega$. Also $\mathbf{C}[p,q]$ is isomorphic to $\oplus_{j=0}^\infty A_\omega^{j+1}/A_\omega^j$. With the Poisson bracket $\{F,G\} = \frac{\partial F}{\partial q}\frac{\partial G}{\partial p} - \frac{\partial F}{\partial p}\frac{\partial G}{\partial q}$, it is straightforward to calculate the δ of (9.1.7), as in

$$\delta(Fdp) = -\frac{\partial F}{\partial q}, \quad \delta(Fdq) = \frac{\partial F}{\partial p}, \quad \delta(Fdqdq) = -dF \qquad (F \in \mathbf{C}[p,q]); \tag{9.2.4}$$

hence one can compute the homology of the Weyl algebra as in [110, p. 247]. In particular, by [68, p. 94] and [109, p. 209] the cyclic homology satisfies $HC_j(A_\omega) = \mathbf{C}$ for $j = 2, 4, \dots$, and is zero otherwise. Also, the Hochschild homology lives in degree two, so that $HH_j(A_\omega) = \mathbf{C}$ for $j = 2$ and is zero otherwise, since $HH_j(A_\omega) \cong H_{DR}^{2-j}(\mathbf{C}[p,q])$ as in Exercise 2.1.5.

(i) The Weyl algebra has the following generalization. Let V be as above, and R a complex algebra with centre Z. Let $\omega\colon \bigwedge^2 V \to R$, and

$$\mathcal{A}_{R,\omega} = S \otimes T(V)/([v_1, v_2] - \omega(v_1, v_2)). \tag{9.2.5}$$

(ii) The universal case arises from the symmetric algebra on $\bigwedge^2 V$ and is $R^{univ} = R(\wedge^2 V)$. Then there is a canonical homomorphism $\mathcal{A}_{R(\wedge^2 V),\omega} \to \mathcal{A}_{R,\omega}$.

(iii) Let $R = (\bigwedge V)$ be the exterior algebra and let ω be the obvious inclusion map $\bigwedge^2 V \subseteq \bigwedge V$, so $\omega(v,w) = v \wedge w$; then let

$$\mathcal{A}_{\wedge V} = \bigwedge V \otimes T(V)/([v_1, v_2] - v_1 \wedge v_2). \tag{9.2.6}$$

We let $S(V)$ be the symmetric tensors of V, which we can regard as polynomial functions on $W = V^*$ as in Lemma 1.4.2. Then $\mathcal{A}_{\wedge V}$ is isomorphic to the algebra of Kähler differential forms $\Omega_{S(V)}$ from Section 4.1 with the Fedosov product

$$\xi \star \eta = \xi\eta + cd\xi d\eta, \tag{9.2.7}$$

where $c = 1/2$. Now $\bigwedge V$ is a subalgebra of $(\Omega_{S(V)}, \star)$ consisting of closed forms. We map

$$\begin{aligned}[v_1, v_2] &\mapsto v_1 \star v_2 - v_2 \star v_1 \\ &= v_1 v_2 + c dv_1 dv_2 - v_2 v_1 - c dv_2 dv_1 \\ &= 2c dv_1 dv_2, \end{aligned} \tag{9.2.8}$$

so with $c = 1/2$, we have $[v_1, v_2] = dv_1 dv_2$. Then we have

$$\mathcal{A}_{\wedge V} = \Omega_{S(V)} \cong S(V) \otimes \wedge V, \tag{9.2.9}$$

where $S(V)$ is generated by $v \in V$ and $\wedge V$ is generated by the exact differentials dv.

Exercise 9.2.4 (Wintner and Wielandt's Theorem) [46, pr. 230]. For H, a non-zero complex Hilbert space, the canonical anti-commutation relation

$$PQ - QP = -iI \tag{9.2.10}$$

does not have a solution with $P, Q \in \mathcal{L}(H)$.

(i) Let $P, Q \in \mathcal{L}(H)$. By considering the geometric series for $(\lambda I - PQ)^{-1}$ for $\lambda \in \sigma(PQ) \setminus \{0\}$, show that

$$\sigma(PQ) \setminus \{0\} = \sigma(QP) \setminus \{0\}. \tag{9.2.11}$$

(ii) Suppose further that P and Q satisfy (9.2.10). Deduce that $\sigma(PQ) \setminus \{0\} = (\sigma(PQ) \setminus \{0\}) + i$, in the sense $(\sigma(PQ) \setminus \{0\}) + i = \{\lambda + i : \lambda \in \sigma(PQ); \lambda \neq 0\}$.

(iii) By iterating this, show that $\sigma(PQ)$ is not compact, contrary to Theorem 3.1.2.

By Exercise 9.2.4, to realize the Weyl relations from operators in H, we need to consider unbounded operators, and to address the issues about domains of definition, the most effective approach is to consider unitary groups of operators and their generators. Let $\mathcal{U}(H)$ be the space of unitary operators on H, and suppose that $u : \mathbf{R} \to \mathcal{U}(H)$ is a unitary representation, so that:

(i) $u(t + t) = u(s)u(t)$;

(ii) $u(s) = u(-s)^*$;

(iii) $u(0) = 1$ and $u(s)\xi \to \xi$ in H as $s \to 0$ for all $\xi \in H$, so the representation is continuous in the strong operator topology.

Theorem 9.2.5 (Stone) *There exists a densely defined self-adjoint operator X in H such that $s^{-1}(U(s)\xi - \xi) \to iX\xi$ as $s \to 0$, for all ξ in the domain of X, and so the unitary group is $u(s) = e^{isX}$. Conversely, any densely defined self-adjoint operator X gives rise to a unitary representation in this way.*

Proof See [41]. □

Suppose further that there exists a pair (u, v) of such unitary representations on H, with $u(s) = e^{isX}$, $v(s) = e^{isY}$, and that (u, v) have a common core, so there exists a dense linear subspace $\mathcal{S}$ on H such that $X\mathcal{S} \subseteq \mathcal{S}$ and $Y\mathcal{S} \subseteq \mathcal{S}$; here we use the notation $\mathcal{S}$ to suggest the Schwartz space. Then the commutator of X and Y is densely defined and

$$u(s)v(s)u(-s)v(-s) = 1 - s^2[X, Y] + O(s^3) \tag{9.2.12}$$

as $s \to 0$. To make this precise, one introduces $w(s) = u(s)v(s)u(-s)v(-s)$ and applies the formula

$$w(t) - w(0) = tw'(0) + \int_0^t (t - s)w''(s)\,ds \tag{9.2.13}$$

on $\mathcal{S}$. Let SAC be the space of self-adjoint operators in H that have core $\mathcal{S}$. Then SAC is a real Lie algebra for

$$(A, B) \mapsto i[A, B], \tag{9.2.14}$$

and $e^{i\cdot} : A \to \mathcal{U}(H) : A \to e^{iA}$ is the imaginary exponential, in the sense that $\frac{d}{dt}e^{itA} = iAe^{itA}$ expresses iA as the generator or tangent of the group.

Now consider a pair of unitary representations U, V of $\mathbf{R}^n$ on H, again with a common core $\mathcal{S}$. Then for $p, q \in \mathbf{R}^n$ there are self-adjoint operators $q \cdot X$ and $p \cdot Y$ such that

$$U(tq)V(tp)U(-tq)V(-tp) = 1 - t^2[q \cdot X, p \cdot Y] + O(t^3) \tag{9.2.15}$$

on $\mathcal{S}$. We consider the case in which

$$i\omega(q, p) = -[q \cdot X, p \cdot Y] \tag{9.2.16}$$

is given by imaginary multiples of the identity operator on H. This style of expressing the anti-commutation relations is often attributed to Weyl, and the following result realizes such a solution of the anti-commutation relations.

Proposition 9.2.6 *Let $\omega(p,q) = p \cdot q$.*

(i) *There exists a pair (U, V) of unitary representations of $\mathbf{R}^n$ on $L^2(\mathbf{R}^n)$ such that*

$$U(q)V(p) = e^{i\omega(q,p)}V(p)U(q). \tag{9.2.17}$$

(ii) *With $U(q) = e^{iq\cdot X}$ and $V(p) = e^{ip\cdot D}$ the product formula holds*

$$e^{i(q\cdot X + p\cdot D)} = e^{iq\cdot p/2}e^{iq\cdot X}e^{ip\cdot D}. \tag{9.2.18}$$

Proof For the standard Schrödinger representation, let

$$e^{iq\cdot X}u(x) = e^{iq\cdot x}u(x), \tag{9.2.19}$$

$$e^{ip\cdot D}u(x) = u(x+p) \tag{9.2.20}$$

be the multiplication and translation groups on $\mathbf{R}^n$. These unitary groups have common core $\mathcal{S} = \mathcal{S}(\mathbf{R}^n;\mathbf{C})$, the Schwartz class of functions on $\mathbf{R}^n$ and generators

$$iq\cdot X = \sum_{j=1}^{n} iq_j x_j, \qquad ip\cdot D = \sum_{j=1}^{n} p_j \frac{\partial}{\partial x_j}, \tag{9.2.21}$$

which satisfy Heisenberg's anti-commutation relation

$$[q\cdot X, p\cdot D] = i\sum_{j=1}^{n} q_j p_j, \tag{9.2.22}$$

which gives $\omega(q,p) = p\cdot q$.

(ii) Now consider the n-fold product

$$\left(e^{iq\cdot X/n}e^{ip\cdot D/n}\right)^n = \left(e^{iq\cdot X/n}e^{ip\cdot D/n}\right)\dots\left(e^{iq\cdot X/n}e^{ip\cdot D/n}\right)$$

and move the first factor of $e^{iq\cdot X/n}$ through $n-1$ factors of $e^{ip\cdot D/n}$, thereby picking up a multiplicative factor of $e^{i(n-1)p\cdot q/n^2}$; then move the second factor of $e^{iq\cdot X/n}$ through $n-2$ factors of $e^{ip\cdot D/n}$, thereby picking up a multiplicative factor of $e^{i(n-2)p\cdot q/n^2}$; and so on. We deduce that

$$\left(e^{iq\cdot X/n}e^{ip\cdot D/n}\right)^n = e^{i(n-1)p\cdot q/2n}e^{iq\cdot X}e^{ip\cdot D}; \tag{9.2.23}$$

by taking $n\to\infty$ and applying Chernoff's generator formula [41], we obtain (9.2.18). □

Conversely, this representation is unique up to unitary equivalence.

Theorem 9.2.7 (Stone–von Neumann) *Let $\omega_+ : \mathbf{R}^n \times \mathbf{R}^n \to \mathbf{R}$ be a non-degenerate bilinear form and (U,V) a pair of unitary representations on $\mathbf{R}^n$ on H such that*

$$U(x)V(y) = e^{i\omega_+(x,y)}V(y)U(x) \qquad (x,y\in\mathbf{R}^n). \tag{9.2.24}$$

Then (U,V) is unitarily equivalent to some multiple of the standard Schrödinger representation on $L^2(\mathbf{R}^n)$.

Proof See [84] for a simple proof. □

Example 9.2.8 (Commuting unitary groups) This example deals with the degenerate case $\omega = 0$, and describes unitary representations of the plane $\mathbf{R}^2$, or more precisely, the one-point compactification S^2 of $\mathbf{R}^2$. Let $\mathbf{R}^2$ have the usual vector space structure, so that $(x,t) + (p,q) = (x + p, t + q)$, and let $W = \mathbf{R}^2$ be the dual group, so that $(\xi,\eta) \in W$ gives a character $\mathbf{R}^2 \to S^1 : (x,t) \mapsto e^{ix\xi+i\eta t}$. Each $f \in \mathcal{S}(\mathbf{R}^2)$ has a Fourier transform

$$\hat{f}(\xi,\eta) = \int_{\mathbf{R}^2} f(x,t)e^{-ix\xi-i\eta t}\, dxdt \qquad ((\xi,\eta) \in W),$$

such that $\hat{f} \in C_0(\mathbf{R}^2;\mathbf{C})$. The usual convolution operation on $\mathcal{S}(\mathbf{R}^2)$ corresponds to pointwise multiplication on $\mathcal{S}(W)$, as in

$$f * g(x,t) = \frac{1}{(2\pi)^2}\int_W \hat{f}(\xi,\eta)\hat{g}(\xi,\eta)e^{ix\xi+i\eta t}\, d\xi d\eta;$$

there are unitary groups $(U_p)_{p\in\mathbf{R}}$ and $(V_q)_{q\in\mathbf{R}}$ such that $U_pV_q = V_qU_p$ and

$$U_pV_qf(x,t) = \frac{1}{(2\pi)^2}\int_W \hat{f}(\xi,\eta)e^{ix(\xi+p)+it(\eta+q)}\, d\xi d\eta,$$

so there is a group representation $\mathbf{R}^2 \to \mathcal{U}(L^2(W)) : (p,q) \mapsto U_pV_q$.

We can take $A_0 = \mathbf{R}$, $A_1 = \mathbf{R}^2$ and introduce $A_0 \oplus A_1$ with the multiplication $(\lambda, x, t) \circ (\lambda', x', t') = (\lambda\lambda', \lambda' x + \lambda' t + \lambda x' + \lambda t')$. In the next section, we consider projective representations of $\mathbf{R}^2$.

Let $\mathcal{A}_0 = \mathcal{S}(\mathbf{R}^2)$ with the standard convolution structure. Given any unitary groups $(e^{-iXp})_{p\in\mathbf{R}}$ and $(e^{-iYq})_{q\in\mathbf{R}}$ which commute in the sense that $e^{-iXp}e^{-iqY} = e^{-iqY}e^{-ipX}$, there exists a homomorphism of $*$-algebras $\mathcal{A}_0 \to \mathcal{L}(H)$

$$f \mapsto \hat{f}(X,Y) = \int_{\mathbf{R}^2} f(p,q)e^{-ipX}e^{-iqY}\, dpdq$$

such that $\|\hat{f}(X,Y)\| \leq \sup\{|\hat{f}(\xi,\eta)| : \xi,\eta \in \mathbf{R}\}$. We can therefore regard $C_0(\mathbf{R}^2;\mathbf{C})$ as the C^*-algebra generated by unitary representations of $(\mathcal{A}_0, *)$.

9.3 Representations of the Heisenberg Group

In this section we show that solutions of the Weyl canonical anti-commutation relations correspond to unitary representations of the Heisenberg group. Then, we extend unitary representations of the Heisenberg group to suitable algebraic structures on the Schwartz class. The advantage of this way of expressing the anti-commutation relations is that the formulas become more symmetrical in P and Q, and the generalization from $\mathbf{R}^2$ to $\mathbf{R}^{2n}$ is straightforward. We begin with the simplest case of the Heisenberg group and the Lie algebra, which can be constructed from upper triangular 3×3 matrices.

Exercise 9.3.1 (Heisenberg's group) (i) In $GL_3(\mathbf{R})$ introduce the matrices

$$P = \begin{bmatrix} 1 & p & 0 \\ 0 & 1 & q \\ 0 & 0 & 1 \end{bmatrix}, \qquad R = \begin{bmatrix} 1 & r & 0 \\ 0 & 1 & s \\ 0 & 0 & 1 \end{bmatrix}. \tag{9.3.1}$$

Show that the group commutator

$$PRP^{-1}R^{-1} = \begin{bmatrix} 1 & 0 & ps - rq \\ 0 & 1 & 0 \\ 0 & 0 & 1 \end{bmatrix}, \tag{9.3.2}$$

where $(p,q) \wedge (r,s) = ps - rq$ is the signed area of the parallelogram in $\mathbf{R}^2$ that is generated by (p,q) and (r,s).

(ii) Let $\mathcal{H}_1$ be the subgroup of $GL_3(\mathbf{R})$ given by

$$\mathcal{H}_n = \left\{ \begin{bmatrix} 1 & p & t \\ 0 & 1 & q \\ 0 & 0 & 1 \end{bmatrix} : p,q,t \in \mathbf{R} \right\}. \tag{9.3.3}$$

Use (i) to compute the Lie algebra of $\mathcal{H}_1$ in terms of

$$X = \begin{bmatrix} 0 & 1 & 0 \\ 0 & 0 & 0 \\ 0 & 0 & 0 \end{bmatrix}, \quad Y = \begin{bmatrix} 0 & 0 & 0 \\ 0 & 0 & 1 \\ 0 & 0 & 0 \end{bmatrix}, \quad Z = \begin{bmatrix} 0 & 0 & 1 \\ 0 & 0 & 0 \\ 0 & 0 & 0 \end{bmatrix}. \tag{9.3.4}$$

One can show that there exists a unique three-dimensional real Lie algebra $\mathcal{H}$ with basis $\{X, Y, Z\}$ such that $[X, Y] = Z$ and Z is in the centre of $\mathcal{H}$.

(iii) Let $H_1 = \{(u, \xi) : u \in S^1, \xi \in \mathbf{R}^2\}$ have the product

$$(u,\xi) \circ (v,\eta) = (uve^{i\xi\wedge\eta}, \xi + \eta) \qquad (u,\xi),(v,\eta) \in H. \tag{9.3.5}$$

Show that H_1 is a group that contains S^1 as a normal subgroup, and $H_1/S^1 \cong \mathbf{R}^2$, so there is an exact sequence of groups

$$S^1 \longrightarrow H_1 \longrightarrow \mathbf{R}^2.$$

(iv) Show that

$$\mathcal{Z} = \left\{ \begin{bmatrix} 1 & 0 & z \\ 0 & 1 & 0 \\ 0 & 0 & 1 \end{bmatrix} : z \in \mathbf{Z} \right\} \tag{9.3.6}$$

is a normal subgroup such that $\mathcal{H}_1/\mathcal{Z}$ is isomorphic to H_1. Note that H_1 is not a closed subgroup of $GL_3(\mathbf{R})$; instead it is a quotient of two closed subgroups of $GL_3(\mathbf{R})$.

(v) With $\mathcal{S}(\mathbf{R}^2)$, the Schwartz class of rapidly decreasing smooth functions on $\mathbf{R}^2$, show that

$$f *_\hbar g(p,q) = \int_{\mathbf{R}^2} e^{i\hbar(ps-qr)} f(p-r,q-s)g(r,s)drds \tag{9.3.7}$$

defines an associative product on $\mathcal{S}(\mathbf{R}^2)$, so $(f *_\hbar g) *_\hbar h = f *_\hbar (g *_\hbar h)$ and $f *_\hbar g \to f * g$ as $\hbar \to 0+$.

We are interested in unitary representations of H_1 on a separable Hilbert space H such that $u \in S^1$ acts by scalar multiplication $u\colon \xi \mapsto u\xi$ for all $u \in S^1$ and $\xi \in H$. The group H_1 is locally compact and has a countable neighbourhood basis. The following disintegration result appears in [81] and enables us to express continuous unitary representations in terms of irreducibles.

Proposition 9.3.2 *For any continuous unitary representation $\pi\colon H_1 \to \mathcal{U}(H)$, there exists a compact space T and a positive measure μ on T such that H disintegrates as a direct integral*

$$H = \int_T H(t)\mu(dt), \tag{9.3.8}$$

where $(H(t))_{t\in T}$ is a measurable field of Hilbert spaces and for almost all $t \in H$ there exists a continuous and irreducible unitary representation $\pi_t\colon H_1 \to \mathcal{U}(H_t)$ such that $\pi(g) = (\pi_t(g))_{t\in T}$ is diagonalized with respect to this disintegration.

The pair of representations $\mathbf{R}^n \to \mathcal{U}(H)$ given by $U(p) = e^{ip\cdot D}$ and $V(q) = e^{iq\cdot D}$ are linked by the anti-commutation relations and together give a projective representation $\mathbf{R}^{2n} \to \mathcal{U}(H)$. Equivalently, this may be expressed as a representation of the Heisenberg group, which we define as follows.

Definition 9.3.3 (Heisenberg group) Let $\mathcal{H}_n$ be the subgroup of $GL_{n+2}(\mathbf{R})$ given by

$$\mathcal{H}_n = \left\{ \begin{bmatrix} 1 & p & t \\ 0 & I_n & q \\ 0 & 0 & 1 \end{bmatrix} : p \in \mathbf{R}^{1\times n}, q \in \mathbf{R}^{n\times 1}, t \in \mathbf{R} \right\}. \tag{9.3.9}$$

The standard symplectic form on $\mathbf{R}^{2n}$ is the 2-form $\sum_{j=1}^n dp_j \wedge dq_j$ which is associated with the skew-symmetric matrix

$$\left\langle \begin{bmatrix} 0 & -I_n \\ I_n & 0 \end{bmatrix} \begin{bmatrix} p \\ q \end{bmatrix} \mid \begin{bmatrix} r \\ s \end{bmatrix} \right\rangle = p \cdot s - q \cdot r \tag{9.3.10}$$

which defines a non-degenerate bilinear form $\mathbf{R}^{2n} \times \mathbf{R}^{2n} \to \mathbf{R}$.

Corollary 9.3.4 *Suppose that $U(p) = e^{ip \cdot D}$ and $V(q) = e^{iq \cdot D}$ satisfy the anti-commutation relations. Then there is a group representation $\pi : \mathcal{H}_n \to \mathcal{U}(H)$:*

$$\begin{bmatrix} 1 & p & t \\ 0 & I_n & q \\ 0 & 0 & 1 \end{bmatrix} \mapsto e^{it} e^{iq \cdot X} e^{ip \cdot D} \tag{9.3.11}$$

such that $\pi(\exp(2\pi Z)) = I$.

Proof This follows from Proposition 9.2.6. □

Also, $e^{iq \cdot X + ip \cdot D} = e^{iq \cdot P} e^{iq \cdot X} e^{ip \cdot D}$, so we can write the representation equivalently as

$$e^{it} e^{iq \cdot X} e^{ip \cdot D} = e^{it - ip \cdot q/2 + iq \cdot X + ip \cdot D}, \tag{9.3.12}$$

since

$$\begin{aligned} &(1/2) p \cdot q + (1/2) p' \cdot q' + p \cdot q' - (1/2)(p + p') \cdot (q + q') \\ &= (1/2)(p \cdot q' - p' \cdot q). \end{aligned} \tag{9.3.13}$$

9.4 Quantum Trace Formula

Now we consider the partition function for the quantum mechanical model. Let $\hbar > 0$ be Planck's constant, and let q_j be the position coordinate which is expressed as multiplication by x_j in $L^2(\mathbf{R}^n)$; let p_j be the momentum coordinates so $p_j \leftrightarrow -i\hbar \partial / \partial x_j$ in $L^2(\mathbf{R}^n)$. Let $\Delta = -\sum_{j=1}^{n} \frac{\partial^2}{\partial x_j^2}$, and consider the quantum Hamiltonian

$$\frac{1}{2} \sum_{j=1}^{n} p_j^2 + U(q) = \frac{\hbar^2}{2} \Delta + U(q). \tag{9.4.1}$$

Then the quantum partition function is $Z(\hbar) = \mathrm{trace}(e^{-\hbar^2 \Delta/2 - U})$.

Example 9.4.1 (Quantum harmonic oscillator) Consider the Hamiltonian $L = -\beta^{-2} d^2/dx^2 + \beta x^2$ in $L^2(\mathbf{R})$ of the one-dimensional quantum harmonic oscillator, as in Proposition 5.8.2, so that $e^{-tL/2}$ corresponds to the integral operator with $\theta = e^{-t}$

$$M(x,y;t) = \frac{\beta}{\sqrt{\pi(1-\theta^2)}} \times \exp\left(-\frac{\beta^2\theta^2 y^2 - 2\beta^2\theta xy + \beta^2\theta^2 x^2}{1-\theta^2} - \frac{\beta^2 x^2 + \beta^2 y^2}{2} \right) \tag{9.4.2}$$

with $\mathrm{trace}[e^{-tL}] = 1/(1-\theta)$. One can likewise obtain an expression for $\beta^{-2}\Delta + \beta^2\|x\|^2$ in $L^2(\mathbf{R}^n)$.

The standard approach to calculating $Z(\hbar)$ involves deriving an asymptotic expansion of the kernel

$$e^{-\hbar^2\Delta/2 - U(q)}(x,y) \leftrightarrow \frac{1}{\hbar^n}\sum_{j=0}^{\infty} a_j(x,y)\hbar^j \tag{9.4.3}$$

in terms of functions $a_j : \mathbf{R}^n \times \mathbf{R}^n \to \mathbf{R}$ as in [104, p. 270], and then one can integrate along the diagonal $x = y$ to compute the trace. We can assign, in a natural way, Schwartz class functions on the cotangent space coordinates $(p,q) \in \mathbf{R}^n \times \mathbf{R}^n$ to operators on the Hilbert space $L^2(\mathbf{R}^n)$ which are of trace class depending upon the parameter $\hbar$ so that

$$\begin{aligned}\mathrm{trace}\big[e^{-\hbar^2\Delta/2 - U(q)}\big] &\asymp \frac{1}{(2\pi)^n}\int_{\mathbf{R}^{2n}} e^{-\sum_{j=1}^n \hbar^2 p_j^2/2 - U(q)} dp_1 \dots dp_n dq_1 \dots dq_n \\ &= \left(\frac{1}{\hbar\sqrt{2\pi}}\right)^n \int_{\mathbf{R}^n} e^{-U(q)} dq_1 \dots dq_n \end{aligned} \tag{9.4.4}$$

as $\hbar \to 0+$. Thus the quantum partition function reduces to a classical partition function.

The leading term in (9.4.4) is justified as follows. By Chernoff's product formula, we have

$$e^{-\hbar^2\Delta/2 - U} = \lim_{m\to\infty}\left(e^{-\hbar^2\Delta/2m} e^{-U/m}\right)^m, \tag{9.4.5}$$

where the product on the right-hand side has factors with Hilbert–Schmidt kernel

$$e^{-\hbar^2\Delta/2m} e^{-U/m} \leftrightarrow \sqrt{\frac{m}{(2\pi\hbar^2)^n}}\, e^{-m\|x-y\|^2/(2\hbar^2) - U(y)/m} \qquad (x, y \in \mathbf{R}^n); \tag{9.4.6}$$

hence the product is trace class.

Example 9.4.2 (Weyl's asymptotic formula) Let Ω be a bounded domain in $\mathbf{R}^n$, let $\Omega^c = \mathbf{R}^n \setminus \Omega$ and for each $k \in \mathbf{N}$ let Ω_k be the subdomain consisting of those $x \in \Omega$ such that $d(x, \Omega^c) > 2^{-k}$. Let $U_k \in C_c^\infty(\mathbf{R}^n; \mathbf{R})$ be a non-negative function such that $U_k(x) = 0$ for all $x \in \Omega_k$ and $U_k(x) = 2^k \|x\|^2$ for all $x \in \Omega_{k+1}^c$; also $U_k \leq U_{k+1}$. It is clear from (9.4.6) that the kernel of $e^{-\hbar^2\Delta/2-U_k}$ is monotone decreasing in k. The limiting operator Δ is the Laplacian in $L^2(\Omega; \mathbf{C})$ with Dirichlet boundary conditions. Hence

$$\operatorname{trace}\left[e^{-\hbar^2\Delta/2}\right] \asymp \left(\frac{1}{\hbar\sqrt{2\pi}}\right)^n \operatorname{vol}(\Omega) \qquad (\hbar \to 0) \tag{9.4.7}$$

by the monotone convergence theorem. In 17.5 of [56], Hörmander uses wave equation methods to give a more precise version of this result, with sharp bounds on the error term.

Of course, the terms in (9.4.3) become increasingly complicated and difficult to interpret geometrically. In the context of Riemannian manifolds, McKean and Singer [74] compute the first three terms in the asymptotic expansion of the heat kernel $e^{-t\Delta}(x, y)$ in terms of local geometrical quantities such as the curvature tensor; however, they noted that the task of describing all the terms in the asymptotic expansion seemed pretty hopeless. Atiyah, Bott and Patodi subsequently overcame these difficulties [7, theorem E.III].

The following approach is more algebraic, and seeks to use cocycles defined by the commutator products to compute the traces of the leading order terms.

Let W be the cotangent space of $\mathbf{R}^{2n}$ and $V = W^*$ the space of linear functions on W. Let $\omega \colon \wedge^2 V \to \mathbf{C}$ be the natural symplectic form on W, and write $\omega = \omega_+ - \omega_-$, where $\omega_\pm$ are rank n bilinear forms on complementary subspaces. We can take the south-west corner of the matrix for ω_+, and the north-east corner for ω_-; results of [57, p. 333] show one can always do this. Let $\mathcal{S}(W)$ be the Schwartz class functions on W, which are C^∞ smooth and of rapid decay. For $\hbar > 0$, we introduce a group law on $\mathcal{H}_\hbar = \mathbf{R} \times V$ by

$$(t_1, v_1) \circ (t_2, v_2) = (t_1 + t_2 + \hbar\omega(v_1, v_2)/2, v_1 + v_2). \tag{9.4.8}$$

Proposition 9.4.3 *Given unitary groups $(e^{ip\cdot D})_{p\in\mathbf{R}^n}$ and $(e^{iq\cdot X})_{q\in\mathbf{R}^n}$ on a Hilbert space such that*

$$e^{ip\cdot D}e^{iq\cdot X} = e^{i\hbar\omega_+(p,q)}e^{iq\cdot X}e^{ip\cdot D}, \tag{9.4.9}$$

then with $v = (q, p)$, there is a unitary representation of $\pi_\hbar \colon \mathcal{H}_\hbar \to \mathcal{U}(H)$ by

$$\pi_\hbar(t, q, p) = e^{it - i\hbar\omega_+(q,p)/2 + iq\cdot X + ip\cdot D}. \tag{9.4.10}$$

Then $\pi_\hbar$ extends to $\mathcal{S}(W) \to \mathcal{L}(H)$ by

$$\pi_\hbar(f) = \int_V \hat{f}(q,p)\pi_\hbar(0,q,p)\,dqdp. \tag{9.4.11}$$

9.5 The Poisson Bracket and Symbols

We begin by introducing pseudo-differential operators within the formalism of Hörmander's general theory.

Definition 9.5.1 (Pseudo-differential operator) For $u \in L^1 \cap L^2(\mathbf{R}^n)$, let

$$\hat{u}(\xi) = \int_{\mathbf{R}^n} e^{-i\langle \xi, x\rangle} u(x)\,dx$$

be the Fourier transform of u. Let $D_x = -i\nabla$ be differentiation

$$D_x u = \left(\frac{\partial u}{i\partial x_1}, \dots, \frac{\partial u}{i\partial x_n}\right), \tag{9.5.1}$$

which is represented by $\hat{u}(\xi) \mapsto (\xi_j \hat{u}(\xi))$. Then for $m \in \mathbf{R}$, let $\mathcal{S}^m = \mathcal{S}^m(\mathbf{R}^n \times \mathbf{R}^n; \mathbf{C})$ be the set of all $p \in C^\infty(\mathbf{R}^n \times \mathbf{R}^n; \mathbf{C})$ such that for all multi-indices α and β there exists $C_{\alpha,\beta} > 0$ such that

$$\left|\frac{\partial^\alpha}{\partial \xi^\alpha}\frac{\partial^\beta}{\partial x^\beta} p(x,\xi)\right| \le C_{\alpha,\beta}(1 + |\xi|)^{m-|\alpha|}; \tag{9.5.2}$$

such a p is a symbol of order m. The corresponding integral operator is $P = \Psi(p)$ where

$$Pu(x) = \frac{1}{(2\pi)^n}\int_{\mathbf{R}^n} e^{i\langle x,\xi\rangle} p(x,\xi)\hat{u}(\xi)\,d\xi, \tag{9.5.3}$$

and we say that P has order m, and $\sigma(P) = p$. The integrand involves both x and ξ variables and can be simplified using the formula

$$e^{iD_\xi \cdot D_y} a(x,y,\xi)|_{y=x} = \frac{1}{(2\pi)^n}\int_{\mathbf{R}^n \times \mathbf{R}^n} e^{i\langle x-y,\eta-\xi\rangle} a(x,y,\xi)\,dyd\eta \tag{9.5.4}$$

for $a \in C_c^\infty(\mathbf{R}^n \times \mathbf{R}^n \times \mathbf{R}^n; \mathbf{R})$. We write $\Psi DO^m = \{\Psi(f) : f \in \mathcal{S}^m(\mathbf{R}^n \times \mathbf{R}^n)\}$ for the pseudo-differential operators of order m on $L^2(\mathbf{R}^n; \mathbf{C})$. The notion of a pseudo-differential operator combines two basic examples of operators from Fourier analysis, described in the following Lemma.

Lemma 9.5.2 (i) *Let $p(x,\xi) = U(x)$ where $U : \mathbf{R}^n \to \mathbf{C}$ is bounded and measurable. Then P is the bounded linear operator of multiplication by $U(x)$.*

(ii) *Let* $p(x,\xi) = p_0(\xi)$, *where* $p_0 \in L^1$ *has* $\hat{p}_0 \in L^1(\mathbf{R}^n;\mathbf{C})$. *Then* P *is a Fourier multiplier, which may be written as the convolution* $Pu(x) = \int \hat{p}_0(x-y)u(y)\,dy$, *or equivalently as* $P = p_0(-i\frac{\partial}{\partial x_1},\dots,-i\frac{\partial}{\partial x_n})$, *and* P *is bounded on* $L^q(\mathbf{R}^n;\mathbf{C})$ *for* $1 \le q \le \infty$.

(iii) *Let* $U \in L^2(\mathbf{R}^n;\mathbf{C})$, $p_0 \in L^2(\mathbf{R}^n;\mathbf{C})$ *and* $p(x,\xi) = U(x)p_0(\xi)$. *Then* $U(x)p_0(-i\hbar\partial/\partial x)$ *is the Hilbert–Schmidt operator on* $L^2(\mathbf{R}^n;\mathbf{C})$ *with kernel* $U(x)\hat{p}_0(x-y)$.

(iv) *In particular, (iii) applies to* $p_0(x) = (1+|x|^2)^{-s/2}$ *for* $s > n/2$.

Proof (ii) This is Young's convolution inequality, as in

$$\|u * \hat{p}_0\|_{L^q(\mathbf{R}^n;\mathbf{C})} \le \|u\|_{L^q(\mathbf{R}^n;\mathbf{C})}\|\hat{p}_0\|_{L^1(\mathbf{R}^n;\mathbf{C})}.$$

(iii) Here $U(x)\hat{p}_0(x-y)$ is square integrable over $\mathbf{R}^2$ since

$$\iint_{\mathbf{R}^2} |U(x)|^2|\hat{p}_0(x-y)|^2 dxdy = 2\pi\int_{\mathbf{R}} |U(x)|^2 dx \int_{\mathbf{R}} |p_0(y)|^2 dy$$

by Plancherel's theorem, hence giving a Hilbert–Schmidt kernel.

(iv) We verify the particular choice $(1+|x|^2)^{-s/2} \in L^2(\mathbf{R}^n;\mathbf{C})$ for $s > n/2$ by using polar coordinates. This example may be compared with Exercise 3.2.11. □

Proposition 9.5.3 *The pseudo-differential operators on* $L^2(\mathbf{R}^n;\mathbf{C})$ *give a graded algebra, in the following sense.*

(i) *Let* $p \in \mathcal{S}^m(\mathbf{R}^n \times \mathbf{R}^n;\mathbf{C})$, *where* $m \le 0$. *Then the corresponding* $P = \Psi(p)$ *belongs to* $\mathcal{L}(L^2(\mathbf{R}^n))$.

(ii) *Let* $p \in \mathcal{S}^m$ *and* $q \in \mathcal{S}^\mu$ *for* $m,\mu \in \mathbf{Z}$. *Then the operators* $P = \Psi(p)$ *and* $Q = \Psi(q)$ *have product* $PQ = R$, *where* R *is a pseudo-differential operator of order* $m+\mu$, *and the commutator* $[P,Q]$ *is a pseudo-differential operator of order* $m+\mu-1$.

(iii) *The pseudo-differential operators of order zero form a subalgebra* ΨDO^0 *of* $\mathcal{L}(L^2(\mathbf{R}^n))$.

(iv) *Suppose that* P *is trace class and* p *is integrable where* $P = \Psi(p)$. *Then*

$$\mathrm{trace}(P) = \frac{1}{(2\pi)^n}\iint_{\mathbf{R}^n\times\mathbf{R}^n} p(x,\xi)dxd\xi. \tag{9.5.5}$$

Proof (i) This is a variant of Schur's Lemma on integral operators [56, 18.8.17].

(ii) As in [56, 1.8.18]. The symbol of R is

$$r(x,\eta) = \frac{1}{(2\pi)^n}\iint_{\mathbf{R}^n\times\mathbf{R}^n} e^{-i\langle x-y,\xi-\eta\rangle}p(x,\eta)q(y,\xi)\,dyd\eta, \tag{9.5.6}$$

which reduces to

$$r(x,\eta) = e^{iD_\xi \cdot D_y} p(x,\eta) q(y,\xi)|_{y=x,\eta=\xi}. \tag{9.5.7}$$

Hence $\Psi DO^m \circ \Psi DO^\mu \subseteq \Psi DO^{m+\mu}$, where $\circ$ denotes operator composition.

Let P_1 and P_2 be pseudo-differential operators of order zero. Then P_1 and P_2 define bounded linear operators on $H = L^2(\mathbf{R}^n)$ such that $[P_1, P_2]$ is a pseudo-differential operator with symbol

$$\begin{aligned}\sigma([P_1,P_2])(x,\eta) &= \frac{1}{(2\pi)^n}\iint_{\mathbf{R}^n\times\mathbf{R}^n} e^{-i\langle x-y,\xi-\eta\rangle}(p_1(x,\eta)p_2(y,\xi) \\ &\quad - p_2(x,\eta)p_1(y,\xi))\,dyd\eta \\ &= e^{iD_\xi\cdot D_y}\big(p_1(x,\eta)p_2(y,\xi) - p_1(y,\xi)p_2(x,\eta)\big)|_{y=x,\eta=\xi}.\end{aligned} \tag{9.5.8}$$

(iii) This combines (i) and (ii).

(iv) This is a variant of Mercer's trace formula. □

Next we link up the Poisson structure on $\mathbf{R}^{2n}$, as discussed in Section 9.1, with the group representation considered in Section 9.3. The first step is to extend representations of the Heisenberg group to the Schwartz class. The Schwartz class has pointwise multiplication and differentiation, hence we can consider the Poisson bracket $\{\cdot,\cdot\}$ of Section 9.1. Also, where possible, we extend the definition to Hörmander's symbol classes S^m via Proposition 9.5.3. We recall the Fourier transform

$$\hat{f}(q,p) = \iint_{\mathbf{R}^n\times\mathbf{R}^n} f(x,\xi)e^{-i\langle q,x\rangle - i\langle p,\xi\rangle}\,dxd\xi$$

for smooth functions f that are of rapid decay, which enables us to define the pseudo-differential operator $\pi_1(f)$ where

$$\pi_1(f)u = f(X,D)u = \frac{1}{(2\pi)^{2n}}\iint_{\mathbf{R}^n\times\mathbf{R}^n} \hat{f}(q,p)e^{i(q\cdot X+p\cdot D)}u\,dqdp. \tag{9.5.9}$$

We use the subscript to indicate the implicit choice $\hbar = 1$. This formula differs from (3.7.12) in two respects: first, the operators X and D are unbounded on $L^2(\mathbf{R}^n)$; second, we have the exponential of $i(q\cdot X + p\cdot D)$ instead of a Wick-ordered product of exponentials.

Proposition 9.5.4 (Weyl calculus) (i) *Let $f \in S^m(\mathbf{R}^n\times\mathbf{R}^n)$, $g \in S^\mu(\mathbf{R}^n\times\mathbf{R}^n)$. Then their usual Poisson bracket $\{f,g\}$ satisfies*

$$\pi_1(\{f,g\}) = -i[\pi_1(f),\pi_1(g)] \tag{9.5.10}$$

modulo, a pseudo-differential operator with symbol in $S^{m+\mu-2}(\mathbf{R}^n\times\mathbf{R}^n)$.

(ii) *Hörmander's class* $(\mathcal{S}^0, \{\cdot,\cdot\})$ *is a Lie algebra, and* $\pi_1 : (\mathcal{S}^0, \{\cdot,\cdot\}) \to (\mathcal{L}(H), -i[\cdot,\cdot])$ *is a Lie algebra homomorphism modulo pseudo-differential operators of order* -2.

(iii) *Suppose that* $f \in \mathcal{S}$ *gives* $\pi_1(f) \in \mathcal{L}^1(H)$. *Then the trace satisfies*

$$\mathrm{trace}(\pi_1(f)) = \frac{1}{(2\pi)^n} \iint_{\mathbf{R}^n \times \mathbf{R}^n} f(x,\xi)\, dx d\xi. \tag{9.5.11}$$

(iv) *The adjoint on* $\mathcal{L}(H)$ *satisfies* $\pi_1(\bar{f}) = \pi_1(f)^*$, *where* $\bar{f}$ *denotes a complex conjugate.*

Proof (i) We introduce a special convolution for $f \in \mathcal{S}^m$ and $g \in \mathcal{S}^\mu$ such that $\pi_1(f)\pi_1(g) = \pi_1(f *_1 g)$; then by page 62 of [104],

$$(f *_1 g)(x,\xi) = f(x,\xi)g(x,\xi) + (i/2)\{f,g\} \qquad \mathrm{mod} \quad \mathcal{S}^{m+\mu-2}.$$

(ii) Here $\pi_1(f) \in \mathcal{L}(H)$ for all $f \in \mathcal{S}^0$ by theorem 18.1.11 of [56] and we can apply (i). In particular ΨDO^{-2} forms an ideal in ΨDO^0, so $\mathcal{E} = \Psi DO^0/\Psi DO^{-2}$ is an algebra and we have a natural map $\mathcal{S}^0 \to \mathcal{E}$ arising from π_1. We can pass to the commutator quotient spaces so

$$\mathcal{S}^0/\{\mathcal{S}^0,\mathcal{S}^0\} \to \mathcal{E}/[\mathcal{E},\mathcal{E}].$$

(iii) The kernel is given by

$$\pi_1(f)(x,y) \leftrightarrow \int_{\mathbf{R}^n} f\left(\frac{x+y}{2},\xi\right) e^{i(x-y)\cdot\xi} d\xi \tag{9.5.12}$$

since

$$\pi_1(f)u(x) = \frac{1}{(2\pi)^n} \iint_{\mathbf{R}^n \times \mathbf{R}^n} f\left(\frac{x+y}{2},\xi\right) e^{i(x-y)\cdot\xi} u(y) dy d\xi; \tag{9.5.13}$$

so when we integrate along $x = y$, we obtain the stated trace formula.

(iv) We can take the adjoint of (9.5.9). □

The results in this section have been formulated for Euclidean space. In Section 9.6, we give versions that hold for Riemannian manifolds. For a compact manifold M and $H = L^2(M)$, the grading on ΨDO matches better with the grading on the $\mathcal{L}^p(H)$.

Exercise 9.5.5 (i) Show that there is an increasing sequence of ideals in the algebra ΨDO^0

$$\dots \subseteq \Psi DO^{-2} \subseteq \Psi DO^{-1} \subseteq \Psi DO^0.$$

(ii) By reviewing Corollary 3.7.2, show that there is a commutative $*$-algebra C_0 that makes the following sequence exact:

$$0 \longrightarrow \Psi DO^{-1} \longrightarrow \Psi DO^{0} \longrightarrow C_0 \to 0.$$

(iii) Using Section 3.7, describe the principal symbol of $P \in \Psi DO^{0}$, namely the highest order term.

Exercise 9.5.6 (i) By taking $f(q,p) = e^{-t(p^2+q^2)}$ in (9.5.12), recover Mehler's kernel of Proposition 5.8.2.

(ii) With ϕ_n as in (5.8.8), compute $\pi(f)$ for $f(p,q) = \phi_n(p)\phi_m(q)$.

Let $\mathcal{S}(W)$ be the space of Schwartz class functions on $W = V^*$, which has various products defined on it through the results of Section 9.2. Let $\bigwedge^+ V$ be the alternating tensors on V that have even degree. Observe that $\bigwedge^+ V$ is commutative, since $\omega \wedge \eta = (-1)^{|\eta|}\eta \wedge \omega = \eta \wedge \omega$ for tensors of even degree. When V has dimension $2n$ over $\mathbf{C}$, we have an exact sequence of commutative algebras

$$0 \longrightarrow \oplus_{j=0}^{n-1} \wedge^{2j} V \longrightarrow \oplus_{j=0}^{n} \wedge^{2j} V \longrightarrow \mathbf{C} \longrightarrow 0, \tag{9.5.14}$$

via the determinant, so $\mathcal{S}(W) \otimes \wedge^+ V$ is a nilpotent extension of $\mathcal{S}(W)$. By Proposition 9.5.4, we can realize $\mathcal{S}$ as pseudo-differential operators on Hilbert space via π_1. If the ΨDO are trace class, then the trace is expressed as integration over the phase space. There are natural maps on the K-groups associated with the trace

$$K_0\big(\mathcal{S}(W) \otimes \wedge^+ V\big) \longrightarrow K_0\big(\mathcal{S}(W)\big) \xrightarrow{\int} \mathbf{C}. \tag{9.5.15}$$

We now return to the situation of Proposition 9.5.4 with phase space $W \cong \mathbf{R}^{2n}$, and recall the maps $\pi_\hbar : \mathcal{S}(W) \to \mathcal{L}(H)$, which are a deformation family including the map $\pi_1 : \mathcal{S}(W) \to \mathcal{L}(H)$ of Proposition 9.5.4, and π_0 which corresponds to pointwise multiplication on $\mathcal{S}$. We can attempt to introduce a new multiplication operation on $\mathcal{S}$ by using $\pi_\hbar$, operator composition in $\mathcal{L}(H)$, and the symbol map σ, as in the following array (9.5.16). We can regard $\mathcal{S}_\hbar$ as a quantized version of $\mathcal{S}$, where the modified multiplication has the Poisson bracket as the first-order correction term. The trace is taken to be common to all the $\mathcal{S}_\hbar$ and given by the integration over phase space as in (9.5.5) and (9.5.11).

Let $\hbar^{-n}\mathbf{C}[[\hbar]]$ be the space of formal Laurent series in $\hbar$ that have a pole of order at most n at zero. As in Example 9.4.2 and [7, theorem E.III], Fedosov [35] observed that for a bounded open set $\Omega \subset \mathbf{R}^{2n}$ and $H = L^2(\Omega; \mathbf{C})$, the Dirichlet Laplacian Δ gives rise to a heat kernel $e^{-\hbar^2\Delta/2} \in \mathcal{L}^1(H)$, such that the trace has an asymptotic expansion

$$\operatorname{trace} e^{-\hbar^2\Delta/2} \asymp \sum_{j=-n}^{\infty} c_j \hbar^j \qquad (\hbar \to 0+),$$

hence we should therefore interpret the trace as a linear map $\mathcal{S}_\hbar \to \hbar^{-n}\mathbf{C}[[\hbar]]$. This set-up is the algebra of quantum observables on $\mathbf{R}^{2n}$.

$$\begin{array}{cccccc} & \mathcal{S}_\hbar \otimes \mathcal{S}_\hbar & \overset{*_\hbar}{\longrightarrow} & \mathcal{S}_\hbar & \\ & \downarrow & & \uparrow & \\ & \Psi DO^0 \otimes \Psi DO^0 & \longrightarrow & A_0 & \\ \pi_\hbar \otimes \pi_\hbar & \downarrow & & \uparrow & \sigma \\ & \mathcal{L}(H) \otimes \mathcal{L}(H) & \longrightarrow & \mathcal{L}(H) & \end{array} \qquad (9.5.16)$$

Claim 9.5.7 Suppose that σ is the symbol map from the pseudo-differential operators of order zero on $L^2(W)$ to $A_0 \subset \mathcal{S}^0$ such that:

(i) $\sigma \circ \pi_\hbar(f) = f$ for all $\hbar > 0$ and $f \in \mathcal{S}^0$;

(ii) $\operatorname{trace} \pi_\hbar(f) = \int_W f(w)dw$ for all $f \in \mathcal{S}$ such that $\pi_\hbar(f) \in \mathcal{L}^1(H)$);

(iii) $f *_\hbar g = \sigma(\pi_\hbar(f)\pi_\hbar(g))$ defines an associative multiplication on $\mathcal{S}$, giving a family of algebras $(\mathcal{S}_\hbar, *_\hbar)$ for $\hbar > 0$.

Then the K-theory of $\mathcal{S}_\hbar$ is a homotopy invariant of algebras as $\hbar \to 0$.

This outline of a theorem has been realized in certain specific circumstances, including the following.

(i) In Section 9.6, we consider the Poisson structure $\{f_1, f_2\}$ and associated commutators $[\rho(f_1), \rho(f_2)]$ of Toeplitz operators, and recall some results of Helton and Howe. By passing to the case of compact manifolds M, we ensure that the grading on products of the pseudo-differential operators in ΨDO_M^{-1} matches the grading on $\mathcal{L}(L^2(M))$ associated with the Schatten ideal $I = \mathcal{L}^m(L^2(M))$ and its powers $I, I^2, \dots, I^m$, so we can define traces and generalized determinants for the Toeplitz operators.

(ii) By Proposition 5.2.2, the heat kernel $e^{-\hbar\Delta/2}$ over a compact manifold has smoothing properties, which enable us to define Fredholm modules with cyclic cocyles as in (5.6.7). In Section 11.12, we use these to state a more sophisticated type of index theorem.

(iii) In Theorem 8.7.1, we showed that a quasi-free algebra A is a subalgebra of the (completed) algebra of even differential forms $\widehat{\Omega}^{ev}A$, where $\Omega^{ev}A$ has the Fedosov product $\star$. In Section 12.2 we introduce $RA = \oplus_{k=0}^{\infty}\Omega^{2k}A$ with ideal $IA = \sum_{k=1}^{\infty}\Omega^{2k}A$ so $A = RA/IA$. We then introduce complex $X(A), X(R/I^n)$ and use the homology of the X complex to compute the periodic cyclic homology of A.

9.6 Cocycles Generated by Commutator Products

In this section we extend the results of Section 2.4, using ideas from [51, 52]. In [18], part II proposition 7, Connes incorporates the fundamental trace form for commutative algebras into his more general cyclic theory. Let $a_j \in \mathcal{L}(H)$ and S_m be the symmetric group of permutations of $\{1, \dots, m\}$. Then by analogy with the definition of the determinant, we define

$$[a_1, \dots, a_m] = \sum_{\varpi \in S_m} \operatorname{sgn}(\varpi) a_{\varpi(1)} \dots a_{\varpi(m)}. \tag{9.6.1}$$

Proposition 9.6.1 *Let A be a commutative algebra and I an ideal in $\mathcal{L}(H)$ such that $I^m \subseteq \mathcal{L}^1(H)$. Suppose that $\rho\colon A \to \mathcal{L}(H)$ is a homomorphism modulo I. Then the following define bounded multilinear functionals*

(i) $$\psi_{2m}(f_1, \dots, f_{2m}) = \operatorname{trace}\big(\omega(f_1, f_2)\omega(f_3, f_4)\dots\omega(f_{2m-1}, f_{2m})\big)$$

(ii) $$\varphi_{2m}(f_1, \dots, f_{2n}) = \operatorname{trace}\big([\rho(f_1), \rho(f_2)][\rho(f_3), \rho(f_4)]\dots [\rho(f_{2m-1}), \rho(f_{2m})]\big),$$

(iii) $$\varphi_{2m+1}(f_0, f_1, \dots, f_{2m}) = \operatorname{trace}\big(\rho(f_0)[\rho(f_1), \rho(f_2)] [\rho(f_3), \rho(f_4)]\dots [\rho(f_{2m-1}), \rho(f_{2m})]\big),$$

(iv) $$\chi(f_1, \dots, f_{2m}) = \operatorname{trace}[\rho(f_1), \dots, \rho(f_{2m})], \tag{9.6.2}$$

where ψ_{2m} in (i) is a cyclic cocycle.

Proof (i) We have $\omega(f_{2j-1}, f_{2j}) \in I$, and can apply Proposition 2.8.2.

(ii) We observe that since A is commutative

$$[\rho(f_1), \rho(f_2)] = \big(\rho(f_1)\rho(f_2) - \rho(f_1 f_2)\big) - \big(\rho(f_2)\rho(f_1) - \rho(f_1 f_2)\big) \in I, \tag{9.6.3}$$

so

$$[\rho(f_1), \rho(f_2)][\rho(f_3), \rho(f_4)]\dots [\rho(f_{2m-1}), \rho(f_{2m})] \in I^m \subset \mathcal{L}^1(H). \tag{9.6.4}$$

Hence φ_{2m} is defined, as in (i).

(iii) This follows from (ii).

(iv) We prove that $[a_1, \dots, a_{2m}]$ is a sum of the product of m simple commutators $[a_j, a_k]$. Let T be the subgroup of S_m generated by the transpositions $(2j-1, 2j)$, and $S_m = \sqcup_{\ell=1}^{L} M\varepsilon_\ell$ a decomposition into cosets. Then

$$\sum_{\varpi \in T} \operatorname{sgn}(\varpi\varepsilon_\ell) a_{\varpi\varepsilon_\ell(1)} a_{\varpi\varepsilon_\ell(2)} \dots a_{\varpi\varepsilon_\ell(2m)} = \operatorname{sgn}(\varepsilon_\ell)[a_{\varepsilon_\ell(1)}, a_{\varepsilon_\ell(2)}]\dots [a_{\varepsilon_\ell(2m-1)}, a_{\varepsilon_\ell(2m)}], \tag{9.6.5}$$

so that

$$[a_1, \dots, a_{2m}] = \sum_{\ell=1}^{L} \sum_{\varpi \in T} \operatorname{sgn}(\varpi \varepsilon_\ell) a_{\varpi \varepsilon_\ell(1)} a_{\varpi \varepsilon_\ell(2)} \dots a_{\varpi \varepsilon_\ell(2m)}$$

$$= \sum_{\ell=1}^{L} [a_{\varepsilon_\ell(1)}, a_{\varepsilon_\ell(2)}] \dots [a_{\varepsilon_\ell(2m-1)}, a_{\varepsilon_\ell(2m)}]. \tag{9.6.6}$$

With $a_j = \rho(f_j)$, we obtain $[\rho(f_1), \dots, \rho(f_{2m})] \in I^m$, so the trace is defined. □

We can regard φ_{2m+1} as a linear functional $\Omega^{2m} A \to \mathbf{C}$

$$f_0 df_1 \dots df_{2m} \mapsto \operatorname{trace}\big(\rho(f_0)[\rho(f_1), \rho(f_2)][\rho(f_3), \rho(f_4)] \dots [\rho(f_{2m-1}), \rho(f_{2m})]\big)$$

on the noncommutative differential forms.

Corollary 9.6.2 *Let M be a smooth compact Riemannian manifold of dimension n, let $A = \mathcal{S}^0(M)$ and let $I = \Psi DO_M^{-1}$. Then $I^m \subseteq \mathcal{L}^1(L^2(M))$ for all $m > n + 1$ and Proposition 9.6.1 applies to give cocycles on A.*

Proof Let U be an open connected subset of M. By Proposition 9.5.4, $[P_1, P_2]$ has an asymptotic expansion in which the leading symbol is the multiple of the Poisson bracket $-i\{p_1, p_2\}$, which has order -1. Hence the range of $[P_1, P_2]$ is contained in $\{f \in L^2(U) : \nabla f \in L^2(U)\}$.

The map $\rho : f \to P_f$, where $\sigma(P_f) = f$, satisfies $\sigma \circ \rho = id$. Also, $T = \rho(fg) - \rho(f)\rho(g)$ is a pseudo-differential operator of order -1, hence $(T^*T)^n$ is pseudo-differential of order $-2n$, hence Hilbert–Schmidt by Lemma 9.5.2(iii), so T gives a compact operator on H by Lemma 3.2.5(iv).

We prove that $[\rho(f_1), \dots, \rho(f_{2n+4})] \in \mathcal{L}^1(H)$ for all $f_j \in C^\infty(U; \mathbf{C})$, and refer the reader to [51, 52] for the rest of the proof. Given an even n, let $n = 2\ell$, and form the product

$$[\rho(f_1), \rho(f_2)][\rho(f_3), \rho(f_4)] \dots [\rho(f_{2\ell+1}), \rho(f_{2\ell+2})], \tag{9.6.7}$$

which is a pseudo-differential operator of order $-(\ell + 1) < -n/2$, hence Hilbert–Schmidt by Lemma 9.5.2(iii), and likewise

$$[\rho(f_{2\ell+3}), \rho(f_{2\ell+4})][\rho(f_{2\ell+5}), \rho(f_{2\ell+6})] \dots [\rho(f_{2n+3}), \rho(f_{2n+4})] \tag{9.6.8}$$

is a pseudo-differential operator of order $-n + 2 + (\ell + 1) < -n/2$, hence Hilbert–Schmidt by Lemma 9.5.2(iii), so

$$[\rho(f_1), \dots, \rho(f_{2n+4})] \tag{9.6.9}$$

is a sum of products of Hilbert–Schmidt operators, hence is trace class. A similar result holds for odd n when we take $n = 2\ell + 1$.

$$\begin{array}{ccccc} & \mathcal{S}^0 \otimes \mathcal{S}^0 & \overset{\pi_1 \otimes \pi_1}{\longrightarrow} \quad \longrightarrow & \mathcal{L}(H) \otimes \mathcal{L}(H) & \\ -i\{\cdot,\cdot\} & \downarrow & & \downarrow & [\cdot,\cdot] \\ & \mathcal{S}^{-1} & \underset{\pi_1}{\longrightarrow} \quad \mathcal{L}^{n+2}(H) & \subset \mathcal{L}(H) & \end{array} \tag{9.6.10}$$

□

Helton and Howe [52] proceed to evaluate the traces introduced in Corollary 9.6.2. The cotangent bundle T^*M has fibres linearly homeomorphic to $\mathbf{R}^n$, so we introduce the unit co-sphere bundle S^*M with fibres $S_x^*M = \{\xi \in T_x^*M : \|\xi\| = 1\}$ and the disc bundle D^*M with fibres $D_x^*M = \{\xi \in T_x^*M : \|\xi\| \leq 1\}$, so D^*M is a compact $2n$-dimensional manifold; also the boundary of D^*M is S^*M. Given a continuous function $f : S_x^*M \to \mathbf{C}$, we extend f to $f : D_x^*M \to \mathbf{C}$ by solving the Dirichlet problem on D_x^*M with boundary condition $f | S_x^*M$. In particular, let $U \subset \mathbf{R}^n$ be open and relatively compact. A pseudo-differential operator can be defined by patching together operators defined locally as in (9.5.3). Given a coordinate patch u, a symbol p restricts to $\sigma(P)(x,\xi) = p(x,\xi)$ for $x \in U$ and $|\xi| = 1$, so we extend the symbol to $\sigma(P)(x,\xi)$ for $\|\xi\| \leq 1$; by patching together such symbols we can introduce $\sigma(P)$ on $D^*(M)$. Let $H = L^2(U,dx)$.

The following result concerns the linear functional on the Kähler differentials $\Omega_A^{2m} \to \mathbf{C}$

$$f_0 df_1 \dots df_{2m} \mapsto \int_{D^*(M)} f_0 df_1 \wedge df_2 \wedge \cdots \wedge df_{2n} \tag{9.6.11}$$

and gives a trace formula extending the case of Toeplitz operators over the circle, in the style of Corollary 3.7.6.

Theorem 9.6.3 *There is an extension*

$$0 \longrightarrow \mathcal{K}(H) \overset{\iota}{\longrightarrow} \mathcal{P} \overset{\sigma}{\longrightarrow} C^\infty(D^*M) \longrightarrow 0 \tag{9.6.12}$$

and a linear map $\rho : C^\infty(D^*M;\mathbf{C}) \to \mathcal{L}(H)$ *such that*

(i) $\sigma \circ \rho = id$ *and* ρ *is a homomorphism modulo* $\mathcal{K}(H)$, *so that* $\rho(f)$ *is a pseudo-differential operator with symbol* f;

(ii) $[\rho(f_1), \dots, \rho(f_{2n})] \in \mathcal{L}^1(H)$ *for all* $f_j \in C^\infty(D^*M;\mathbf{C})$,

$$\mathrm{trace}[\rho(f_1), \dots, \rho(f_{2n})] = \frac{n!}{(2\pi i)^n} \int_{D^*(M)} df_1 \wedge df_2 \wedge \cdots \wedge df_{2n}. \tag{9.6.13}$$

Proof The proof requires a refinement of Proposition 9.6.1. Let $\mathcal{A}_0$ be a subalgebra of $\mathcal{L}(H)$, and let $\mathcal{A}_1$ be the ideal in $\mathcal{A}_0$ that is generated by the commutator subspace $[\mathcal{A}_0, \mathcal{A}_0]$. Then we let $\mathcal{A}_2$ be ideal in $\mathcal{A}_0$ that is generated by $\mathcal{A}_1^2$ and the commutator subspace $[\mathcal{A}_0, \mathcal{A}_1]$. Generally we let $\mathcal{A}_{j+1}$ be the ideal generated by $\mathcal{A}_1^{j+1}$ and the commutator subspace $[\mathcal{A}_0, \mathcal{A}_j]$. Thus we produce a decreasing filtration

$$\mathcal{L}(H) \supset \mathcal{A}_0 \supset \mathcal{A}_1 \supseteq \mathcal{A}_2 \supseteq \cdots \tag{9.6.14}$$

as in Exercise 1.6.9. If $\mathcal{A}_1 \subseteq \mathcal{L}^m$, then $\mathcal{A}_m \subseteq \mathcal{L}^1(H)$ and the trace is defined on $\mathcal{A}_m$, and we can apply Proposition 9.6.1 and Corollary 9.6.2. We refer the reader to [52] for the details of how this trace in (9.6.13) is computed in terms of the symbols of the pseudo-differential operators. □

10
Cyclic Cochains

Suppose that we have an exact sequence of algebra homomorphisms

$$0 \longrightarrow B \longrightarrow E \overset{\pi}{\longrightarrow} R \longrightarrow 0 \qquad (10.0.1)$$

and a linear map $\rho \colon R \to E$ that splits the sequence, so $\pi \circ \rho = id$, as vector spaces. When can we split the sequence as algebras? Now the deviation of ρ from being a homomorphism is measured by the curvature $\omega(x, y) = -\rho(xy) + \rho(x)\rho(y)$. If R is a free algebra $R = T(V)$, then we can define a canonical homomorphism $h \colon R \to E$ by $h(v_1, \dots v_n) = \rho(v_1) \dots \rho(v_n)$. In this chapter, we make use of this basic idea.

This chapter considers the tensor algebra and coalgebra associated with a noncommutative associative algebra. The idea is to introduce universal models for derivations, especially the Hochschild b operator, which is associated with the bar construction. There are some subtleties as to how unital and non-unital algebras are treated. Ultimately, we derive the exact sequences that are needed for the Connes's bicomplex, which is considered in the next chapter.

10.1 Traces Modulo Powers of an Ideal

Let A and R be algebras, and I an ideal in R. Recall that a cyclic 0-cocycle on A is precisely a trace on A. Let $\rho \colon A \to R$ be a $\mathbf{k}$-linear map that is an algebra homomorphism modulo I, so the curvature $\omega = \delta\rho + \rho^2$ has $\omega \in C^2(A, I)$, where $\omega(a_1, a_2) = \rho(a_1)\rho(a_2) - \rho(a_1 a_2)$. Observe that I^n is an R-bimodule, so $[R, I^n] \subseteq I$. Let $\tau \colon I^n \to V$ be a $\mathbf{k}$-linear map to a vector space V such that $\tau | [R, I^n] = 0$. For $f \in C^p(A, I^n)$, we define $\mathrm{tr}_\tau(f) = N\tau(f) \in C_\lambda^{p-1}(A, V)$. When $n = 1$ and $V = \mathbf{k}$, we have the situation of Lemma 1.8.3. However, we wish to produce interesting cyclic cocycles on A in the case in which I^n has interesting traces for some $n > 1$, but I and R/I have no interesting traces.

Proposition 10.1.1 (i) $\mathrm{tr}_\tau(\omega^n)$ *is a cyclic* $2n-1$*-cocycle on* A.

(ii) *The cyclic cohomology class depends only upon the induced homomorphism* $\tilde{\rho}\colon A \to R/I$.

Proof (i) By Bianchi's identity, we have

$$b\mathrm{tr}_\tau(\omega^n) = -\mathrm{tr}_\tau(\delta\omega^n) = -\mathrm{tr}_\tau([\rho,\omega^n]) = 0; \tag{10.1.1}$$

also

$$\begin{aligned}\mathrm{tr}_\tau(\omega^n)(a_1,\dots,a_{2n}) &= \tau(\omega(a_1,a_2)\dots\omega(a_{2n-1},a_{2n}))\\ &\quad + \text{cyclic permutations with sign.}\end{aligned} \tag{10.1.2}$$

(ii) Let ρ be a family of such maps, depending differentiably upon a parameter $t \in [0,1]$. Then writing $\dot{}$ for d/dt, we have $\dot{\omega} = \delta\dot{\rho} + \dot{\rho}\rho + \rho\dot{\rho} = \delta\dot{\rho} + [\dot{\rho},\rho]$ for the superbracket. Hence

$$\begin{aligned}\frac{d}{dt}\mathrm{tr}_\tau\omega^n &= \sum_{i=1}^{n}\mathrm{tr}_\tau(\omega^i\dot{\omega}\omega^{n-i-1})\\ &= \sum_{i=1}^{n}\mathrm{tr}_\tau(\omega^i\delta\dot{\rho}\omega^{n-i-1}) + \sum_{i=1}^{n}\mathrm{tr}_\tau(\omega^i[\dot{\rho},\rho]\omega^{n-i-1})\\ &= n\mathrm{tr}_\tau\delta\dot{\rho}\omega^{n-1} + n\mathrm{tr}_\tau[\dot{\rho},\rho]\omega^{n-1}.\end{aligned} \tag{10.1.3}$$

Now $\delta\omega = [\rho,\omega]$, so

$$\begin{aligned}b\mathrm{tr}_\tau(\dot{\rho}\omega^{n-1}) &= \mathrm{tr}_\tau(\delta\dot{\rho}\omega^{n-1}) + \sum_{i=1}^{n}\mathrm{tr}_\tau(\dot{\rho}\omega^{i-1}(\delta\omega)\omega^{n-i-1})\\ &= \mathrm{tr}_\tau(\delta\dot{\rho}\omega^{n-1}) + \sum_{i=1}^{n}\mathrm{tr}_\tau(\dot{\rho}\omega^{i-1}[\rho,\omega]\omega^{n-1-i})\end{aligned} \tag{10.1.4}$$

and

$$0 = \mathrm{tr}_\tau[\rho,\dot{\rho}\omega^{n-1}] = \mathrm{tr}_\tau([\rho,\dot{\rho}]\omega^{n-1}) + \sum_{i=1}^{n-1}\mathrm{tr}_\tau(\dot{\rho}\omega^{i-1}[\rho,\omega]\omega^{n-1-i}), \tag{10.1.5}$$

so

$$\frac{d}{dt}\mathrm{tr}_\tau\omega^n = nb\mathrm{tr}_\tau(\dot{\rho}\omega^{n-1}). \tag{10.1.6}$$

□

Example 10.1.2 (Cocycles for powers of an ideal) Connes [18, proposition 3, p. 72] considered the case of $R = \mathcal{L}(H)$ and $I = \mathcal{L}^{p/2}(H)$, so that $I^n \subseteq \mathcal{L}^1(H)$ for all $n \in \mathbf{N}$ such that $n \geq p/2$. Then the usual trace τ on $\mathcal{L}^1(H)$ satisfies $\tau|[I^n, R] = 0$. Given $\rho\colon A \to R$ so that ρ is a homomorphism modulo I, let $A'' = \{a \in A\colon \rho(a) \in I\}$ and introduce A' to make the sequence

$$0 \longrightarrow A'' \longrightarrow A \longrightarrow A' \longrightarrow 0 \tag{10.1.7}$$

exact. Then we have an exact sequence

$$0 \longrightarrow I \longrightarrow \rho(A) + I \longrightarrow A' \longrightarrow 0. \tag{10.1.8}$$

We can proceed to define

$$\operatorname{trace}\big(\rho(a_0)\omega(a_1, a_2) \dots \omega(a_{2n-1}, a_{2n})\big). \tag{10.1.9}$$

In this discussion we do not assume the existence of traces on I or on R/I. This is an essentially different situation from Lemma 1.8.3, and resembles more closely the results of Section 9.6.

10.2 Coalgebra

We introduce the notion of a coalgebra via a similar diagram by which we introduced a unital associative algebra; see [1]. Given a field $\mathbf{k}$, a coalgebra is a triple (C, Δ, ε) consisting of a vector space C over $\mathbf{k}$, a $\mathbf{k}$-linear map $\Delta\colon C \to C \times C$ and a co-unit ε, which is a $\mathbf{k}$-linear map $\varepsilon\colon C \to \mathbf{k}$ with the following properties. Following a standard shorthand from the theory of Hopf algebras, we write $\Delta(c) = \sum c_{(1)} \otimes c_{(2)}$, for $c_{(1)}, c_{(2)} \in C$, with notation alluding to some suitably chosen set of indices, and we require that the following diagrams commute: first to express co-associativity,

$$\begin{array}{ccc} C \otimes C \otimes C & \longleftarrow & C \otimes C \\ \downarrow & & \uparrow \\ C \otimes C & \longrightarrow & C \end{array}$$

$$\begin{array}{ccc} \sum c_{(1)(1)} \otimes c_{(1)(2)} \otimes c_{(2)} = \sum c_{(1)} \otimes c_{(2)(1)} \otimes c_{(2)(2)} & \longleftarrow & \sum c_{(1)} \otimes c_{(2)} \\ \downarrow & & \uparrow \\ \sum c_{(1)} \otimes c_{(2)} & = & c \end{array} \tag{10.2.1}$$

then to describe the operation of the co-unit on the leftmost tensor:

$$\begin{array}{ccc} C\otimes C & \longleftarrow & C\otimes C \\ \downarrow & & \uparrow \\ \mathbf{k}\otimes C & \longrightarrow & C \end{array} \qquad \begin{array}{ccc} \sum c_{(1)}\otimes c_{(2)} & \longleftarrow & \sum c_{(1)}\otimes c_{(2)} \\ \downarrow & & \uparrow \\ \sum \varepsilon(c_{(1)})c_{(2)} & = & c \end{array} \tag{10.2.2}$$

Then to describe the operation of the co-unit on the right tensor:

$$\begin{array}{ccc} C\otimes C & \longleftarrow & C\otimes C \\ \downarrow & & \uparrow \\ C\otimes \mathbf{k} & \longrightarrow & C \end{array} \qquad \begin{array}{ccc} \sum c_{(1)}\otimes c_{(2)} & \longleftarrow & \sum c_{(1)}\otimes c_{(2)} \\ \downarrow & & \uparrow \\ \sum \varepsilon(c_{(2)})c_{(1)} & = & c \end{array} \tag{10.2.3}$$

Example 10.2.1 (Hopf algebras) (i) Let $C = \mathbf{C}[\mathbf{N}\cup\{0\}]$ be the space of complex-valued and finitely supported functions on the non-negative integers, with basis $\{c_j\}_{j=1}^{\infty}$, where $c_j(n) = 1$ for $n = j$ and $c_j(n) = 0$ for $n \neq j$. Then we define the co-multiplication by taking $\Delta(c_n) = \sum_{j=0}^{n} c_j \otimes c_{n-j}$ and extending linearly; then we define ε by $\varepsilon(c_j) = 1$ for $j = 0$ and $\varepsilon(c_j) = 0$ for all $j \neq 0$, then extending linearly.

Another notation for this Hopf algebra has $\mathbf{C}[x]$ where $\varepsilon(x^j) = 1$ for $j = 0$ and $\varepsilon(x^j) = 0$ for all $j = 1, 2, \dots$; then we write the coproduct as $\Delta(x^n) = \sum_{j=0}^{n} x^j \otimes x^{n-j}$. We also take $\tau(x^n) = \varepsilon$ for $n = 1$ and $\tau(x^n) = 0$ for all $n \neq 1$.

(ii) Let $C = \mathbf{C}[G]$ be the group algebra, as in Proposition 2.9.19. There is additional structure on $\mathbf{C}[G]$ given by a coproduct $\Delta\colon \mathbf{C}[G] \to \mathbf{C}[G]\otimes\mathbf{C}[G]$

$$\Delta\colon \left(\sum_{g\in G} a_g g\right) \mapsto \left(\sum_{g\in G} a_g g \otimes g\right)$$

and a co-unit $\varepsilon\colon \mathbf{C}[G] \to \mathbf{C}$

$$\varepsilon\colon \sum_{g\in G} a_g g \mapsto \sum_{g\in G} a_g.$$

(iii) A coalgebra C over $\mathbf{k}$ has a co-associative coproduct $\Delta\colon C \to C\otimes C$ and $\eta\colon C \to \mathbf{k}$. In particular, let A be a $\mathbf{k}$-vector space with $T(A) = \oplus_{n=0}^{\infty} A^{\otimes n}$; then $C = T(A)$ is a coalgebra with $\Delta\colon C \to C\otimes C$ given by

$$\Delta(a_0, \dots, a_n) = \sum_{i=0}^{n} (a_0, \dots, a_i) \otimes (a_{i+1}, \dots, a_n). \tag{10.2.4}$$

Here we have

$$(a_0, \dots, a_i) \otimes (a_{i+1}, \dots, a_n) = a_0 \otimes a_1 \otimes \cdots \otimes a_n,$$

but our notation makes the grouping of terms in the coproduct more clear. We have diagrams

$$C \overset{\Delta}{\longrightarrow} C^{\otimes 2} \overset{\Delta\otimes id - id\otimes\Delta}{\longrightarrow} C^{\otimes 3} \tag{10.2.5}$$

Let σ be the forward shift operator $\sigma(a_1,\dots,a_n) = (a_n,a_1,\dots,a_{n-1})$; then let

$$I(a_0\otimes(a_1,\dots,a_n)) = \sum_{0\leq j\leq k\leq n} (a_{k+1},\dots,a_n,a_0,a_1,\dots,a_j)\otimes(a_{j+1},\dots,a_k) \tag{10.2.6}$$

and define

$$J = I\otimes\Delta - \Delta\otimes I + \sigma(I\otimes\Delta). \tag{10.2.7}$$

We use the mixture of tensor and bracket notation so that the arguments of the operators are more transparent.

10.3 Quotienting by the Commutator Subspace

Lemma 10.3.1 *The following sequence is exact:*

$$0 \longrightarrow A\otimes C \overset{I}{\longrightarrow} C^{\otimes 2} \overset{J}{\longrightarrow} C^{\otimes 3} \tag{10.3.1}$$

Proof We consider the case of a tensor algebra $R = T(V)$ on a $\mathbf{k}$-vector space V, so

$$R^{\otimes 4} \overset{b'}{\longrightarrow} R^{\otimes 3} \overset{p}{\longrightarrow} R\otimes V\otimes R \tag{10.3.2}$$

is exact for maps

$$b'(x,y,z,w) = (xy,z,w) - (x,yz,w) + (x,y,zw) \tag{10.3.3}$$

and

$$p(x\otimes v_1,\dots,v_n\otimes y) = \sum_{i=1}^{n}(xv_1,\dots,v_{i-1})\otimes v_i\otimes(v_{i+1},\dots,v_n y). \tag{10.3.4}$$

For any algebra R and R-bimodule M, we have

$$\mathrm{Hom}_{R\text{-}bimod}(\mathrm{Coker}(b'),M) = \mathrm{Der}(R,M) \tag{10.3.5}$$

and for $R = T(V)$, we have $\mathrm{Der}(R, M) = \mathrm{Hom}_{\mathbf{k}}(V, M)$ by the universal property of $T(V)$. Also

$$\mathrm{Hom}_{\mathbf{k}}(V, M) = \mathrm{Hom}_{R\text{-}bimod}(R \otimes V \otimes R, M), \tag{10.3.6}$$

with $D \in \mathrm{Hom}_{\mathbf{k}}(V, M)$ producing

$$D(v_1, \ldots, v_n) = \sum_{i=1}^{n} v_1 \ldots v_{i-1}(Dv_i)v_{i+1} \ldots v_n \tag{10.3.7}$$

on the subspace $V^n \subset R$. Now we quotient each module M by $[R, M]$ to produce $M \mapsto M \otimes_R$

$$\begin{array}{ccccccc} R^{\otimes 4} & \overset{b'}{\longrightarrow} & R^{\otimes 3} & \overset{p}{\longrightarrow} & R \otimes V \otimes R & \longrightarrow & 0 \\ \downarrow & & \downarrow & & \downarrow & & \\ R^{\otimes 3} & \overset{b}{\longrightarrow} & R^{\otimes 2} & \overset{\bar{p}}{\longrightarrow} & R \otimes V & \longrightarrow & 0 \end{array} \tag{10.3.8}$$

so $b(x, y, x) = (xy, z) - (x, yz) + (zx, y)$ and

$$\bar{p}(x \otimes v_1 \ldots v_n) = \sum_{i=1}^{n} v_{i+1} \ldots v_n x v_1 \ldots v_{i-1} \otimes v_i. \tag{10.3.9}$$

The next step uses the diagram

$$\begin{array}{ccccccc} R^{\otimes 3} & \overset{b}{\longrightarrow} & R^{\otimes 2} & \overset{p}{\longrightarrow} & R \otimes V & \longrightarrow & 0 \\ \downarrow \sigma & & \downarrow \sigma & & \downarrow \sigma & & \\ R^{\otimes 3} & \overset{\tilde{b}}{\longrightarrow} & R^{\otimes 2} & \overset{\tilde{p}}{\longrightarrow} & R \otimes V & \longrightarrow & 0 \end{array} \tag{10.3.10}$$

where

$$\tilde{b} = \sigma b \sigma : (x, y, z) \mapsto (y, zx) - (xy, z) + (x, yz), \tag{10.3.11}$$

$$\tilde{p} : (v_1 \ldots v_n \otimes x) \mapsto \sum_{i=1}^{n} v_i \otimes v_{i+1} \ldots v_n x v_1 \ldots v_{i-1}, \tag{10.3.12}$$

hence the bottom row is exact.

To prove the Lemma in the case where $\dim A < \infty$, we let $V = A^*$ so $C_n = A^{\otimes n}$ has dual $C_n^* = V^{\otimes n}$. The dual sequence is exact, and we check that the duals map to I and J are $\tilde{p}$ and $\tilde{b}$. Indeed, with $\mu : R \otimes R \to R$ the multiplication $\mu(x, y) = xy$, we have

$$\tilde{b} = (1 \otimes \mu)\sigma^{-1} - \mu \otimes 1 + 1 \otimes \mu \tag{10.3.13}$$

with dual

$$J = \sigma(1 \otimes \Delta) - \Delta \otimes 1 + 1 \otimes \Delta \tag{10.3.14}$$

and

$$\Big\langle I\big(a_0 \otimes (a_1, \dots, a_n)\big), (v_0 \dots v_p) \otimes (v_{p+1} \dots v_n)\Big\rangle$$
$$= \Big\langle a_0 \otimes (a_1, \dots, a_n), \tilde{p}((v_1 \dots v_p) \otimes (v_{p+1}, \dots, v_n)\Big\rangle. \tag{10.3.15}$$

To see this, expand the left-hand side as

$$\Bigg\langle \sum_{0 \le j \le k \le n} (a_{k+1}, \dots, a_n, a_0, \dots, a_j) \otimes (a_{j+1}, \dots a_k), (v_0 \dots v_p) \otimes (v_{p+1} \dots v_n) \Bigg\rangle \tag{10.3.16}$$

in which the summands contribute zero unless $k - j = n - p$. Meanwhile, the right-hand side is

$$\Bigg\langle a_0(a_1, \dots, a_n), \sum_{i=0}^{p} v_i \otimes v_{i+1} \dots v_p(v_{p+1} \dots v_n)v_0 \dots v_{i-1} \Bigg\rangle \tag{10.3.17}$$

$$= \sum_{i=0}^{p} \langle a_0, v_i \rangle \langle a_1, v_{i+1} \rangle \dots \langle a_{p-i}, v_p \rangle \dots \langle a_n, v_{i-1} \rangle \tag{10.3.18}$$

$$= \sum \Big\langle (a_{n-i-1}, \dots, a_n, a_0, \dots, a_{p-1}) \otimes (a_{p-i-1}, \dots, a_n),$$
$$(v_0, \dots, v_p) \otimes (v_{p+1}, \dots, v_n) \Big\rangle. \tag{10.3.19}$$

To obtain the general statement of the Lemma for infinite-dimensional V, we take the inductive limits over finite-dimensional subspaces. □

10.4 Bar Construction

We carry out constructions with values in a tensor algebra since this makes the definitions simplest. Given the results for the tensor algebra, we subsequently use universal properties to deduce results for other algebras. Let A be a non-unital algebra over $\mathbf{k}$ and let $A^+ = \mathbf{k} \oplus A$ be the augmented algebra, so A^+ is unital. We locate A as the degree one component $A[1]$ of $C = T(A)$, so $C = \otimes_{n=0}^{\infty} C_n$ with $C_0 = \mathbf{k}$ and $C_1 = A[1]$. Let $\Delta(a_0, \dots, a_n)$ be as above,

$$J = 1 \otimes \Delta - \Delta \otimes 1 + \sigma^{-1}(1 \otimes \Delta) \tag{10.4.1}$$

$$I(a_0 \otimes (a_1, \dots, a_n)) = \sum_{0 \le j \le k \le n} (-1)^{n(k+1)} (a_{k+1}, \dots, a_n, a_0, a_1, \dots, a_j)$$
$$\otimes (a_{j+1}, \dots, a_k). \tag{10.4.2}$$

Definition 10.4.1 (Bar construction) The bar construction of A^+ is the differential graded algebra $C = T(A)$ with b'. (We don't need to quotient out by an ideal since A is non-unital.)

Theorem 10.4.2 *There is an exact sequence of complexes*

$$0 \longrightarrow A[1] \otimes C \xrightarrow{I} C \otimes C \xrightarrow{J} C^{\otimes 3} \tag{10.4.3}$$

where $A[1] \otimes C$ is equipped with the Hochschild differential b.

Proof We use the lemma as follows. Let $\nu \colon C \to A$ be the projection onto the component $A[1]$. Then I is injective since $(\nu \otimes I)I = id$; indeed

$$I(a_0 \otimes (a_1, \dots, a_n)) = \sum_{0 \le j \le k \le n} (-1)^{n(k+1)} (a_{k+1}, \dots, a_n, a_0, \dots, a_{j-1}) \otimes (a_j, \dots, a_k) \tag{10.4.4}$$

has

$$Ib = (b' \otimes id + id \otimes b')I. \tag{10.4.5}$$

Indeed,

$$\begin{aligned}(b' \otimes id + id \otimes b')I(a_0 \otimes (a_1, \dots, a_n)) \\ = \sum (-1)^{n(k+1)} b'(a_{k+1}, \dots, a_n, a_0, \dots, a_{j-1}) \otimes (a_j, \dots, a_k) \\ + (a_{k+1}, \dots, a_n, a_0, \dots, a_j) \otimes b'(a_{j+1}, \dots, a_k)\end{aligned} \tag{10.4.6}$$

and so

$$\begin{aligned}(\nu \otimes id)(b' \otimes id + id \otimes b')(a_0 \otimes (a_1, \dots, a_n)) \\ = \sum_{0 \le j \le k \le n} (-1)^{n(k+1)} \nu b'(a_{k+1}, \dots, a_n, a_0, \dots, a_j) \otimes (a_{j+1}, \dots, a_k) \\ + (-1)^{(j+1)nk} \nu(a_{k+1}, \dots, a_n, a_0, \dots, a_j) \otimes b'(a_{j+1}, \dots, a_k).\end{aligned} \tag{10.4.7}$$

To give a non-zero contribution, ν should operate on a tensor in $A[1]$. The final term contributes only if $j = 0$ and $k = n$; the first term in the sum contributes when $k = n - 1$ and $j = 0$; or $k = n$ and $j = 1$; so

$$
\begin{aligned}
&(\nu \otimes id)(b' \otimes id + id \otimes b')I(a_0 \otimes (a_1, \dots, a_n)) \\
&\quad = -a_0 \otimes b'(a_1, \dots, a_n) + (-1)^n b'(a_n, a_0) \otimes (a_1, \dots, a_{n-1}) \\
&\qquad + b'(a_0) \otimes (a_1, \dots, a_n) \\
&\quad = -a_0 \otimes \sum_{i=1}^{n} (-1)^{i-1} (a_1, \dots, a_i a_{i+1}, \dots, a_n) \\
&\qquad + (-1)^n a_n a_0 \otimes (a_1, \dots, a_{n-1}) + a_0 a_1 \otimes (a_2, \dots, a_n) \\
&\quad = b(a_0 \otimes (a_1, \dots, a_n)).
\end{aligned} \tag{10.4.8}
$$

This suffices to prove (10.4.3) since $A \otimes C$ has no terms in degree zero. □

Definition 10.4.3 (Hochschild complex) $(A \otimes C, b)$ is called a Hochschild complex.

When A is unital, $H_*(A \otimes C; b) = H_*(A, A)$. Also

$$
H_n(C) = \begin{cases} \mathbf{k}, & \text{for } n = 0; \\ 0, & \text{for } n \neq 0. \end{cases} \tag{10.4.9}
$$

The tensor algebra has a co-unit $\eta\colon C \to \mathbf{k}$, so η picks out the component in degree zero.

$$
A \otimes C \xrightarrow{I} C \otimes C \xrightarrow{id \otimes \eta} C \otimes \mathbf{k} = C. \tag{10.4.10}
$$

Recall the cyclic permutation with sign λ and the symmetrizing operator $N = \sum_{i=0}^{n} \lambda^i$ on $A^{\otimes n+1}$. Then on $A^{\otimes n+1}$ in $A[1] \otimes C$,

$$
\begin{aligned}
N(a_0, \dots, a_n) &= \sum_{j=0}^{n} (-1)^{j(n+1)} (a_{j+1}, \dots, a_n, a_0, \dots, a_j) \\
&= (1 \otimes \eta) I(a_0, \dots, a_n).
\end{aligned} \tag{10.4.11}
$$

To avoid confusion, we write this as $N = \bar{\partial}$ and conclude that $b'\bar{\partial} = \bar{\partial} b$.

The map $\Delta - \sigma\Delta\colon C \to C \otimes C$ is dual to $R \otimes R \to R\colon (x, y) \mapsto [x, y]$. Also $J(\Delta - \sigma\Delta) = 0$ since $b^2 = 0$. With $u = (\nu \otimes id)(\Delta - \sigma\Delta)$ has $u\colon C \to A[1] \otimes C : u(a_1, \dots, a_n) = (a_1, \dots, a_n) - (-1)^{n-1}(a_n, a_1, \dots, a_{n-1})$ so $u = 1 - \lambda$ on $A^{\otimes n}$, hence $b(1 - \lambda) = (1 - \lambda)b'$. We have a commuting diagram:

$$
\begin{array}{ccc}
C & & \\
u \downarrow & \searrow_{\Delta - \sigma\Delta} & \\
A[1] \otimes C & \rightarrow & C \otimes C
\end{array} \tag{10.4.12}
$$

Thus we have the chain version of Connes's bicomplex:

$$\begin{array}{ccccc} A^{\otimes 3} & \overset{N}{\longleftarrow} & A^{\otimes 3} & \overset{1-\lambda}{\longleftarrow} & A^{\otimes 3} \\ & & \downarrow b & & \downarrow b' \\ A^{\otimes 2} & \overset{N}{\longleftarrow} & A^{\otimes 2} & \overset{1-\lambda}{\longleftarrow} & A^{\otimes 2} \\ & & \downarrow b & & \downarrow b' \\ A & \overset{N}{\longleftarrow} & A & \overset{1-\lambda}{\longleftarrow} & A \end{array} \tag{10.4.13}$$

Definition 10.4.4 (Cyclic complex) The cyclic complex of A is $CC(A) = \operatorname{Im}\{\bar{\partial}\colon A \otimes C \to C\}$.

From the exact sequence

$$0 \longleftarrow CC(A) \longleftarrow A \otimes C \overset{1-\lambda}{\longleftarrow} C \longleftarrow CC(A) \longleftarrow 0 \tag{10.4.14}$$

we have a long exact sequence

$$HC_n(A) \overset{I}{\longrightarrow} H_{n+1}(A, A) \overset{B}{\longrightarrow} HC_{n+1}(A) \overset{S}{\longrightarrow} HC_{n-1}(A) \overset{B}{\longrightarrow} \dots. \tag{10.4.15}$$

In Lemma 2.1.2, we asserted that the connecting map exists via a diagram chase that involves selecting representatives for homology classes. In Section 11.2, we will define the connecting map S by specifying how these representatives can found explicitly.

10.5 Cochains with Values in an Algebra

So far we have carried out computations with cochains that take values in the tensor algebra $T(A)$. To bring the discussion down to earth, we wish to follow through similar calculations for cochains with values in an algebra such as $\mathcal{L}(H)$.

Let A and its tensor algebra $C = T(A)$ be as above, with $A[1]$ in degree one, and let L be an algebra over $\mathbf{k}$ with multiplication $\mu\colon L \times L \to L$. Let $C(A, L) = \operatorname{Hom}_{\mathbf{k}}(C, L)$, which is differential graded algebra with the product defined by

$$fg = \mu(f \otimes g)\Delta$$

$$C \overset{\Delta}{\longrightarrow} C \otimes C \overset{f \otimes g}{\longrightarrow} L \otimes L \overset{\mu}{\longrightarrow} L. \tag{10.5.1}$$

More explicitly, we have for $f \in C^p(A,L)$ and $g \in C^q(A,L)$, the signed cup product

$$fg(a_1,\dots,a_{p+q}) = (-1)^{pq} f(a_1,\dots,a_p)g(a_{p+1},\dots,a_{p+q}). \qquad (10.5.2)$$

The differential is $\delta f = -(-1)^{deg(f)} f \circ b'$ so

$$\begin{aligned}\delta f(a_1,\dots,a_{p+1}) &= (-1)^{p+1} f(b'(a_1,\dots,a_{p+1}))\\ &= (-1)^{p+1}\sum_{i=0}^{p-1} f(a_1,\dots,a_{i+1}a_{i+2},\dots,a_{p+1}). \qquad (10.5.3)\end{aligned}$$

We define a pairing on the bar cochains with values in the Hochschild cochains

$$C((A\otimes[1]\otimes B,\mathbf{k}) \to C(A)\colon (f,h) \mapsto \tau((\partial f)h). \qquad (10.5.4)$$

Let M be an L-bimodule with $\mu\colon L\times M \to M$ the left multiplication; also let $\tau\colon M \to \mathbf{k}$ be a trace, namely a linear functional such that $\tau|[L,M] = 0$. More generally, one can consider $\mathbf{k}$-linear maps from M into a vector space.

A Hochschild cochain is an element of $\mathrm{Hom}_{\mathbf{k}}(A\otimes C,\mathbf{k})$ of the form $\tau((\delta f)h)$ as defined for $f \in C(A,L)$ and $h \in C(A,M)$ by

$$A\otimes C \overset{I}{\longrightarrow} C\otimes C \overset{f\otimes h}{\longrightarrow} L\otimes M \overset{\mu}{\longrightarrow} M \overset{\tau}{\longrightarrow} \mathbf{k}. \qquad (10.5.5)$$

Proposition 10.5.1 $\mathrm{Hom}_{\mathbf{k}}(A\otimes L,\mathbf{k})$ *is a complex with differential* δ *where*

$$\delta\varphi = -(-1)^{deg(\varphi)}\varphi\circ b.$$

The differential satisfies (i) $\delta\tau((\partial f)h) = \tau((\partial(\delta f))h) + (-1)^{deg(f)}\tau((\partial f)\delta h)$;
(ii) *for* $f,g \in C(A,L)$ *and* $h \in C(A,M)$,

$$\tau(\partial(fg)h) = \tau((\partial f)gh) + (-1)^{deg(f)(deg(g)+deg(h))}\tau((\partial g)hf). \qquad (10.5.6)$$

Proof (i) When $f \in C^p(A,L)$ and $h \in C^q(A,M)$ with $n = p+q$, the right-hand side is

$$\begin{aligned}
&\tau\big(\partial(\delta f)h)(a_0 \otimes (a_1,\dots,a_{p+q}) + (-1)^{p+1}\tau(\partial f\delta h)(a_0 \otimes (a_1,\dots,a_{p+q})\big)\\
&\quad = \tau\Big(\mu(\delta f \otimes h)\Big(\sum_{0\le j\le k\le n}(a_{k+1},\dots,a_n a_0,\dots,a_j)\otimes(a_{j+1},\dots,a_k)\Big)\\
&\qquad + (-1)^{p+1}\tau\Big(\mu\partial f\delta h\sum_{0\le j\le k\le n}(a_{k+1},\dots,a_n,a_0,\dots,a_j)\otimes(a_{j+1},\dots,a_k)\Big)\\
&\quad = \tau\sum_j f(b'(a_{q+j+1},\dots,a_j)h(a_{j+1},\dots,a_{j+q}))\\
&\qquad + (-1)^{p+1}\tau\sum_j f(a_{j+q+1},\dots,a_{p+q},a_0,\dots,a_{j-1})h(b'(a_j,\dots,a_{j+q}))
\end{aligned} \tag{10.5.7}$$

and the left-hand side is

$$\begin{aligned}
&(-1)^{deg(\varphi)+1}\varphi\circ b(a_0,\dots,a_{p+q})\\
&\quad = (-1)^{p+q}\tau\mu(f\otimes h)I(b(a_0,\dots,a_{p+q}))\\
&\quad = (-1)^{p+q}\tau\mu(f\otimes h)(b'\otimes id + id\otimes b')I(a_0,\dots,a_{p+q})\\
&\quad = (-1)^{p+q}\tau\sum_j f(b'(a_{j+q+1},\dots,a_{p+q},a_0,\dots,a_j)h(a_{j+1},\dots,a_{j+q})\\
&\qquad + (-1)^{p+q}\tau\sum_j f(a_{j+q+1},\dots,a_{p+q},a_0,\dots,a_{j+1})h(b'(a_j,\dots,a_{j+q})).
\end{aligned} \tag{10.5.8}$$

(ii) Basically, this follows from $JI = 0$:

$$\begin{array}{ccccccc}
A\otimes C & \overset{I}{\longrightarrow} & C\otimes C & \overset{f.g\otimes h}{\longrightarrow} & L\otimes M & \overset{\tau\mu}{\longrightarrow} & \mathbf{k}\\
 & \Delta\otimes id & \downarrow & & \uparrow & \mu\otimes id & \\
 & & C\otimes C\otimes C & \longrightarrow & L\otimes L\otimes M & &
\end{array} \tag{10.5.9}$$

The cochains in (ii) are

$$\begin{aligned}
\tau(\partial(fg)h) &= \tau\mu(\mu\otimes id)(f\otimes g\otimes h)(\Delta\otimes id)I,\\
\tau(\partial(f)gh) &= \tau\mu(id\otimes\mu)(f\otimes g\otimes h)(\Delta\otimes id)I,\\
\tau((\partial g)hf) &= \tau\mu(\mu\otimes id)\sigma^{-1}(f\otimes g\otimes h)\sigma(\Delta\otimes id)I;
\end{aligned} \tag{10.5.10}$$

the hypotheses that $\tau|[L,M] = 0$ implies that the maps

$$\tau\mu(\mu\otimes id),\quad \tau\mu(id\otimes\mu),\quad \tau\mu(1\otimes\mu)\sigma^{-1} \tag{10.5.11}$$

are the same. For instance,

$$\tau\mu(id\otimes\mu)\sigma^{-1}(x,y,z) = \tau\mu(y,mx) = \tau(ymx) = \tau(xym)$$
$$= \tau\mu(id\otimes\mu)(x,y,m). \tag{10.5.12}$$

So we take the alternating sum and use the identity $JI = 0$. □

10.6 Analogue of $\Omega^1 R$ for the Bar Construction

For a unital algebra A, we work with $A\otimes\bar{A}^{\otimes n}$; whereas for a non-unital algebra A, we consider $A[1]\otimes A^{\otimes n}$.

The bar construction of the augmented algebra $A^+ = \mathbf{k}\oplus A$ is the differential graded coalgebra $\mathcal{B} = \oplus_{n=0}^{\infty} A^{\otimes n}$ with differential $d = b'$ and comultiplication

$$\Delta(a_1,\dots,a_n) = \sum_{i=1}^{n}(a_1,\dots,a_i)\otimes(a_{i+1},\dots,a_n). \tag{10.6.1}$$

With $C^n(A,L) = \mathrm{Hom}(\mathcal{B}_n, L)$, we have $C(A,L) = \oplus C^n(A,L)$ and inside $C(A,L)$, we can do curvature calculations. With $\rho\in C^1(A,L)$ we can build $\omega = \delta\rho+\rho^2$ so that $\delta\omega = -ad(\rho)(\omega^n)$. If τ is a trace on L, then tr_τ is a trace on $C(A,L)$, and $\mathrm{tr}_\tau(\gamma) = N\tau\gamma\in C_\lambda(A)$ is a cyclic cochain such that $\delta\mathrm{tr}_\tau(\omega^n) = 0$. In Corollary 11.2.2 we achieve our goal by defining Connes's S operator such that

$$S\left[\mathrm{tr}_\tau\left(\frac{\omega^n}{n!}\right)\right] = \left[\mathrm{tr}_\tau\left(\frac{\omega^{n+1}}{(n+1)!}\right)\right], \tag{10.6.2}$$

where the S operator has degree 2 and is a periodicity map, as in Section 7.7.

Proposition 10.6.1 *Let $R = C(A,L)$. Then there exists an R-bimodule $\Omega^1 R$ such that*

$$f\partial(gh)k = f(\partial g)hk + fg(\partial h)k. \tag{10.6.3}$$

Proof The first step is to take $R = C(A,L)$ and produce $\Omega^1 R$. The differentials relate to bar cochains

$$R^{\otimes 4}\xrightarrow{b'} R^{\otimes 3}\xrightarrow{b'} R^{\otimes 2}\xrightarrow{b'} \tag{10.6.4}$$

With $\mathcal{B}$ the tensor algebra as above, $\mathcal{B}$ is co-unital, so we have an exact sequence

$$\begin{array}{ccccccc} \mathcal{B} & \xrightarrow{\Delta} & \mathcal{B}^{\otimes 2} & \xrightarrow{\Delta\otimes id - id\otimes} & \mathcal{B} & \xrightarrow{\Delta\otimes id\otimes id - id\otimes\Delta\otimes id + id\otimes id\otimes\Delta} & \mathcal{B}^{\otimes 4} \\ & & P\downarrow & \nearrow_I & & & \\ & & \mathcal{B}\otimes A[1]\otimes\mathcal{B} & & & & \end{array} \tag{10.6.5}$$

In this diagram, we take I to distribute two tensor factors into a sum of three tensor factors

$$I((a_1,\ldots,a_p)\otimes(a_{p+1},\ldots,a_n)) = \sum_{0\le j\le p<k\le n}(a_1,\ldots,a_j)\otimes(a_{j+1},\ldots,a_k)$$
$$\otimes(a_{k+1},\ldots,a_n) \tag{10.6.6}$$

and P builds three tensor factors by picking out the difference between the last term in the first tensor factor and the first term in the last tensor factor

$$P((a_1,\ldots,a_p)\otimes(a_{p+1},\ldots,a_n)) \tag{10.6.7}$$

$$\begin{aligned} &= (a_1,\ldots,a_{p-1})\otimes a_p\otimes(a_{p+1},\ldots,a_n) \\ &\quad - (a_1,\ldots,a_p)\otimes a_{p+1}\otimes(a_{p+2},\ldots,a_n), \end{aligned} \tag{10.6.8}$$

so $IP = \Delta\otimes id - id\otimes\Delta$.

There is a projection $\mathcal{B}\to A$ giving a retraction $\mathcal{B}\otimes\mathcal{B}\otimes\mathcal{B}\to B\otimes A[1]\otimes B$, and hence an induced differential on $B\otimes A[1]\otimes B$, namely

$$\begin{aligned} &d(a_1,\ldots,a_p)\otimes a_{p+1}\otimes(a_{p+2},\ldots,a_n) \\ &\quad = b'(a_1,\ldots,a_p)\otimes a_{p+1}\otimes(a_{p+2},\ldots,a_n) \\ &\qquad + (-1)^{p-1}(a_1,\ldots,a_{p-1})\otimes a_pa_{p+1}\otimes(a_{p+2},\ldots,a_n) \\ &\qquad + (-1)^{p}(a_1,\ldots,a_p)\otimes a_{p+1}a_{p+2}\otimes(a_{p+3},\ldots,a_n) \\ &\qquad + (-1)^{p+1}(a_1,\ldots,a_p)\otimes a_{p+1}\otimes b'(a_{p+2},\ldots,a_n). \end{aligned} \tag{10.6.9}$$

For $f,g,h\in C(A,L)$, we define $f(\partial g)h\in \mathrm{Hom}(\mathcal{B}\otimes A[1]\otimes\mathcal{B},L)$ as the composition

$$\mathcal{B}\otimes A[1]\otimes\mathcal{B}\xrightarrow{I}\mathcal{B}^{\otimes 3}\xrightarrow{f\otimes g\otimes h}L^{\otimes 3}\xrightarrow{\mu}L. \tag{10.6.10}$$

Now $\mathrm{Hom}_{\mathbf{k}}(\mathcal{B}\otimes A[1]\otimes\mathcal{B},L)$ is a bimodule over $C(A,L)$, and $f(\partial g)h$ is consistent with the bimodule operations, as follows. With $\eta\colon\mathcal{B}\to\mathbf{k}$ the co-unit, we can write $\partial g = \mu(\eta \otimes g \otimes \eta)$, so $f(\partial g)h$ is as above. Moreover ∂ acts as a derivation for the $C(A,L)$ bimodule in the sense that

$$f\partial(gh)k = f(\partial g)hk + fg(\partial h)k. \tag{10.6.11}$$

□

10.7 Traces Modulo an Ideal

We extend this to traces modulo an ideal. The ultimate aim is to deal with the trace class operators $\mathcal{L}^1(H)$ in $\mathcal{L}(H)$. So we let L be an algebra, K an ideal in L and $\tau : K \to \mathbf{k}$ a trace, so $\tau | [L,K] = 0$. For $\alpha \in \mathrm{Hom}(\mathcal{B} \otimes A[1] \otimes, K)$, we define

$$\tau_\sharp(\alpha) = \tau\alpha\bar{I} \in \mathrm{Hom}_{\mathbf{k}}(A[1] \otimes \mathcal{B}, \mathbf{k}). \tag{10.7.1}$$

The sign conventions dictate that

$$\tau_\sharp(f\alpha) = (-1)^{deg(f)deg(\alpha)}\tau_\sharp(\alpha f). \tag{10.7.2}$$

Suppose that $f,g,h \in C(A,L)$ and one or more of f,g,h takes values in K. Then $f(\partial g)h \in \mathrm{Hom}(\mathcal{B} \otimes A[1] \otimes \mathcal{B}, K)$. Then we can form

$$\tau_\sharp(f(\partial g)h) \in \mathrm{Hom}(A[1] \otimes \mathcal{B}, \mathbf{k})$$

and we can identify this space $\mathrm{Hom}(A[1] \otimes \mathcal{B}, \mathbf{k})$ with $C(A, A^*)$.

Corollary 10.7.1 *$\tau_\sharp(f(\partial g)h)$ gives an element of $C(A,A^*)$ such that*

(i) $$\tau_\sharp(\partial(fg)h) = \tau_\sharp(\partial(f)gh) + (-1)^{deg(f)(deg(g)+deg(h))}\tau_\sharp(\partial(g)hf). \tag{10.7.3}$$

(ii) $$\delta\tau_\sharp(\partial(f)g) = \tau_\sharp(\partial(\delta f)g) + (-1)^{deg(f)}\tau_\sharp(\partial(f)\delta g). \tag{10.7.4}$$

10.8 Analogue of the Quotient by Commutators

Recall that for an R-module M, we have commutator quotient space $M \otimes_R = M/[R,M]$. The next step is an analogue for the bar construction of $\Omega^1 R \otimes_R$.

Definition 10.8.1 (Analogue of bar construction) Let M be a C-bimodule. There exist a left comodule map $\Delta_\ell : M \to C \otimes M$ and a right comodule map $\Delta_r : M \to M \otimes C$ which commute. With $\sigma : M \otimes C \to C \otimes M$ the flip map, define

$$M \sqcup^C = \mathrm{Ker}\{\Delta_\ell - \sigma\Delta_r : M \to C \otimes M\}. \tag{10.8.1}$$

Lemma 10.8.2 *For* $C = T(V)$*, the map* $\sigma(id\otimes\Delta)\colon V\otimes C \to (C\otimes V\otimes C)\sqcup^{C}$ *is an isomorphism.*

Proof Let $\eta\colon C \to \mathbf{k}$ be the co-unit and $\sigma : V \otimes C \otimes C \to C \otimes V \otimes C$ the flip map. Then $\Delta\colon C \to C \otimes C$ is a comultiplication on $C = T(V)$:

$$0 \longrightarrow V\otimes C \;\underset{\eta\otimes id\otimes id}{\overset{\sigma(id\otimes\Delta)}{\rightleftarrows}}\; C\otimes V\otimes V \;\underset{id\otimes\eta\otimes id\otimes id}{\overset{\Delta\otimes id\otimes id-\sigma(id\otimes\Delta\otimes id)}{\rightleftarrows}}\; C\otimes C\otimes V\otimes C. \tag{10.8.2}$$

The sequence is exact.

$$\begin{array}{ccccccc} \mathcal{B}\otimes\mathcal{B} & \overset{P}{\longrightarrow} & \mathcal{B}\otimes A[1]\otimes\mathcal{B} & \overset{I}{\longrightarrow} & \mathcal{B}\otimes\mathcal{B}\otimes\mathcal{B} & \overset{\Delta\otimes id\otimes id-id\otimes\Delta\otimes id-id\otimes id\otimes\Delta}{\longrightarrow} & \mathcal{B}^{\otimes 4} \\ \sigma\Delta\uparrow & & \uparrow\sigma(id\otimes\Delta) & & \uparrow\sigma(id\otimes\Delta) & & \\ \mathcal{B} & \underset{\bar{P}}{\longrightarrow} & A[1]\otimes\mathcal{B} & \underset{\bar{I}}{\longrightarrow} & \mathcal{B}\otimes\mathcal{B} & \underset{id\otimes\Delta-\Delta\otimes id+\sigma(id\otimes\Delta)}{\longrightarrow} & \mathcal{B}^{\otimes 3} \end{array} \tag{10.8.3}$$

The top row is exact and hence the bottom row is also exact for the maps

$$\bar{I}(a_0\otimes(a_1,\ldots,a_n) = \sum_{0\le i\le k\le n}(-1)^{n(k+1)}(a_{k+1},\ldots,a_n,a_0,\ldots,a_i) \otimes(a_{i+1},\ldots,a_k) \tag{10.8.4}$$

and

$$\bar{P}(a_0\otimes(a_1,\ldots,a_n)) = -a_0\otimes(a_1,\ldots,a_n)+(-1)^{n-1}a_n\otimes(a_0,\ldots,a_{n-1}). \tag{10.8.5}$$

The induced differential is $d\colon A[1]\otimes\mathcal{B}\to A[1]\otimes\mathcal{B}$ on $A[1]\otimes A^{\otimes n}$

$$d(a_0\otimes(a_1,\ldots,a_n)) = a_0a_1\otimes(a_2,\ldots,a_n) - a_0\otimes b'(a_1,\ldots,a_n) + (-1)^n(a_na_0)\otimes(a_1,\ldots,a_{n-1}). \tag{10.8.6}$$

When we identify $A[1]\otimes A^{\otimes n}$ with $A^{\otimes n+1}$, we have $d = b$ the Hochschild differential and $\bar{P} = \lambda - 1$, where λ is the cyclic permutation with sign. □

Now we seek a sequence of complexes and maps ∂ and β which have properties analogous to (10.4.13). Let $C^n(A) = (A^{\otimes n})^*$ and $C^n(A,A^*) = (A^{\otimes n})^*$, then $C(A) = \mathrm{Hom}_{\mathbf{k}}(\mathcal{B},\mathbf{k})$ are the bar cochains, and $C(A,A^*) = \mathrm{Hom}_{\mathbf{k}}(A[1]\otimes\mathcal{B},\mathbf{k})$ are the cyclic bar cochains; also $(A[1]\otimes\mathcal{B},b)$ is the cyclic bar construction. We identify the following spaces

$$C^n(A,A^*) = \mathrm{Hom}((A[1]\otimes\mathcal{B})_{n+1},\mathbf{k}) = (A\otimes A^{\otimes n})^* = C_{\sharp}^{n+1}(A),$$

where $(A\otimes A^{\otimes n})^*$ has the differential b.

The operator $\bar{I}$ gives rise to $N = (id \otimes \eta)I$, where $N\colon A[1]\otimes\mathcal{B} \to \mathcal{B}\colon N = \sum_{i=0}^{n} \lambda^i$ is the cyclic symmetrization operator in degree n. This gives rise to ∂. The operator $\bar{P}$ gives rise to $\beta = \lambda - 1$.

In the following complex, the rows are exact and the image of $\partial = N$ is equal to the kernel of β, namely $C_\lambda(A)$, the cyclic cochain complex of A. We write

$$C_\lambda^n(A) = \mathrm{Im}\{\partial\colon C^{n+1}(A) \to C^n(A, A^*)\}.$$

$$\begin{array}{ccccccccc}
C^3(A) & \xrightarrow{\partial} & C^2(A,A^*) & \xrightarrow{\beta} & C^3(A) & \xrightarrow{\partial} & C^2(A,A^*) & \longrightarrow & \\
 & & b\uparrow & & \uparrow -b' & & \uparrow b & & \\
C^2(A) & \xrightarrow{\partial} & C^1(A,A^*) & \xrightarrow{\beta} & C^2(A) & \xrightarrow{\partial} & C^1(A,A^*) & \longrightarrow & (10.8.7) \\
 & & b\uparrow & & \uparrow -b' & & \uparrow b & & \\
C^1(A) & \overset{\partial}{\hookrightarrow} & C^0(A,A^*) & \xrightarrow{\beta} & C^1(A) & \xrightarrow{\partial} & C^0(A,A^*) & \longrightarrow &
\end{array}$$

10.9 Connes's Chain and Cochain Bicomplexes

The chain version of Connes's bicomplex is given below. The third column is the cyclic bar construction $A[1] \otimes \mathcal{B}$, while the fourth column is $\mathcal{B} = \oplus_{n=0}^{\infty} A^{\otimes n}$. This is the bar construction of $\bar{A}$ up to renorming $\mathbf{k} = A^{\otimes 0}$ and shifting dimensions:

$$\begin{array}{ccccccccc}
0 & \longleftarrow & A_\lambda^{\otimes 3} & \longleftarrow & A^{\otimes 3} & \xleftarrow{1-\lambda} & A^{\otimes 3} & \xleftarrow{N} & A^{\otimes 3} \\
 & & & & \downarrow b & & \downarrow -b' & & \downarrow b \\
0 & \longleftarrow & A_\lambda^{\otimes 2} & \longleftarrow & A^{\otimes 2} & \xleftarrow{1-\lambda} & A^{\otimes 2} & \xleftarrow{N} & A^{\otimes 2} \\
 & & & & \downarrow b & & \downarrow -b' & & \downarrow b \\
0 & \longleftarrow & A_\lambda & \longleftarrow & A & \xleftarrow{1-\lambda} & A & \xleftarrow{N} & A
\end{array} \qquad (10.9.1)$$

Taking the duals, we obtain the cochain complex.

$$\begin{array}{ccccccccc}
C_\lambda^2(A) & \hookrightarrow & C^2(A,A^*) & \xrightarrow{\beta} & C^3(A) & \xrightarrow{\partial} & C^2(A,A^*) & \longrightarrow & \\
 & & b\uparrow & & \uparrow -b' & & \uparrow b & & \\
C_\lambda^1(A) & \hookrightarrow & C^1(A,A^*) & \xrightarrow{\beta} & C^2(A) & \xrightarrow{\partial} & C^1(A,A^*) & \longrightarrow & (10.9.2) \\
 & & b\uparrow & & \uparrow -b' & & \uparrow b & & \\
C_\lambda^0(A) & \hookrightarrow & C^0(A,A^*) & \xrightarrow{\beta} & C^1(A) & \xrightarrow{\partial} & C^0(A,A^*) & \longrightarrow &
\end{array}$$

The cyclic homology of A is defined by $(\Omega A, b, B)$:

- $HH_n(A)$ Hochschild homology $H_n(\Omega A, b)$;
- $HC_n(A)$ cyclic homology $H_n(\Omega A, b + B)$;
- $HP(A)$ periodic cyclic homology, as in Section 12.3.

11
Cyclic Cohomology

Connes's S operator emerged in Section 7.7 as a connecting operator in the exact homology for the periodic complex. The S operator is analogous to the suspension operation in K-theory, which gives the Bott periodicity theorem, as in Theorem 3.3.7. In this chapter we obtain a more explicit form for S. The chapter begins with Connes's bicomplex, which is the ultimate product of the computations in the preceding Chapter 10. The bicomplex exhibits a periodicity which is realized through Connes's S operator, as discussed in terms of the results from Chapter 10. The chapter includes some homotopy formulas for cyclic cocycles. Finally, there are examples of cyclic cocycles arising from Dirac operators on manifolds, and Dirac operators twisted by connections. These results use methods from Chapters 5 and 9.

11.1 Connes's Double Cochain Complex

The cyclic cochain complex of A is the following, in which the image of ∂ equals the kernel of β, namely the cyclic cochain complex of $C_\lambda(A)$ of A. The differentials are δ on $C^n(A)$, where $\delta\varphi = -(-1)^{deg(\varphi)}b'\varphi$, and δ on $C^n(A, A^*)$ is $\delta\psi = -(-1)^{deg(\psi)}b\psi$:

$$\begin{array}{ccccccccc} C_\lambda^2 & \hookrightarrow & C^2(A,A^*) & \xrightarrow{\beta} & C^3(A) & \xrightarrow{\partial} & C^2(A,A^*) & \longrightarrow & \\ & & b\uparrow & & \uparrow -b' & & \uparrow b & & \\ C_\lambda^1 & \hookrightarrow & C^1(A,A^*) & \xrightarrow{\beta} & C^2(A) & \xrightarrow{\partial} & C^1(A,A^*) & \longrightarrow & \\ & & b\uparrow & & \uparrow -b' & & \uparrow b & & \\ C_\lambda^0 & \hookrightarrow & C^0(A,A^*) & \xrightarrow{\beta} & C^1(A) & \xrightarrow{\partial} & C^0(A,A^*) & \longrightarrow & \end{array} \qquad (11.1.1)$$

In the next section, we carry out the following diagram chase:

$$
\begin{array}{ccccccc}
\varphi & & & & & & \\
\uparrow & & & & & & \\
* & \xrightarrow{N} & \psi & \xrightarrow{1-\lambda} & * & & \\
\uparrow b' & & \uparrow b & & \uparrow b' & & \\
* & \xrightarrow{N} & * & \xrightarrow{1-\lambda} & \varphi & \longrightarrow & * \\
 & & & & \uparrow & & \uparrow \\
 & & & & & & \psi
\end{array}
\tag{11.1.2}
$$

We pass from the chain complex to the cochain complex by taking duals. The operators are identified by formulas such as

$$bf(a_1, \dots, a_n) = f(b(a_1, \dots, a_n)), \tag{11.1.3}$$

which make sense for multilinear maps $f \in C^n(A, V)$.

11.2 Connes's S Operator

Let $HC^n(A) = H^n(C_\lambda(A))$. The operator $S\colon HC^n(A) \to HC^{n+2}(A)$ is induced by embedding the double cochain complex in itself by moving two steps to the right. This exploits the periodicity of the array. The explicit formula for S at the level of cochains is found as follows. Suppose that $\varphi \in C^n(A, A^*)$ satisfies $b\varphi = 0$ and $(1 - \lambda)\varphi = 0$. Then the rows are exact, so we can choose $\psi \in C^{n+1}(A)$ and $\psi' \in C^{n+1}(A, A^*)$ such that

$$\varphi = N\psi \tag{11.2.1}$$

then $N(-b')\psi = bN\psi = b\varphi = 0$, so there exists $\psi' \in C^{n+1}(A, A^*)$ such that

$$b'\psi = (1 - \lambda)\psi'. \tag{11.2.2}$$

Then $S[\varphi]$ is represented by $b\psi'$. We define $S[\varphi] = -[b\psi']$. The statement of Corollary 11.2.2 shows how this relates to some of the fundamental cocycles that we have been using.

Let $\rho\colon A \to L$ be a linear map that is a homomorphism modulo an ideal K of L, and $\tau\colon K^m \to \mathbf{k}$ a trace on K^m. Then $\tau|[L, K^m] = 0$, and $\tau\colon [K^i, K^j] = 0$ for $i + j \geq m$. As usual, let $\omega = \delta\rho + \rho^2$ be the curvature. We use from Section 10.7 the notation $\tau_\sharp$ to refer to traces modulo an ideal.

Proposition 11.2.1 (i) *For $n \geq m$, $\tau(\omega^n) \in C^{2n}(A)$ and $\tau_\sharp((\partial\rho)\omega^n) \in C^{2n}(A, A^*)$;*

$$\text{(ii)} \qquad \delta\tau(\omega^n) = \beta\tau_\sharp((\partial\rho)\omega^n); \tag{11.2.3}$$

$$\text{(iii)} \qquad \delta\tau_\sharp\left((\partial\rho)\frac{\omega^n}{n!}\right) = \partial\tau\left(\frac{\omega^{n+1}}{(n+1)!}\right). \tag{11.2.4}$$

Proof (i) We have $\omega \in C^2(A, K)$, and taking products we deduce that $\omega^n \in C^{2n}(A, K^n)$, so $\tau(\omega^n) \in C^{2n}(A)$. Likewise, $\tau_\sharp((\partial\rho)\omega^n) \in C^{2n+1}(A)$, which we identify with $C^{2n}(A, A^*)$.

(ii) By Bianchi's identity

$$\begin{aligned}\delta\tau(\omega^n) &= \tau(\delta\omega^n)\\ &= \tau(-\rho\omega^n + \omega^n\rho)\\ &= \beta\tau_\sharp((\partial\rho)\omega^n).\end{aligned} \tag{11.2.5}$$

(iii) Likewise, $\delta\omega = \partial\delta\rho + (\delta\rho)\rho + \rho(\delta\rho)$, so

$$\begin{aligned}\delta\tau_\sharp((\partial\rho)\omega^n) &= \tau((\partial(\delta\rho)\omega^n - (\partial\rho)\delta\omega^n)\\ &= \tau_\sharp(\partial(\delta\rho)\omega^n + \partial\rho(\rho\omega^n - \omega^n\rho))\\ &= \tau_\sharp(\partial(\delta\rho) + \rho\partial\rho + (\partial\rho)\rho)\omega^n)\\ &= \tau_\sharp((\partial\omega)\omega^n),\end{aligned} \tag{11.2.6}$$

which turns out to be

$$\begin{aligned}\partial\tau_\sharp(\omega^{n+1}) &= \tau_\sharp\left(\sum_{i=0}^{n}\omega^i(\partial\omega)\omega^{n-i}\right)\\ &= (n+1)\tau_\sharp((\partial\omega)\omega^n).\end{aligned} \tag{11.2.7}$$

□

The cocycles $\varphi_{2n} = \tau(\omega^n)$ and $\psi_{2n+1} = \tau_\sharp((\partial\rho)\omega^n)$ have a special role in the theory. We have

$$\begin{aligned}\varphi_{2n}(a_1, \dots, a_{2n}) &= \tau(\omega(a_1, a_2)\dots\omega(a_{2n-1}, a_{2n})),\\ \psi_{2n+1}(a_0, \dots, a_{2n}) &= \tau(\rho(a_0)\omega(a_1, a_2)\dots\omega(a_{2n-1}, a_{2n})),\end{aligned} \tag{11.2.8}$$

which satisfy

$$b'\varphi_{2n} = (1 - \lambda)\psi_{2n+1}, \tag{11.2.9}$$

$$b\psi_{2n+1} = \frac{1}{n+1}N\varphi_{2n+2}, \tag{11.2.10}$$

as one can verify by direct computation. □

Corollary 11.2.2 *For $m \geq n$, $N\tau(\omega^n)$ is a cyclic cocycle of degree $2n-1$ and so determines a cyclic cocycle class $[N\tau(\omega^n)] \in HC^{2n-1}(A)$ such that*

$$S\left[N\tau\left(\frac{\omega^n}{n!}\right)\right] = \left[N\tau\left(\frac{\omega^{n+1}}{(n+1)!}\right)\right]. \tag{11.2.11}$$

Proof We let $\varphi = N\tau(\omega^n/n!)$, so $\varphi = N\psi$ where $\psi = \tau(\omega^n/n!)$ and

$$-b'\psi = -(1-\lambda)\tau_\sharp\left((\partial\rho)\frac{\omega^n}{n!}\right); \tag{11.2.12}$$

so let

$$\psi' = \tau_\sharp\left((\partial\rho)\frac{\omega^n}{n!}\right)$$

and then

$$-b\psi' = N\tau\left(\frac{\omega^{n+1}}{(n+1)!}\right). \tag{11.2.13}$$

Then $S[\varphi] = -[b\psi']$ gives (11.2.11). Also, $\lambda^2\tau(\omega^n) = \tau(\omega^n)$, so

$$\begin{aligned} N\tau(\omega^n) &= \sum_{j=0}^{n-1}\lambda^j\tau(\omega^n) \\ &= n\tau(\omega^n)(1+\lambda)(a_1,\dots,a_{2n}) \\ &= n\tau(\omega(a_1,a_2)\dots\omega(a_{2n-1},a_{2n})). \end{aligned} \tag{11.2.14}$$

□

11.3 Connes's Long Exact Sequence

The following sequence is exact:

$$HC^{n-1}(A) \overset{S}{\longrightarrow} HC^{n+1}(A) \overset{I}{\longrightarrow} H^{n+1}(A,A^*) \overset{\beta}{\longrightarrow} HC^n(A) \tag{11.3.1}$$

The columns of the bicomplex (11.1.1) give

$$0 \longrightarrow C_\lambda \longrightarrow C \longrightarrow C/C_\lambda \longrightarrow 0, \tag{11.3.2}$$

so by the five lemma [33, p. 641] we have a long exact sequence in homology:

$$\begin{array}{ccccccccc} HC^n(C_\lambda) & \longrightarrow & H^n(C) & \longrightarrow & H^n(C/C_\lambda) & \longrightarrow & H^{n+1}(C_\lambda) & \longrightarrow & H^{n+1}(C) \\ \downarrow = & & \downarrow = & & & & \downarrow B & & \downarrow = \\ HC^n(A) & \overset{I}{\longrightarrow} & H^n(A,A^*) & \overset{B}{\longrightarrow} & HC^n(A) & \longrightarrow & H^{n+1}(A) & \longrightarrow & H^{n+1}(A,A^*) \end{array}$$

11.4 A Homotopy Formula for Cocycles Associated with Traces

Let A be a non-unital associative algebra over $\mathbf{k}$, and $R = RA$ where $RA = \oplus_{n=0}^{\infty} A^{\otimes n+1}$ with augmented algebra $R^+ = R + \mathbf{k}1$. We have

$$\Omega^1 R \cong R^+ \otimes A \otimes R^+, \tag{11.4.1}$$

$$\Omega^1 R/[R, \Omega^1 R] \cong R^+ \otimes A. \tag{11.4.2}$$

The terminology used suggests applications to homotopy of cocycles which we discussed in Section 10.5. The formulas themselves resemble the calculations of Section 6.5.

Definition 11.4.1 (Homotopy of cocycles)

(i) A trace τ on R is null homotopic if $\tau = \tau' \circ d$, for some trace τ' on the R-bimodule $\Omega^1 R$ that is given by cochains $\omega^{n-1} d\rho$ and $-\rho\omega^{n-1} d\rho$.

(ii) The g-cochain of τ' is the sequence

$$g_{2n-1} = \tau'(\omega^{n-1} d\rho), \tag{11.4.3}$$

$$g_{2n} = \tau'(-\rho\omega^{n-1} d\rho). \tag{11.4.4}$$

(iii) Let $\mu_{2n-1} = \sum_{i=1}^{n} \omega^{i-1}(d\rho)\omega^{n-i}$, so

$$\mu_{2n-1}(a_0, \dots, a_{2n-1}) = \sum_{i=1}^{n} \omega(a_1, a_2) \dots \omega(a_{2i-3}, a_{2i-2}) \times (d\rho(a_{2i-1}))\omega(a_{2i}, a_{2i+1}) \dots \omega(a_{2n-2}, a_{2n-1}). \tag{11.4.5}$$

Then the h-cochain of τ' is

$$h_{2n-1} = \tau'(\mu_{2n-1}), \tag{11.4.6}$$

$$h_{2n} = \tau'(-\rho\mu_{2n-1}). \tag{11.4.7}$$

Theorem 11.4.2 *If $\tau = \tau' \circ d$, then the cocycle of τ is the h-cochain associated with τ'.*

Proof We require to prove that

$$\text{(i)} \qquad \tau' \circ d(\omega^n) = b' h_{2n-1} + (\lambda - 1) h_{2n}, \tag{11.4.8}$$

$$\text{(ii)} \qquad \tau' \circ d(\rho\omega^n) = -b h_{2n-1} + \frac{N}{n+1} h_{2n+1}. \tag{11.4.9}$$

(i) We start with $\omega = \delta\rho + \rho^2 \colon A^{\otimes 2} \to R$ and compute

$$d\omega = \delta d\rho + \rho d\rho + (d\rho)\rho = (\delta + ad(\rho))(d\rho), \tag{11.4.10}$$

so $(\delta + ad(\rho))\omega = 0$ and

$$\begin{aligned} d\omega^n &= \sum_{i=1}^{n} \omega^{i-1}(d\omega)\omega^{n-i} \\ &= \sum_{i=1}^{n} \omega^{i-1}(\delta + ad(\rho))(d\rho)\omega^{n-i} \\ &= (\delta + ad(\rho)) \sum_{i=1}^{n} \omega^{i-1}(d\rho)\omega^{n-i} \\ &= (\delta + ad(\rho))\mu_{2n-1}. \end{aligned} \tag{11.4.11}$$

We deduce that

$$\begin{aligned} b'\mu_{2n-1}(a_1, \dots, a_{2n}) &= \delta\mu_{2n-1}(a_1, \dots, a_{2n}) \\ &= (-\rho\mu_{2n-1} - \mu_{2n-1}\rho + d\omega^n)(a_1, \dots, a_{2n}) \\ &= -\rho(a_1)\mu_{2n-1}(a_1, \dots, a_{2n-1}) \\ &\quad - \mu_{2n-1}(a_1, \dots, a_{2n-1})\rho(a_{2n}) + d(\omega^n)(a_1, \dots, a_{2n}) \end{aligned} \tag{11.4.12}$$

and we apply τ' to this expression where τ' commutes with b', so

$$\begin{aligned} b'h_{2n-1}(a_1, \dots, a_{2n}) &= h_{2n}(a_1, \dots, a_{2n}) + h_{2n}(a_{2n}, a_1, \dots, a_{2n-1}) \\ &\quad + \tau' d(\omega^n)(a_1, \dots, a_{2n}), \end{aligned} \tag{11.4.13}$$

which is (i).

(ii) We compute $h_{2n} = -\tau'(\rho\mu_{2n-1})$ and

$$\begin{aligned} bh_{2n}(a_0, \dots, a_{2n}) &= \tau'\big(\rho(a_0a_1)\mu_{2n-1}(a_2, \dots, a_{2n})\big) \\ &\quad - \tau'\big(\rho(a_0)b'\mu_{2n-1}(a_1, \dots, a_{2n})\big) \\ &\quad + \tau'\big(\rho(a_{2n}a_0)\mu_{2n-1}(a_1, \dots, a_{2n-1})\big), \end{aligned} \tag{11.4.14}$$

and we use the formula from (i) to eliminate the term involving b' and obtain

$$\begin{aligned}
&bh_{2n}(a_0,\dots,a_{2n})\\
&\quad=\tau'\Bigg(\sum_{i=1}^{n}\omega^i(a_0,\dots,a_{2i-1})d\rho(a_{2i})\omega^{n-i}(a_{2i-1},\dots,a_{2n})\\
&\qquad+\sum_{i=1}^{n}\omega^i(a_{2n},a_0,\dots,a_{2i-2})d\rho(a_{2i-1})\omega^{n-i}(a_{2i},\dots,a_{2n-1})\\
&\qquad-\rho(a_0)d(\omega^n)(a_1,\dots,a_{2n})\Bigg)\\
&\quad=\tau'\Bigg(\sum_{i=1}^{n}d\rho(a_{2i})\omega^{n-i}(a_{2i-1},\dots,a_{2n})\omega^i(a_0,\dots,a_{2i-1})\\
&\qquad+\sum_{i=1}^{n}d\rho(a_{2i-1})\omega^n(a_{2i},\dots,a_{2i-2})-\rho(a_0)d(\omega^n)(a_1,\dots,a_{2n})\Bigg)\\
&\quad=\sum_{j=0}^{n}\tau'(d\rho(a_j)\omega^n(a_{j+1},\dots,a_{j-1})-\tau'd(\rho\omega^n)(a_0,\dots,a_{2n})\\
&\quad=(n+1)N\tau'((d\rho)\omega^n)(a_0,\dots,a_{2n})-\tau'\big(d(\rho\omega^n)\big)(a_0,\dots,a_{2n}),
\end{aligned}\tag{11.4.15}$$

where we have used the trace property to rearrange terms. Hence (ii) follows. □

This proof admits of variations which are significant in homotopy theory, as discussed in Section 10.5.

11.5 Universal Graded and Ungraded Cocycles

Suppose that $F\in\mathcal{L}(H)$ satisfies $F=F^*$ and $F^2=1$, and there is a $*$-homomorphism $\pi: A\to\mathcal{L}(H)$ such that $[F,a]\in\mathcal{L}^m(H)$ for all $a\in A$. We say that (H,π,F) is an m-summable Fredholm module. In this section, we use universal properties of QA to produce cocycles. Note that m-summability is the type of condition involved in Corollary 9.6.2, and is a much more restrictive condition than theta summability, as introduced in Section 5.6.

Then by the universal property of QA, there is a homomorphism $QA \to \mathcal{L}(H)$: $\iota(a) \mapsto a$ and $\bar{\iota}(a) \mapsto FaF$ such that

$$a^+ \mapsto (1/2)(a + FaF),$$
$$a^- \mapsto (1/2)(a - FaF) = (1/2)F[F,a] \in \mathcal{L}^m(H).$$

Let $J = \mathrm{Ker}(\mathrm{fold}\colon A \star A \to A\}$ be the ideal in QA, which is the kernel of the folding map, so that under the correspondence with the reduced tensor algebra $J^m \cong \oplus_{n=m}^{\infty} A \otimes (\bar{A})^{\otimes n}$

$$a^+ a_1^- \dots a_n^- \longleftrightarrow (a_0, a_1, \dots, a_n). \tag{11.5.1}$$

Since $a_1^- \dots a_n^- \in \mathcal{L}^m(H)$ for $n \geq m$, we have $J^m \subseteq \mathcal{L}^1(H)$, and we can apply the usual trace to J^m. We also use the superbracket and adjoint $ad(\rho)$ such that

$$ad(\rho)(\alpha) = [\rho, \alpha] = \rho\alpha - (-1)^{deg(\rho)deg(\alpha)}\alpha\rho. \tag{11.5.2}$$

The discussion splits into the ungraded and the graded cases.

(1) In the ungraded case, we let $\tau\colon J^m \to \mathbf{C}$ be $\tau(x) = \mathrm{trace}(Fx)$. Then

$$\tau(xy) = (-1)^{deg(x)deg(y)}\tau(yx) \qquad (x \in J^i, y \in J^{m-i}) \tag{11.5.3}$$

and $\tau(xy) = \tau(yFxF)$ since

$$\begin{aligned}\tau(xy) &= \mathrm{trace}(Fxy) = \mathrm{trace}(yFx) = \mathrm{trace}(yFxF^2)\\ &= \mathrm{trace}(FyFxF) = \tau(yFxF).\end{aligned} \tag{11.5.4}$$

We introduce an even cocycle

$$\begin{aligned}f_n(a_0, \dots, a_{2n}) &= \mathrm{trace}(Fa_0^+ a_1^- \dots a_{2n}^-)\\ &= c_n \mathrm{trace}(Fa_0^+[F,a_1]\dots[F,a_{2n}])\end{aligned} \tag{11.5.5}$$

for come constant c_n.

(2) In the graded case, let $\varepsilon \in \mathcal{L}(H)$ such that $\varepsilon = \varepsilon^*$ and $\varepsilon^2 = 1$, with $\varepsilon a = a\varepsilon$ for all $a \in A$ and $\varepsilon F = -F\varepsilon$. The pair (ε, F) gives a $\mathbf{Z}/(2)$ grading of H into the $\pm$ eigenspaces of ε, so $H = H_+ \oplus H_-$. Then we define

$$\tau(x) = \mathrm{trace}(\varepsilon x) \qquad (x \in J^m), \tag{11.5.6}$$

so $\tau(FxF) = -\tau(x)$. Then we introduce a cocycle

$$\begin{aligned}f_{2n-1}(a_0, \dots, a_{2n-1}) &= \mathrm{trace}(\varepsilon a_0^+ a_1^- \dots a_{2n}^-)\\ &= c_n' \mathrm{trace}(\varepsilon F a_0[F,a_1]\dots[F,a_{2n-1}])\end{aligned} \tag{11.5.7}$$

for some constant c'_n. These cocycles are κ invariant and are normalized Hochschild cochains such that

$$bf_n = \frac{2}{n+2}Bf_{n+2} \qquad \kappa f_n = f_n. \tag{11.5.8}$$

These facts follows from the next lemma. We recall $d\colon \Omega^n A \to \Omega^{n+1}A$ given by

$$d(a_0da_1\dots da_n) = da_0da_1\dots da_n \tag{11.5.9}$$

can otherwise be expressed as the $s\colon A\otimes(\bar{A})^{\otimes n} \to (\bar{A})^{\otimes n+1}\colon (a_0,\dots,a_n)\mapsto (1,a_0,\dots,a_n)$.

Lemma 11.5.1 *Let $f_n(a_0,\dots,a_n) = \tau(a_0^+a_1^-\dots a_n^-)$. Then τ vanishes on $[J,J^{m+1}]$ if and only if*

$$\begin{aligned} bf_n &= \frac{2}{n+2}Bf_{n+2},\\ \kappa f_n &= f_n \qquad (n\geq m). \end{aligned} \tag{11.5.10}$$

Proof We compute

$$\begin{aligned} b(pq^n)(a_0,\dots,a_n) &= (a_0a_1)^+a_2^-\dots a_{n+1}^- - a_0^+b'(q^n)(a_1,\dots,a_{n+1})\\ &\quad + (-1)^{n+1}(a_{n+1}a_0)^+a_1^-\dots a_n^-\\ &= (a_0^-a_1^-)a_2^-\dots a_{n+1}^- + (-1)^n a_0^+a_1^-\dots a_n^-a_{n+1}^+\\ &\quad + (-1)^{n+1}(a_{n+1}a_0)^+a_1^-\dots a_n^-\\ &= a_0^-\dots a_{n+1}^- + (-1)^{n+1}a_{n+1}^-a_0^-\dots a_n^-\\ &\quad + (-1)^{n+1}[a_0^+a_1^-\dots a_n^-,a_{n+1}^+]\\ &= (1+\lambda)q^{n+2}(a_0,\dots,a_{n+1})\\ &\quad + (-1)^{n+1}[a_0^+a_1^-\dots a_n^-,a_{n+1}^+] \end{aligned} \tag{11.5.11}$$

and also

$$\begin{aligned} (1-\kappa)\rho q^n)(a_0,\dots,a_n) &= a_0^+a_1^-\dots a_n^- - \kappa(a_0^+a_1^-\dots a_n^-)\\ &= a_0^+a_1^+\dots a_n^- - (-1)^n a_n^-a_0^+a_1^-\dots a_{n-1}^-\\ &= [a_0^+a_1^-\dots a_{n-1}^-,a_n^-]. \end{aligned} \tag{11.5.12}$$

Hence

$$bf_n - (1+\lambda)df_{n+2})(a_0\dots,a_{n+1}) = \tau[a_0^+a_1^-\dots a_n^-,a_{n+1}^+]$$

and

$$(1-\kappa)f_n(a_0,\dots,a_n) = \tau[a_0^+a_1^-\dots a_{n-1}^-,a_n^-]. \tag{11.5.13}$$

If τ is a supertrace, then the right-hand sides are zero, so

$$\kappa f_n = f_n,$$

$$df_n = d\kappa f_n = \kappa df_n = \lambda df_n,$$

and

$$(1+\lambda)df_n = 2df_n = \frac{2}{n+2}Ndf_{n+2} = \frac{2}{n+2}Bf_{n+2}. \tag{11.5.14}$$

The converse is similar. □

11.6 Deformations of Fredholm Modules

The goal is to obtain a homotopy formula for a family of Fredholm modules (H, π_t, F_t) over an algebra A, stating that if F_t is deformed smoothly, then the corresponding cocycles are cohomologous. In Section 8.3, we considered the case of connections over a commutative algebra and showed that the Chern class is independent of the choice of connection. Here, our ultimate aim is to have invariance of cyclic cohomology classes.

Lemma 11.6.1 *Let $\{F_t : t \in \mathbf{R}\}$ be a continuously differentiable family $\mathcal{L}(H)$ such that $F_t^* = F_t$ and $F_t^2 = 1$. Then there exists a continuously differentiable family of unitary operators $\{U_t : t \in \mathbf{R}\}$ such that $U_t F_t U_t^* = F_0$ is constant.*

Proof Using a dot to indicate a derivative with respect to t, we have $\dot{F}_t F_t + F_t \dot{F}_t = 0$ and $\dot{F}_t = \dot{F}_t^*$, so $F_t \dot{F}_t \in \mathcal{L}(H)$ is a skew operator. Hence for all $t_0 > 0$, the initial value problem

$$\begin{aligned} \dot{U}_t &= (1/2)U_t F_t \dot{F}_t \\ U_0 &= I \end{aligned} \tag{11.6.1}$$

has a unique solution U_t on $[-t_0, t_0]$ such that $U_t U_t^* = 1 = U_t^* U_t$ by the existence theory for linear ordinary differential equations. Now

$$\begin{aligned} (U_t F_t U_t^*)^{\cdot} &= \dot{U}_t F_t U_t^* + U_t \dot{F}_t U_t^* - U_t F_t U_t^* \dot{U}_t U_t^* \\ &= (1/2)U_t F_t \dot{F}_t F_t U_t^* + U_t \dot{F}_t U_t^* - (1/2)U_t F_t F_t \dot{F}_t U_t^* \\ &= (1/2)U_t (F_t \dot{F}_t F_t + \dot{F}_t F_t) U_t^* \\ &= 0, \end{aligned} \tag{11.6.2}$$

so $U_t F_t U_t^*$ is constant, with value F_0. □

We include the original $*$-homomorphism $\pi: A \to \mathcal{L}(H)$ in the family of $*$-representations $\pi_t: A \to \mathcal{L}(H): a \mapsto U_t a U_t^*$ and then extend to $QA \to \mathcal{L}(H): \iota(a) = U_t a U_t^*$ and $\tilde{\iota}(a) = F_0 U_t a U_t^* F_0$. Now F_t anti-commutes with $(1/2)F_t \dot{F}_t = U_t^* \dot{U}_t$, so $F_0 = U_t F_t U_t^*$ anti-commutes with $L = \dot{U}_t U_t^*$. Hence we have

$$(\iota(a))^{\cdot} = (U_t a U_t^*)^{\cdot} = [L, \iota(a)] \tag{11.6.3}$$

and

$$(\tilde{\iota}a)^{\cdot} = F_0[L, \iota(a)]F_0 = -[L, \tilde{\iota}(a)], \tag{11.6.4}$$

hence

$$(a^+)^{\cdot} = [L, a^-]; \qquad (a^-)^{\cdot} = [L, a^+]. \tag{11.6.5}$$

Let QLQ be the free super bimodule over QA generated by L, so $QLA \cong Q \otimes Q$.

Lemma 11.6.2 *There is a unique even derivation of degree zero $D: QA \to QLA$ such that*

$$D(a^+) = La^- - a^- L, \quad D(a^-) = La^+ - a^+ L. \tag{11.6.6}$$

Proof This follows from the lemma. □

We extend this to the bar construction as follows. Let $\mathcal{B}(A) = \oplus_{n=0}^{\infty} A^{\otimes n}$ so that $\mathcal{B}(A)$ has differential b' in degree n. Also $\mathcal{B}(A)$ is a differential graded coalgebra with

$$\Delta(a_1, \dots, a_n) = \sum_{i=0}^{n} (a_1, \dots, a_i) \otimes (a_{i+1}, \dots, a_n).$$

The space $\mathrm{Hom}(\mathcal{B}(A), Q)$ is the set of all inhomogeneous cochains on A with values in $Q = QA$. We introduce a $\mathbf{Z}/(2)$ grading on $\mathrm{Hom}(\mathcal{B}(A), Q)$ by

$$\begin{aligned}
\mathrm{Hom}(\mathcal{B}(A), Q)^+ &= \mathrm{Hom}(\mathcal{B}(A)^+, Q^+) \oplus \mathrm{Hom}(\mathcal{B}(A)^-, Q^-), \\
\mathrm{Hom}(\mathcal{B}(A), Q)^- &= \mathrm{Hom}(\mathcal{B}(A)^+, Q^-) \oplus \mathrm{Hom}(\mathcal{B}(A)^-, Q^+).
\end{aligned} \tag{11.6.7}$$

There is a product defined on $\mathrm{Hom}(\mathcal{B}(A), Q)$ by $fg = \mu_Q(f \otimes g)\Delta$, so

$$fg(a_1, \dots, a_n) = \sum_{i=0}^{n} (-1)^{i\, deg(g)} f(a_1, \dots, a_i) g(a_{i+1}, \dots, a_n) \tag{11.6.8}$$

and a derivation $\delta f = (-1)^{deg(f)} f \circ b'$.

The cocycles that are particularly related to homotopy arise as follows. Let $\rho(a) = a^+$ and $q(a) = a^-$, then let $(\rho q^n)(a_0, \dots, a_n) = a_0^+ a_1^- \dots a_n^-$. Also, let $\delta L = 0$ and

$$Dq = L\rho + \rho L = (\delta + ad(\rho))(L), \tag{11.6.9}$$

$$D\rho = Lq + qL.$$

Let $\mu_n = \sum_{i=0}^{n-1} q^i L q^{n-i-1}$, so

$$\mu_n(a_1, \dots, a_n) = \sum_{i=0}^{n-1} (-1)^i a_1^- \dots a_i^- L a_{i+1}^- \dots a_n^-. \tag{11.6.10}$$

The following theorem contains the basic homotopy formula for cocycles.

Theorem 11.6.3 *Let τ' be a supertrace on QLQ. Then*

$$\tau' D(pq^n) = -b\tau'(-\rho\mu_{n-1}) + \frac{2}{n+2} B\tau'(-\rho\mu_{n+1}), \tag{11.6.11}$$

where the right-hand side is the coboundary of the cochain $\tau'(-\rho\mu_n)$.

Proof We have $(\delta + ad(\rho))(q) = 0$ since

$$-q(a_0a_1) + \rho(a_0)q(a_1) + q(a_0)\rho(a_1) = -(a_0a_1)^- + a_0^+ a_1^- + a_0^- a_1^+ = 0. \tag{11.6.12}$$

Hence we have

$$\begin{aligned} Dq^{n+1} &= \sum_{i=0}^{n} q^i (Dq) q^{n-i} \\ &= \sum_{i=0}^{n} q^i ((\delta + ad(\rho)) q^{n-i} \\ &= (\delta + ad(\rho))\mu_n; \end{aligned} \tag{11.6.13}$$

so applying τ' to $-\rho\mu_n$ gives a sum

$$\begin{aligned} \tau'(-\rho q^i L q^{n-i})(a_0, \dots, a_n) &= (-1)^i \tau'\big(a_0^+ a_1^- \dots a_i^- L a_{i+1}^- \dots a_n^-\big) \\ &= (-1)^i (-1)^{(n-i+1)i} \tau'\big(L a_{i+1}^- \dots a_0^+ a_1 \dots a_i\big), \end{aligned} \tag{11.6.14}$$

where $K^{n-i}(a_0^+ a_1^- \dots a_n^-) = (-1)^{(n-i)i} a_{i+1}^- \dots a_n^- a_0^+ a_1^- \dots a_i^-$, so

$$\tau'(-\rho q^i L q^{n-i}) = \kappa^{n-i} \tau'\big(L\rho q^n\big) \tag{11.6.15}$$

and hence

$$\tau'(-\rho\mu_n) = \sum_{i=0}^{n} \kappa^{n-i}\tau'(L\rho q^n). \tag{11.6.16}$$

One can now use similar arguments to the proof of Proposition 10.1.1 and Theorem 11.4.2 to conclude. □

11.7 Homotopy Formulas

Let $\mathbf{C}[t]$ be the algebra of complex polynomials with the usual formal derivative $d/dt = \dot{}$, and let dt be the infinitesimal such that $(dt)^2 = 0$; then consider the algebra $\mathbf{C}[t,dt] = \{f(t) + g(t)dt \colon f(t), g(t) \in \mathbf{C}[t]\}$ with the derivative $\partial_t((f(t) + g(t)dt) = \dot{f}(t) + \dot{g}(t)dt$, and the grading associated with powers of t. Then for any algebra L with ideal K, $\mathbf{C}[t,dt] \otimes L$ is a graded algebra.

Let $\rho \colon A \to \mathbf{C}[t,dt] \otimes L$ be $\mathbf{C}$-linear and consider

$$C(A, \mathbf{C}[t,dt] \otimes L) = \mathrm{Hom}_{\mathbf{C}}(\mathcal{B}, \mathbf{C}[t,dt] \otimes L), \tag{11.7.1}$$

which forms a differential graded algebra with the bigrading

$$C^{p,q} = \mathrm{Hom}(\mathcal{B}_q, (\mathbf{C}[t,dt])_p) \tag{11.7.2}$$

for tensors of order q and coefficients of t^p. Let $\xi \in C^{p,q}$ and $\eta \in C^{p',q'}$, the product is $\xi.\eta = \mu(\xi \otimes \eta)\Delta$, so

$$\xi\eta(a_1, \dots, a_{p+q+1}) = (-1)^{q(p'+q')}\xi(a_1, \dots, a_q)\eta(a_{q+1}, \dots, a_{q+q'}) \tag{11.7.3}$$

and the total derivative acting on ξ is

$$d_{total}\xi = (dt)\partial_t\xi + \delta\xi, \tag{11.7.4}$$

where the differential δ is $\delta\xi = (-1)^{p+q+1}\xi \circ b'$.

In particular, for $\rho \in C^{0,1}$, we compute the total curvature $\tilde{\omega}$, which involves the curvature $\delta\rho + \rho^2$ and the time differential $\dot{\rho}dt$, as in

$$\begin{aligned} \tilde{\omega} &= (d_{total}\rho + \rho^2) \\ &= dt\dot{\rho} + (\delta\rho + \rho^2) \\ &= (dt)\dot{\rho} + \omega. \end{aligned} \tag{11.7.5}$$

On account of $(dt)^2 = 0$, we have

$$\begin{aligned}(\tilde{\omega})^n &= (dt\,\dot{\rho} + \omega)^n \\ &= \omega^n + (dt)\sum_{i=1}^{n} \omega^{i-1}\dot{\rho}\omega^{n-i} \\ &= \omega^n + (dt)\mu_n, \end{aligned} \tag{11.7.6}$$

where $\dot{\rho} \in C^1(A, L[t]) = C^{0,1}$ and $\mu_n \in C^{0,2n-1}$.

Let $\tau\colon K^m \to \mathbf{C}$ be a $\mathbf{C}$-linear trace such that $\tau|[K^m, L] = 0$. Then τ extends to $\mathbf{C}[t, dt] \otimes K^m \to \mathbf{C}[t, dt]$ and then to

$$C(A, \mathbf{C}[t, dt] \otimes K^m) \to C(A, \mathbf{C}[t, dt]) \tag{11.7.7}$$

as a map of complexes. Then we have

$$\tau(\tilde{\omega}^n) = \tau(\omega^n) + (dt)\tau(\mu_n), \tag{11.7.8}$$

$$\tau_\sharp((\partial\rho)\tilde{\omega}^n) = \tau_\sharp((\partial\rho)\omega^n - (dt)\tau_\sharp((\partial\rho)\mu_n), \tag{11.7.9}$$

where $\tau(\mu_n) \in C^{2n+1}(A, L[t])$.

Lemma 11.7.1 *Then*

$$((dt)\partial_t + \delta)\tau(\tilde{\omega}^n) = \beta\tau_\sharp((\partial\rho)\omega^n)), \tag{11.7.10}$$

$$((dt)\partial_t + \delta)\tau_\sharp(\partial\tilde{\omega}^n) = \partial\tau(\tilde{\omega}^{n+1}/(n+1)). \tag{11.7.11}$$

□

Proposition 11.7.2 (i) *Suppose that $\dot{\rho} \in C^1(A, \mathbf{C}[t] \otimes K)$. Then*

$$[N\tau(\omega^n)] \in HC^{2n-1}(A)[t] \tag{11.7.12}$$

is independent of t for all $n \geq m$.

(ii) *If $\dot{\rho} \in C^1(A, \mathbf{C} \otimes L)$, then the cyclic cohomology classes for $n > m$ are homotopy invariant.*

(iii) *If τ is defined on L itself, then $[N\tau(\omega^n)] = 0$ for all n.*

Proof (i) The hypothesis implies that $\rho\colon A \to (L/K)[t]$ is constant. Now $\mu_n \in C^{2n-1}(A, K^n)$, so $\tau(\mu_n)$ and $\tau_\sharp((\partial\rho)\mu_n)$ are defined for $n \geq m$. Also

$$N\tau(\omega^n)^{\cdot} = \delta N\tau(\omega^n), \tag{11.7.13}$$

so $[N\tau(\omega^n)] \in \mathrm{Hom}(HC_{2n-1}(A), \mathbf{C}[t])$ is independent for t for all $n \geq m$.

(ii) More generally, we have $\mu_n \in C^{2n-1}(A, \mathbf{C}[t] \otimes K^{n-1})$, so $\tau(\mu_n)$ and $\tau_\sharp((\partial\rho)\mu_n)$ are defined for $n > m$.

(iii) In this case, we have a homotopy from ρ to the zero homomorphism given by the path $\rho_t = t\rho \in C^1(A, L[t])$, such that $\dot{\rho}_t = \rho \in C(A, L)$, such that

$$\omega_t = \delta\rho_t + \rho_t^2 = t\delta\rho + t^2\rho^2 \tag{11.7.14}$$

and

$$\mu_{n,t} = \sum_{i=1}^{n} \omega_t(\dot{\rho}_t)\omega_t^{n-i} = \sum_{i=1}^{n} \omega_t^{i-1}\rho\omega_t^{n-i}. \tag{11.7.15}$$

For $t = 0$, all the terms vanish, so $\rho_0 = 0$, hence $\omega_0 = 0 = \mu_{n,0}$;
for $t = 1$, we have the original terms $\rho_1 = \rho$ and $\omega_1 = \omega$. By (i), we have

$$\left[\tau\left(\frac{\omega^n}{n!}\right)\right] = 0 = \left[\tau\left(\frac{(\partial\rho)\omega^n}{n!}\right)\right]. \tag{11.7.16}$$

□

11.8 Cyclic Cocycles over the Circle

Example 11.8.1 Let $H = L^2(S^1, d\phi/(2\pi))$ and $X = d/dx$, so e^{tX^2} is the heat semigroup on H, and e^{tX^2} as an integral operator has kernel, given in terms of the elliptic theta function

$$\vartheta_3\left(\frac{\phi_1 - \phi_2}{2\pi}\,\middle|\,\frac{it}{\pi}\right) \qquad (t > 0), \tag{11.8.1}$$

so e^{tX^2} is trace class. Consider $C(A, \mathcal{L}(H))$ and observe that e^{tX^2} is a 0-cochain in $\mathcal{L}^1(H)$, and $\theta \colon A \to \mathcal{L}(H)$ takes $a \mapsto \theta(a)$, where $\theta(a) \colon f \mapsto af$ is the multiplication operator. The curvature is

$$(\delta\theta + \theta^2)(a_1, a_2) = -\theta(a_1a_2) + \theta(a_1)\theta(a_2) = 0, \tag{11.8.2}$$

since θ is a homomorphism. Also $[X, \theta]$ is a 1-cochain with $[X, \theta](a) = \theta(a')$, since $X\theta(a)(f) - \theta(a)Xf = (af)' - af' = a'f$. In calculations, we sometimes abbreviate

$$e^{tX^2}[X, \theta]e^{tX^2}[X, \theta](a_1, a_2) = e^{tX^2}[X, a_1]e^{tX^2}[X, a_2]. \tag{11.8.3}$$

Example 11.8.2 (Interacting particles on circle) Connes considers the following example of a cocycle. Suppose that distinct points P_j for $j = 0, \dots, r$ lie on the circle S^1 and are specified by angles $\phi_j \in [0, 2\pi]$, which are unordered. Suppose that P_j interacts with the point with neighbouring indices P_{j-1} via the difference quotient

$$\frac{a_j(\phi_j) - a_j(\phi_{j-1})}{e^{i\phi_j} - e^{i\phi_{j-1}}}. \tag{11.8.4}$$

Then we consider

$$\psi(a_0, \dots, a_r) = \int_{[0,2\pi]^{r+1}} a_0(\phi_0)\frac{a_1(\phi_1) - a_1(\phi_0)}{e^{i\phi_1} - e^{i\phi_0}} \dots \frac{a_r(\phi_r) - a_r(\phi_{r-1})}{e^{i\phi_r} - e^{i\phi_{r-1}}}\frac{d\phi_0}{2\pi}\dots\frac{d\phi_r}{2\pi}, \tag{11.8.5}$$

and likewise for cocycles with argument $(a_1, \dots, a_n)$.

11.9 Connections over a Compact Manifold

Let E be a vector bundle over M with connection ∇. Let A be the algebra of sections of the endomorphism bundle $A = \Gamma(M, \mathrm{End}(E)) = \Omega^0(M, \mathrm{End}(E))$, which operates on $\Omega(M, E)$ by left multiplication. The connection gives rise to a derivation $ad(\nabla) \in \Omega^1(M, \mathrm{End})$ so that $ad(\nabla)$ operates on $\Omega(M, E)$ via

$$ad(\nabla)(\xi) = \nabla\xi - (-1)^{deg(\xi)}\xi\nabla. \tag{11.9.1}$$

As in (8.1.7), we take $\nabla^2 \in \Omega^2(M, \mathrm{End})$ to be the curvature, so $(ad(\nabla))^2 = ad(\nabla^2)$.

Consider the tensor algebra $\mathcal{B} = T(A)$ and $C(A, \Omega(\mathrm{End}(E))) = \mathrm{Hom}(\mathcal{B}, \Omega(\mathrm{End}(E)))$, which is bigraded with the (p,q)th component

$$C^{p,q}(A, \Omega(\mathrm{End}(E))) = \mathrm{Hom}(\mathcal{B}_q, \Omega^p(\mathrm{End}(E))). \tag{11.9.2}$$

On $C(A, \Omega(\mathrm{End}(E)))$ there are derivations δ of degree $(0,1)$ and $\tilde{\nabla} = ad(\nabla)$ of degree $(1,0)$

$$\delta_{p,q}\xi(a_0, \dots, a_q) = (-1)^{p+q+1}\xi(b'(a_0, \dots, a_q)) \tag{11.9.3}$$

and

$$(ad(\nabla))\xi(a_1, \dots, a_p) = [\nabla, \xi(a_1, \dots, a_p)], \tag{11.9.4}$$

such that δ and $ad(\nabla)$ anti-commute. Let $\theta\colon A \to \mathcal{B}$ be the canonical inclusion, which is a homomorphism, and let

$$K = \nabla^2 + [\nabla, \theta] \in C^{2,0}(A, \Omega(\mathrm{End}(E))) + C^{1,1}(A, \Omega(\mathrm{End}(E))), \tag{11.9.5}$$

which we regard as a unified curvature.

Proposition 11.9.1 *The operator $D = (\delta + ad(\theta) + ad(\nabla))$ is a derivation and Bianchi's identity holds, so*

$$(\delta + ad(\theta) + ad(\nabla))K^n = 0 \qquad (n = 1, 2, \dots). \tag{11.9.6}$$

Proof We observe that $\theta \in \mathrm{Hom}(A, \Omega^0(\mathrm{End}(E)))$ hence $\delta\theta + \theta^2 = 0$, as above. Let $[\nabla, \theta] = (\tilde{\nabla}\theta)$, where δ and $\tilde{\nabla}$ anti-commute, so $\delta\tilde{\theta} = \tilde{\nabla}\theta^2$. Now

$$\begin{aligned} DK &= (\delta + ad(\theta) + \tilde{\nabla})(\nabla^2 + \tilde{\nabla}(\theta)) \\ &= \delta\nabla^2 + \delta(\tilde{\nabla}\theta) + ad(\theta)(\nabla^2) + ad(\theta)(\tilde{\nabla}\theta) + \tilde{\nabla}(\nabla^2) + \tilde{\nabla}(\tilde{\nabla}\theta) \\ &= 0 + \delta(\tilde{\nabla}\theta) + [\theta, \nabla^2] + [\theta, \tilde{\nabla}\theta] + \tilde{\nabla}(\tilde{\nabla}\theta) \\ &= \delta(\tilde{\nabla}(\theta)) + [\theta, \tilde{\nabla}\theta] + [\theta, \nabla^2] + [\nabla^2, \theta] \\ &= \tilde{\nabla}\theta^2 + [\theta, \tilde{\nabla}\theta] \\ &= 0. \end{aligned} \tag{11.9.7}$$

Since $\delta + ad(\theta) + ad(\nabla)$ is a derivation, we deduce that $(\delta + ad(\theta) + ad(\nabla))K^n = 0$ for $n = 2, 3, \dots$. In particular, we have

$$\begin{aligned} \partial\tau(K^n) &= \tau_\sharp(\partial K^n) \\ &= \tau_\sharp\left(\sum_{i=1}^{n} K^{i-1}(\partial K)K^{n-i}\right) \\ &= n\tau_\sharp\left((\partial K)K^{n-1}\right). \end{aligned} \tag{11.9.8}$$

□

By analogy with Definition 8.2.3, we can interpret $\tau(K^n)$ as the nth Chern class, and the next step is to introduce a Chern character, as in $\tau(e^K)$. It is tempting to use the formula

$$e^K = id + K + \frac{K^2}{2!} + \cdots + \frac{K^n}{n!} + \cdots \tag{11.9.9}$$

to infer results about e^K. However, the operators in this series are unbounded, so in Section 11.11 we need to proceed by an indirect route via Duhamel's principle.

11.10 The Trivial Bundle

Let M be a compact n-dimensional manifold with $A = C^\infty(M; \mathbf{C})$. Then $H^n(A, A^*)$ is the space of de Rham currents of dimension n, and $H^*(A)$ is the de Rham homology of M.

Let E be the trivial bundle with fibre $\mathbf{C}$, so the connection ∇ reduces to the de Rham connection $d = \nabla$ with $d^2 = 0$, and $\tau\colon \Omega \to \Omega$ is the identity. Also, $\theta\colon A \to A$ has $\theta(a) = a$ and

$$(d\theta)^r(a_1, \dots, a_r) = da_1 \dots da_r \tag{11.10.1}$$

and

$$(\partial\theta)(d\theta)^r(a_0, \dots, a_r) = a_0 da_1 \dots da_r. \tag{11.10.2}$$

For a closed r-dimensional current γ,

$$\begin{aligned}\psi(a_0, \dots, a_r) &= \int_\gamma \tau_\sharp((\partial\theta)e^{d\theta})(a_0, \dots, a_r)\\ &= \int_\gamma a_0 da_1 \dots da_r\end{aligned} \tag{11.10.3}$$

and

$$\begin{aligned}\varphi &= \int_\gamma \tau(e^{d\theta})\\ &= \int_\gamma \frac{(d\theta)^r}{r!}\\ &= \int_\gamma d\frac{\theta(d\theta)^{r-1}}{r!}\\ &= 0,\end{aligned} \tag{11.10.4}$$

since γ is closed. Now $\beta\psi = \delta\varphi = 0$ and $\delta\psi = \partial\varphi = 0$. Hence ψ is a cyclic cochain and a Hochschild coboundary, so ψ is a cyclic cocycle.

For example, we can take E to be the tangent bundle over the elliptic curve (torus) as in [7, p. 312] and Example 6.1.4. In index theory, there are significant examples of cocycles available when one operates on vector bundles equipped with a grading structure, as we consider in Section 11.11.

11.11 Cocycles Arising from the Connection

We resume the general discussion of Section 11.9. The vector space E gives a standard trace on $\mathrm{End}(E)$ via the usual matrix representation, which does not depend upon the choice of bases. So we can define $\mathrm{tr}_E\colon \Omega(\mathrm{End}) \to \Omega$ by fibrewise operations such that $\mathrm{tr}_E([\xi, \eta]) = 0$. Writing $\tau = \mathrm{tr}_E$, we have $\tau\colon C(A, \Omega(\mathrm{End}(E)) \to C(A, \Omega)$ such that

$$\tau \circ ad(\nabla) = d \circ \tau, \tag{11.11.1}$$

where $d \circ \tau = \mathrm{tr}_E([\nabla, \xi])$. Similarly, τ induces a map

$$\tau_\sharp : \mathrm{End}(\mathcal{B} \otimes A[1] \otimes \mathcal{B}, \Omega(\mathrm{End}(E))) \to \mathrm{End}(A[1] \otimes \mathcal{B}, \Omega(M)), \quad (11.11.2)$$

where the right-hand side is the complex of Hochschild cochains $C_\sharp(A, \Omega(M))$ with values in Ω, such that

$$\delta \tau_\sharp = \tau_\sharp \delta, \quad (11.11.3)$$

$$d\tau_\sharp = \tau_\sharp ad(\nabla). \quad (11.11.4)$$

Let ∂ be defined for the $C(A, \Omega(\mathrm{End}(E))$ bimodule as in (11.9.8).

Theorem 11.11.1 (i) *The elements* $\tau(e^K) \in C(A, \Omega)$ *and* $\tau_\sharp((\partial\theta)e^K) \in C_\sharp(A, \Omega)$ *satisfy*

$$(\delta + d)\tau(e^K) = \beta\tau_\sharp((\partial\theta)e^K) \quad (11.11.5)$$

and

$$(\delta + d)\tau_\sharp((\partial\theta)e^K) = \partial\tau(e^K). \quad (11.11.6)$$

(ii) *Let* γ *be a closed current in* M *of dimension* r*, then*

$$\varphi = \int_\gamma \tau(e^K) \in C(A),$$

$$\psi = \int_\gamma \tau_\sharp((\partial\theta)e^K) \in C_\sharp(A)$$

satisfy

$$\delta\varphi = \beta\psi \quad \text{and} \quad \delta\psi = \partial\varphi. \quad (11.11.7)$$

Proof (i) We consider $L = \nabla^2$ and $K = \nabla^2 + [\nabla, \theta]$, so that $[\nabla, \theta](a) = (\tilde{\nabla}\theta)(a) \in \mathcal{L}(H)$ for all $a \in A$, and e^{tL} is the heat semigroup of Section 5.2. The perturbation to L involves a first-order differential operator, so we are in a similar situation to Example 5.7.2. Hence we can apply Duhamel's formula

$$D\tau(e^K) = \tau(e^K DK) \quad (11.11.8)$$

with $D = \delta + ad(\nabla) + ad(\theta)$, so $DK = 0$ by Proposition 11.9.1; hence

$$\begin{aligned}(\delta + d)\tau(e^K) &= \tau((\delta + ad(\nabla))e^K) \\ &= \tau((-ad(\theta))e^K) \\ &= \beta\tau_\sharp((\partial\theta)e^K). \end{aligned} \quad (11.11.9)$$

Also $K = \nabla^2 + [\nabla, \theta]$, where $\partial(\nabla^2) = 0$, so

$$\begin{aligned}(\delta + d)\tau_\sharp\big((\partial\theta)e^K\big) &= \tau_\sharp((\delta + ad(\nabla))((\partial\theta)e^K))\\ &= \tau_\sharp\big(\partial(\delta\theta + [\nabla,\theta)e^K - \partial\theta(\delta + ad(\nabla))e^K\big)\\ &= \tau_\sharp\big(\partial\theta^2 + \partial[\nabla,\theta])e^K + \partial\theta(\theta e^K - e^K\theta))\\ &= \tau_\sharp\big(-\theta\partial\theta e^K + \partial[\nabla,\theta]e^K - \partial\theta e^K\theta\big)\\ &= \tau_\sharp\big((\partial[\nabla,\theta])e^K\big)\\ &= \tau_\sharp\big(\partial K e^K\big)\\ &= \partial\tau\big(e^K\big). \end{aligned} \tag{11.11.10}$$

(ii) We repeatedly use Lemma 5.7.1 to obtain

$$\begin{aligned} e^{tK} &= e^{tL} + \int_0^t e^{(t-t_1)L}(K-L)e^{t_1L}\,dt_1 \\ &+ \int_0^t\int_0^{t_1} e^{(t-t_1-t_2)L}(K-L)e^{t_1L}(K-L)e^{t_2L}\,dt_2dt_1 \\ &+\dots+ \int_{0\le t_1+\dots+t_n\le t} e^{(t-t_1-\dots-t_n)L}(K-L)e^{t_1L} \\ &\times (K-L)e^{t_2L}\dots(K-L)e^{t_nK}\,dt_1\dots dt_n. \end{aligned} \tag{11.11.11}$$

By induction one proves that

$$\int_{t_1+\dots+t_n\le t}(t-t_1-\dots-t_n)^{i_0}t_1^{i_1}\dots t_n^{i_n}\,dt_1\dots dt_n = \frac{i_0!\dots i_n!\,t^{i_0+\dots+i_n+n}}{(n+i_0+\dots+i_n)!}, \tag{11.11.12}$$

so in particular the volume of the simplex $\{(t_1,\dots,t_n);0\le t_j;t_1+\dots+t_n\le 1\}$ is of order $1/n!$, hence we can take $n\to\infty$ and deduce the convergent series

$$\begin{aligned}\varphi &= \int_\gamma \tau(e^K) + \sum_{n=1}^\infty \int_{0\le t_j;t_1+\dots+t_n\le 1} \\ &\times \tau\Big(e^{(1-t_1-\dots-t_n)\nabla^2}[\nabla,\theta]e^{t_1\nabla^2}[\nabla,\theta]\dots e^{t_n\nabla^2}\Big)dt_1\dots dt_n \end{aligned} \tag{11.11.13}$$

and

$$\varphi = \sum_r \sum_{i_0+\dots+i_n=r}\tau\left(\frac{\nabla^{2i_0}[\nabla,\theta]\nabla^{2i_1}\dots[\nabla,\theta]\nabla^{2i_n}}{(n+i_1+\dots+i_n)!}\right). \tag{11.11.14}$$

□

The periodic cyclic cohomology class represented by the cochains in Theorem 11.10.1 does not depend upon the particular choice of connection.

11.12 Super Connections and Twisted Dirac Operators

One of the most challenging problems in cyclic theory was raised by Jaffe, Lesniewski and Osterwalder [58] in their development of quantum K-theory, and Quillen considered various ways in which this could be described by connections. He regarded the definition of the Chern character for connections as fundamental to the theory and sought to interpret cyclic classes as Chern character forms. In his notebooks on 12 April 1989 he wrote: 'Working with extensions seems difficult and in the wrong direction from the entire theory. So we return to JLO. It seems to me we can link Connes's approach in his entire paper, where he uses traces on the Cuntz algebra, with JLO.'

The following construction is an extension of Section 6.2, where we constructed the Chern character of a vector bundle over a manifold. Let M be a compact and connected Riemannian manifold and E a complex vector bundle over M with connection ∇. We further suppose that X is a skew adjoint operator and $\sigma \in \mathrm{End}(E)$ a linear involution such that $\sigma^2 = id$ on E. We introduce the algebra $\Omega(\mathrm{End}(E))[\sigma] = \Omega(\mathrm{End}(E)) \oplus \Omega(\mathrm{End}(E))[\sigma]$ with product $\sigma\omega = (-1)^{deg(\omega)}\omega\sigma$, which operates on E-valued forms by $\sigma\xi = (-1)^{deg(\xi)}\xi$. This makes sense at a formal level, but goes beyond the context of Section 8.2 since X can be an unbounded operator.

The curvature of the superconnection is the operator $K = (\nabla + \sigma X)^2$ on $\Omega(E)$, which satisfies

$$K = \nabla^2 + \nabla\sigma X + \sigma X\nabla + \sigma X\sigma X = \nabla^2 + [\nabla, X]\sigma + X^2. \tag{11.12.1}$$

To form cocycles, we need a trace, and the theory has two cases.

(i) (Ungraded) Let $\tau \colon \Omega(\mathrm{End}(E)) \to \Omega$ be $\tau(\alpha + \beta\sigma) = tr_E(\beta)$. Then $\tau(e^K)$ is an odd form, which is closed.

(ii) (Graded) Assume that E is a graded vector bundle with grading

$$\varepsilon = \begin{bmatrix} 1 & 0 \\ 0 & -1 \end{bmatrix} \quad \text{on} \quad E = \begin{matrix} E^+ \\ E^- \end{matrix}, \tag{11.12.2}$$

compatible with ∇, so $[\nabla, \varepsilon] = 0$ and $\varepsilon X = -X\varepsilon$, where, in the style of Section 5.6,

$$X = \begin{bmatrix} 0 & -T^* \\ T & 0 \end{bmatrix}. \tag{11.12.3}$$

Here we have a trace $\tau(\alpha + \beta\sigma) = \mathrm{tr}_E(\varepsilon\alpha)$. We say that X is a twisted Dirac operator with twisting given by a vector bundle and connection ∇. The terminology arises from the case in which $X = \sigma_3\partial/\partial x_3$ with σ_3 the Pauli matrix as in (5.5.4); compare X with iD in Proposition 5.6.1.

Proposition 11.12.1 *The cochain*

$$\tau\left(e^{X^2+[\nabla,X]\sigma+\nabla^2}\right) \in \Omega^{ev} \tag{11.12.4}$$

is closed.

Proof Let $K = (\nabla + \sigma X)^2$, so that

$$d\tau(e^K) = \tau([\nabla, K]) = \tau([\nabla + X\sigma, e^K]) = 0. \qquad \square$$

The next idea is to treat $\delta + \theta$ as a sort of connection and combine this with the operator $X\sigma$. So the super connection operator is $\delta + \theta + X\sigma$, with curvature

$$(\delta + \theta + X\sigma)^2 = (X\sigma)^2 + (\delta + \theta)^2 + [\delta + \theta, X\sigma] = X^2 + [\theta, X\sigma]. \tag{11.12.5}$$

By analogy with the results of the previous section, we define the curvature of the superconnection operator to be $K = X^2 + [\theta, X\sigma]$. We have

$$e^K = e^{X^2} + \sum_{n=1}^{\infty} \int_{0\leq t_j;\, t_1+\dots+t_n\leq 1} e^{(1-t_1-\dots-t_n)X^2}[\theta, X\sigma]e^{t_1 X^2} \dots [\theta, X\sigma]e^{t_n X^2} dt_1 \dots dt_n, \tag{11.12.6}$$

which belongs to $\prod C(A, \mathcal{L}(H)[\sigma])$. The analytical situation is different since X is an unbounded operator and cannot be interpreted as a finite matrix.

Theorem 11.12.2 *The cochains $\varphi = \tau(e^K) \in C(A)$ and $\psi = \tau_\sharp((\partial\theta)e^K) \in C_\sharp(A)$ form a double complex such that*

$$\delta\tau(e^K) = \beta\tau_\sharp((\partial\theta)e^K). \tag{11.12.7}$$

Proof The operator e^K is trace class, and we can use Duhamel's formula to compute its derivatives. Bianchi's identity holds in the form

$$(\delta + ad(\theta + X\sigma))K = 0, \tag{11.12.8}$$

so

$$(\delta + ad(\theta + X\sigma))e^K = 0.$$

Also $\tau((ad(X\sigma)e^K) = \tau([X\sigma, e^K]) = 0$ since e^K is trace class, and so

$$\begin{aligned}\delta\tau(e^K) &= \tau\big((\delta + ad(X\sigma))e^K\big)\\ &= \tau\big(-[\theta + X\sigma, e^K]\big)\\ &= \beta\tau_\sharp\big((\partial\theta)e^K\big);\end{aligned} \tag{11.12.9}$$

likewise

$$\begin{aligned}\tau_\sharp\big((\partial\theta)e^K\big) &= \tau_\sharp\big(-\partial(\theta^2)e^K - (\partial\theta)\delta e^K\big)\\ &= \tau_\sharp\big((-(\partial\theta)\theta - \theta(\partial\theta)e^K) + (\partial\theta)[\theta + X\sigma, e^K]\big), \qquad (11.12.10)\end{aligned}$$

where we have used (11.12.4). Then

$$\begin{aligned}\delta\tau_\sharp\big((\partial\theta)e^K\big) &= \tau_\sharp\big((\partial\theta)X\sigma e^K - e^K X\sigma\big)\\ &= \tau_\sharp\big((\partial[\theta, X\sigma])e^K\big), \qquad (11.12.11)\end{aligned}$$

which we compare with

$$\begin{aligned}\partial\tau\big(e^K\big) &= \tau_\sharp\Big(\int_0^1 e^{(1-s)K}\partial K e^{sK}\,ds\Big)\\ &= \tau_\sharp\big((\partial K)e^K\big)\\ &= \tau_\sharp\big((\partial(X^2 + [\theta, X\sigma])e^K\big)\\ &= \tau_\sharp\big((\partial([\theta, X\sigma])e^K\big). \qquad (11.12.12)\end{aligned}$$

We can replace X by $\hbar X$ in the previous computation and consider

$$\begin{aligned}&\tau_\sharp\big((\partial\theta)e^{\hbar^2X^2+\sigma[\hbar X,\theta]}\big)(a_0,\dots,a_n) = \pm\int_{0\le t_j;\,t_1+\dots+t_n\le 1}\\ &\tau\Big(a_0e^{t_0\hbar^2X^2}[a_1, X\sigma]e^{t_1\hbar^2X^2}[a_2, X\sigma]\dots[a_n, X\sigma]e^{t_n\hbar^2X^2}\Big)dt_1\dots dt_n, \qquad (11.12.13)\end{aligned}$$

where we have abbreviated $t_0 = 1 - t_1 - \dots - t_n$ in the first exponent. □

We now consider the connection between the results of Section 11.9, which involve connections, and this section which involve twisted connections. Getzler observed that the quantum harmonic oscillator from Section 5.8 can be used to compute cyclic cocycles such as (11.12.13). In [38, 39] he developed a symbolic calculus based upon the asymptotic expansion of pseudo-differential operators into graded components.

Let M be a compact $2m$-dimensional Riemannian manifold with spin structure, let E be a vector bundle over M with connection ∇ and let A be the algebra of endomorphisms of E. The $\hat{A}$ genus associated with ∇ is defined in a similar way to the Chern character of Section 8.2, namely

$$\widehat{A}(\nabla^2) = \det\left[\frac{\nabla^2/(2\pi i)}{\sin\nabla^2/(2\pi i)}\right].$$

Let X be a twisted Dirac operator, as above. Then the following holds [88]:

$$\lim_{\hbar\to 0} \tau_{\natural}\Big(\partial\theta e^{\hbar^2 X^2+\sigma[\hbar X,\theta]}\Big) = \Big(\frac{i}{2\pi}\Big)^m \int_M \hat{A}(M)\mathrm{tr}^{\natural}_E\Big(\partial\theta e^{\nabla^2+[\nabla,\theta]}\Big)\mu(dx). \tag{11.12.14}$$

On the right-hand side, the first factor involves the $\hat{A}$ genus of the manifold, a geometrical quantity, while the final factor involves the connection ∇. See also Theorem 5.3 of [111].

12
Periodic Cyclic Homology

In this chapter we introduce a complex $X(A)$ and use it to compute the periodic cyclic homology of a quasi-free algebra, obtaining an analogue of Theorem 6.2.9. In Section 12.2, we compute the periodic cyclic homology for a commutative and smooth algebra, thereby resolving the homology theories of Kähler differentials and noncommutative differentials. The process of passing from Theorem 6.2.9 to 6.2.10 involves resolution of singularities of an algebraic variety. In Sections 12.3 and 12.4, we consider a version of this in cyclic theory for noncommutative algebras, and ultimately obtain the desired result in Section 12.4.

12.1 The X Complex and Periodic Cyclic Homology

The operators b and B satisfy

$$b^2 = B^2 = bB + Bb = 0, \tag{12.1.1}$$

so $b + B$ has $(b + B)^2 = 0$. Now let

$$\widehat{\Omega}^{ev} A = \prod_{n=0}^{\infty} \Omega^{2n} A, \quad \widehat{\Omega}^{odd} A = \prod_{n=0}^{\infty} \Omega^{2n+1} A, \tag{12.1.2}$$

so

$$\widehat{\Omega}^{ev} A \overset{b+B}{\longrightarrow} \widehat{\Omega}^{odd} A \overset{b+B}{\longrightarrow} \widehat{\Omega}^{ev} A \overset{b+B}{\longrightarrow} \tag{12.1.3}$$

gives a complex. The difference between this complex and (6.5.13) is that here we allow sequences of chains that have infinitely many non-zero terms, and the hats indicate a process of completion.

Definition 12.1.1 (Periodic cyclic homology) The periodic cyclic homology is the homology of the above complex

$$HP_i = H_i\left(\widehat{\Omega}^{ev} A \overset{b+B}{\rightleftharpoons} \widehat{\Omega}^{odd} A\right) \qquad (i \in \{0, 1\}). \tag{12.1.4}$$

Consider a unital complex algebra A and the space of noncommutative differential forms $(\Omega A, b)$, where $b(\omega da) = (-1)^{deg(\omega)}(\omega a - a\omega)$, so b has degree minus one, and $b\Omega^{n+1} A \subseteq [A, \Omega^n A]$. For an A-bimodule M, we write $M_\natural = M/[A, M]$ and also $\natural: M \to M_\natural$ for the natural quotient map.

Let $X(A)$ be the complex

$$A \underset{d}{\overset{b}{\leftrightarrows}} (\Omega^1 A)_\natural. \tag{12.1.5}$$

The space $X(A)$ is endowed with:

(i) a product $A \otimes A \to A: x \otimes y \mapsto xy$;

(ii) the pairing $A \otimes A \to (\Omega^1 A)_\natural: x \otimes y \mapsto \natural(xdy)$, such that

$$\natural(xd(yz)) = \natural(xydz) + \natural(zxdy), \tag{12.1.6}$$

so $f(x, y) = \natural(xdy)$ satisfies

$$bf(x, y, z) = f(xy, z) - f(x, yz) + f(zx, y) = 0; \tag{12.1.7}$$

(iii) $X(A)$ is a $\mathbf{Z}/(2)$ graded complex with $bB = Bb = 0$.

The following is the analogue of the Lefschetz, Atiyah–Hodge and Grothendieck Theorem 6.2.9, and may be viewed alongside Corollary 7.7.6, which dealt with $HH(A)$ and $HC(A)$.

Theorem 12.1.2 *If A is quasi-free, then*

$$H_i(X(A)) = HP_i(A) \qquad (i = 0, 1). \tag{12.1.8}$$

Proof For quasi-free A, the Hochschild cohomology vanishes in degrees >1 by Definition 6.4.7, so the periodic cyclic homology is calculated by $X(A)$. There is a lifting homomorphism $A \to \widehat{RA}$, or equivalently $A \to \widehat{\Omega}^{ev} A$, as in Theorem 8.7.1. Then the complex

$$\begin{array}{ccccccc} \hat{\Omega}^{ev} A & \overset{b+B}{\longrightarrow} & \hat{\Omega}^{odd} A & \qquad & \hat{\Omega}^{ev} A & \overset{b+B}{\longleftarrow} & \hat{\Omega}^{odd} A \\ \mu \downarrow & & \downarrow \mu & & \mu \downarrow & & \downarrow \mu \\ A & \underset{d}{\longrightarrow} & (\Omega^1 A)_\natural & & A & \underset{b}{\longleftarrow} & (\Omega^1 A)_\natural \end{array} \tag{12.1.9}$$

has μ, a quasi-isomorphism, so the homology of $\hat{\Omega}^{ev} A \rightleftarrows \hat{\Omega}^{odd} A$ is equal to the homology of the $X(A)$ complex $A \rightleftarrows (\Omega^1 A)_\natural$, so we get the stated result. □

12.2 $X(A)$ for Commutative Differential Graded Algebras

We recall Fedosov's construction, with a rescaling. Let

$$\Omega^0 \longrightarrow \Omega^1 \longrightarrow \Omega^2 \longrightarrow \tag{12.2.1}$$

be a differential graded algebra, and for a fixed constant c, define the Fedosov product

$$x \star y = xy + (-1)^{|x|} c\,dx\,dy, \tag{12.2.2}$$

where $|x|$ denotes the degree of x. We introduce the operation $\nu(x) = |x|x$ for homogeneous elements and write

$$[x, y]^\star = x \star y - y \star x. \tag{12.2.3}$$

Let $\Omega = \Omega A$ and let Ω^{ev} be the even part of Ω, and Ω^{odd} be the odd part of Ω.

Example 12.2.1 $(\Omega, \star)$ is isomorphic to the Cuntz algebra QA, and $(\Omega^{ev}, \star)$ is isomorphic to RA via

$$a_0 da_1 \dots da_{2n} \leftrightarrow \rho(a_0)\omega^n(a_1, \dots, a_{2n}). \tag{12.2.4}$$

Let A be a commutative unital algebra over $\mathbf{C}$, so we can compare the various homology theories. From the exact sequence given by multiplication

$$0 \longrightarrow I \longrightarrow A \otimes A \longrightarrow A \longrightarrow 0,$$

we introduce the ideal I and the first-order Kähler differentials $\Omega^1_A = I/I^2$ as in Section 6.1; the differential $d\colon A \to \Omega^1_A\colon da = a \otimes 1 - 1 \otimes a \pmod{I^2}$ gives the basic Kähler differential. Then we let $\Omega_A = \wedge_A \Omega^1_A$, which we split into the even degree terms $\Omega^{ev} = \oplus_{n=0}^{\infty} \Omega^{2n}$ with $\Omega^0_A = A$ and the odd degree terms $\Omega^{odd}_A = \oplus_{n=0}^{\infty} \Omega^{2n+1}_A$. The Kähler differential gives $d\colon \Omega^{ev}_A \to \Omega^{odd}_A$ and $d\colon \Omega^{odd}_A \to \Omega^{ev}_A$. Then we introduce $\mathbf{X} = \mathrm{Hom}_{alg}(A, \mathbf{C})$.

We now specialize to the context of a commutative differential graded algebra Ω, and write $R = (\Omega^{ev}, \star)$. Let $\bar{b}(xdy) = [x, y]^*$ and $\bar{d}x = dx$ modulo the commutator subspace $[R, \Omega^1 R]$. Consider the pair of diagrams:

$$\begin{array}{ccccccc} R & \xrightarrow{\bar{d}} & \Omega^1 R_\natural & \qquad & R & \xleftarrow{\bar{b}} & \Omega^1 R_\natural \\ =\downarrow & & \downarrow \Phi & & =\downarrow & & \downarrow \Phi \\ \Omega^{ev} & \xrightarrow[Nd]{} & \Omega^{odd} & & \Omega^{ev} & \xleftarrow[2cd]{} & \Omega^{odd} \end{array} \tag{12.2.5}$$

Lemma 12.2.2 *Let $\Phi(x, y) = xdy + |y|d(xy)$. Then Φ determines a map on $\Omega^1 R_\natural$, such that both diagrams commute.*

Proof We have $\Phi\colon R \otimes R \to \Omega^{odd}$, so Φ restricts to $\Phi\colon \Omega^1 R \to R$. Using the operator $b\colon R^{\otimes 2} \to R^{\otimes 3}$ for the $\star$ product, we need $b\Phi = 0$ so that Φ gives the zero map on the commutator subspace $[\Omega^1 R, R]^\star$, which hence passes to the quotient $\Omega^1 R_\natural$. So we have to show

$$\Phi(x \star y, z) - \Phi(x, y \star z) + \Phi(z \star x, y) = 0, \tag{12.2.6}$$

where

$$\begin{aligned}\Phi(z \star x, y) &= (1 + |y|)(z \star x)dy + |y|d(z \star x)y \\ &= (1 + |y|)(zx + cdzdx)dy + |y|\big((dz)xy + z(dx)y\big),\end{aligned} \tag{12.2.7}$$

and likewise

$$\Phi(x \star z, y) = (1 + |z|)(xy + cdxdy)dz + |z|\big((dx)yz + x(dy)z\big) \tag{12.2.8}$$

and

$$\begin{aligned}\Phi(x, y \star z,) &= \Phi(x, yz) + c\Phi(x, dydz) \\ &= (1 + |y| + |z|)x\big((dy)z + y(dz)\big) + \big(|y| + |z|\big)(dx)yz \\ &\quad + c\big(1 + |y| + 1 + |z|\big)dxdydz,\end{aligned} \tag{12.2.9}$$

whence the result.

Now $\bar{b}(xd^R y) = [x, y]^\star$, where

$$[x, y]^\star = x \star y - y \star x = xy + cdxdy - yx - cdydx = 2cdxdy, \tag{12.2.10}$$

since the algebra is commutative and x and y are even. Also,

$$d^R y \mapsto \phi(x, y) = xdy + |y|d(xy), \tag{12.2.11}$$

where $2cd(xdy + |y|d(xy)) = 2cdxdy$, so the left-hand diagram commutes.

Also, $\bar{d}\colon y \mapsto d^R y$ has $\Phi\colon d^R y \mapsto dy + |y|dy$, while $\nu(dy) = (1 + |y|)dy$, so the right-hand diagram also commutes. □

Suppose that A is a commutative and unital complex algebra, and from Section 6.2 recall the space of Kähler differential forms Ω_A. A closed current C on A is a linear functional on Ω_A which kills $d\Omega_A$; in dimension k we have

$$\varphi(f_0, \dots, f_k) = C(f_0 df_1 \wedge \dots \wedge f_k) \tag{12.2.12}$$

such that

$$\varphi(1, f_1, \dots, f_k) = C(df_1 \wedge \dots \wedge f_k) = 0. \tag{12.2.13}$$

The following theorem asserts that closed currents determine the periodic cyclic cohomology and that every class is represented in this way. The space $H^i(\Omega_A)$ is a classical geometrical object, which is amenable to calculation, as in Sections 6.2 and 8.2.

Theorem 12.2.3 (Connes) *Let A be a commutative and smooth algebra. Then*

$$HP_i(A) = H^i(\Omega_A) \qquad (i = 0, 1). \tag{12.2.14}$$

There is a canonical homomorphism of differential graded algebras $\mu\colon \Omega A \to \Omega_A$, *and* $b\colon \Omega A \to \Omega A\colon b(\omega da) = (-1)^{|\omega|}(\omega a - a\omega)$. *Then we define b on* Ω_A *to be zero. We write* $N_{\kappa^2} b = \sum_{j=0}^{n-1} \kappa^{2j} b$. *Then* μ *is compatible with b and d, and the following diagrams commute:*

$$\begin{array}{ccccccc} \Omega^{ev} A & \xrightarrow{\bar d} & \Omega^{odd} A & \quad & \Omega^{ev} A & \xleftarrow{\bar b} & \Omega^{odd} A \\ \mu \downarrow & & \downarrow \mu & & \mu \downarrow & & \downarrow \mu \\ \Omega^{ev}_A & \xrightarrow[\nu d]{} & \Omega^{odd}_A & & \Omega^{ev}_A & \xleftarrow[-2d]{} & \Omega^{odd}_A \end{array} \tag{12.2.15}$$

Then

$$1 - \kappa = bd + db = 0; \qquad -N_{\kappa^2} b + B = B = \nu d; \qquad b - (1+\kappa)d = -2d. \tag{12.2.16}$$

Here the Karoubi operator is the identity. This is the case of $c = -1$.

Remarks 12.2.4 In [109] Kassel discusses the circumstances under which complex associative algebras A and A' satisfy

$$HP_*(A \otimes A') = HP_*(A) \otimes HP_*(A'). \tag{12.2.17}$$

In particular, it holds for all A when A' is separable and flat, as in Example 6.4.4. The result also holds for all A when $A' = \mathbf{C}[\mathbf{X}]$, where $\mathbf{X}$ is a smooth complex affine variety, in which case we can use Theorem 12.2.3.

12.3 The Canonical Filtration

In this section, we state an extension of Theorem 12.1.2 to the case of $A = R/I$, where R is quasi-free. Given an inverse system of complexes

$$\cdots \xrightarrow{d} K_{n+1} \xrightarrow{d} K_n \xrightarrow{d} K_{n-1} \xrightarrow{d} \cdots \tag{12.3.1}$$

there exist homology groups $H_i(K_j)$ and linking homomorphisms, hence we can form $H_i(\lim_{\leftarrow n} K_n)$ and $\lim_{\leftarrow n} H_i(K_n)$. See [45] for a discussion of inverse limits. The operation of taking inverse limits does not commute with forming homology groups, so these groups are not necessarily isomorphic; however, they are related by Milnor's exact sequence

$$0 \longrightarrow \lim_{\leftarrow n}{}^{1} H_{i+1}(K_n) \longrightarrow H_i(\lim_{\leftarrow n} K_n) \longrightarrow \lim_{\leftarrow n} H_i(K_n) \longrightarrow 0. \tag{12.3.2}$$

The following is the analogue of the Zariski–Grothendieck Theorem 6.2.10.

Theorem 12.3.1 *If $A = R/I$ where R is quasi-free, then*

$$HP_i(A) = H_i(\varprojlim X(R/I^n)). \tag{12.3.3}$$

In Theorem 12.4.2, we prove that this holds in the case of the universal extension $A = RA/IA$. Also, we prove in Lemma 12.4.4 that $H_i(\lim_{\leftarrow n} X(R/I^n))$ is independent of R when $A = R/I$. Compare this with Exercise 1.1.4.

In this section, we proceed to look at a more general situation. Let $TA = \oplus_{k=0}^{\infty} A^{\otimes k}$ and $RA = T(A)/(1_{T(A)} - 1_A)$ be the free algebra which may be characterized:

(i) *As the universal algebra for the unital linear map $\rho\colon A \to RA$.*

(ii) *The even differential forms $RA = \oplus_{k=0}^{\infty} \Omega^{2k} A$, where RA has the Fedosov product $x \star y = xy - dxdy$, as in Chapter 6. The algebra RA has an ideal $IA = \oplus_{k=1}^{\infty} \Omega^{2k} A$ arising from the homomorphism $dxdy \mapsto 0$, and $(IA)^n = \oplus_{k=n}^{\infty} \Omega^{2k} A$.*

Proposition 12.3.2 *The algebra A is isomorphic to RA/IA, where RA is quasi-free, and there is an inverse system*

$$\longrightarrow X(RA/(IA)^n) \longrightarrow X(RA/(IA)^{n-1}) \longrightarrow \cdots \longrightarrow X(RA/(IA)) \cong X(A). \tag{12.3.4}$$

Proof Let $\delta\colon RA \to \Omega^1 RA$ be the canonical differential, and extend this as in Theorem 6.3.1 to δ on ΩRA. Now for all RA-bimodules M, we have

$$\mathrm{Hom}_{RA \otimes RA^{op}}(\Omega^1 RA, M) = \mathrm{Hom}(\bar{A}, M), \tag{12.3.5}$$

hence $\Omega^1 RA \cong RA \otimes \bar{A} \otimes RA$ via $(x(\delta a)y \leftrightarrow x \otimes \bar{a} \otimes y$. Note that $\Omega^1 RA \cong RA \otimes \bar{A} \otimes RA$ is projective as an RA-bimodule, so RA is quasi-free by 6.4.7. We now take commutators $M \mapsto [M, RA]$, then pass to quotient spaces $M \mapsto M_\natural = M/[M, RA]$, as in $\Omega^1 RA_\sharp \cong RA \otimes \bar{A}$. Abusing notation, we introduce $\delta\colon RA \mapsto (\Omega^1 RA)_\natural$ by $x \mapsto \natural(\delta x)$ and $\beta\colon \natural(x\delta y) = x \star y - y \star x$. The inverse system

$$\longrightarrow R/I^n \longrightarrow R/I^{n-1} \longrightarrow \cdots \longrightarrow R/I^2 \longrightarrow R/I \tag{12.3.6}$$

produces

$$\longrightarrow X(R/I^n) \longrightarrow X(R/I^{n-1}) \longrightarrow \cdots \longrightarrow X(R/I^2) \longrightarrow X(R/I), \tag{12.3.7}$$

in general, and we can take $R = RA$ and $I = IA$ in particular. □

Definition 12.3.3 (Inverse limits of X complex) Given an inverse system (12.3.7), we define

$$\widehat{X}(R,I) = \lim_{\leftarrow n} X(R/I^n). \tag{12.3.8}$$

Corollary 12.3.4 *The noncommutative differential forms* $\Omega A = \Omega^{ev} A \oplus \Omega^{odd} A$ *give decreasing sequences of powers of ideals and spaces*

$$(IA)^n = \bigoplus_{k=n}^{\infty} \Omega^{2k} A, \quad \natural(IA)^n \delta(RA) = \bigoplus_{k=n}^{\infty} \Omega^{2k+1} A \tag{12.3.9}$$

and a complex

$$X(RA/(IA)^n): \quad RA/(IA)^n \underset{\leftarrow}{\overset{\rightarrow}{}} \left(\Omega^1\left(RA/(IA)^n\right)\right)_\natural. \tag{12.3.10}$$

Proof Compare the odd differentials with the products on the left-hand side. We have $\natural(z\delta a) = zda$, and RA is generated by A, so

$$\natural\big(z\delta(a_1 \star \cdots \star a_q)\big) = \natural\left(\sum_{j=1}^{q}\big(a_{j+1} \star \cdots \star a_q \star z \star a_1 \star \cdots \star a_{j-1}\big)\delta a_j\right),$$

so

$$\natural(IA)^n \delta(RA) = \natural(IA)^n \delta A = \bigoplus_{k=n}^{\infty} \Omega^{2k+1} A. \qquad \square$$

The preceding result becomes more useful when we identify more explicitly the operators on the complex $X(RA/(IA)^n)$ in terms of the Karoubi operation. The formulas are similar to those we used for (Ω, b) in the proof of Theorem 7.5.2.

Lemma 12.3.5 *Let* RA *have the Fedosov product* $\star$. *Then* $X(RA)$ *is the complex*

$$\Omega^{ev} A \cong RA \underset{\delta}{\overset{\beta}{\leftrightarrows}} (\Omega^1 RA)_\natural \cong \Omega^{odd} A, \tag{12.3.11}$$

where

(i) $\beta = b - (1+\kappa)d$ *and* $\delta = -\sum_{j=0}^{n-1} \kappa^{2j} b + B$ *on* $\Omega^{2n} A$;

(ii) *the pairing* $RA \times RA \to (\Omega^1 RA)_\natural$ *is*

$$(x,y) \mapsto \natural(x\delta y) = -\sum_{j=0}^{n-1} \kappa^{2j} b(x \star y) + \sum_{j=0}^{2n-1} \kappa^j d(x \star y)$$
$$+ \kappa^{2n}(xdy) \qquad (y \in \Omega^{2n} A). \tag{12.3.12}$$

Proof We have isomorphisms $RA \otimes \bar{A} \cong \Omega^{odd} A$ via $x \otimes \bar{a} \mapsto xda$, and $RA \otimes \bar{A} \cong \Omega^1(RA)_\natural : x \otimes \bar{a} \mapsto \natural(xda)$. (i) The first identity follows from

$$\begin{aligned} x \star y - y \star x &= xy - dxdx - yx + dydx \\ &= [x, y] - dxdy + dydx \\ &= b(xdy) - (1 + \kappa)d(xdy). \end{aligned} \tag{12.3.13}$$

(ii) We prove this by induction on n. For $n = 0$, the identity holds since we have

$$\natural(xda) = xda. \tag{12.3.14}$$

Now suppose the identity holds for $n - 1$ and consider $y \in \Omega^{2n} A$ of the standard form $y = zda_1da_2$, where $z \in \Omega^{2(n-1)} A$. Then

$$\begin{aligned} \natural(x\delta y) &= \natural(x\delta(zda_1da_2)) \\ &= \natural(x\delta(z \star da_1da_2)) \\ &= \natural(x \star z\delta(da_1da_2)) + \natural((da_1da_2 \star x)\delta z) \end{aligned} \tag{12.3.15}$$

and we consider these terms separately. By induction the last term satisfies

$$\begin{aligned} \natural((da_1da_2 \star x)\delta z) = &-\sum_{j=0}^{n-2} \kappa^{2j}(da_1da_2 \star x \star z) + \sum_{j=0}^{2n-3} \kappa^j d(da_1da_2 \star x \star z) \\ &+ \kappa^{2n-2}(da_1da_2 \star x \star z), \end{aligned} \tag{12.3.16}$$

which involves $da_1da_2 \star x \star z = \kappa^2(x \star z \star da_1da_2)$, so induction applies. The other term is

$$\begin{aligned} \natural((x \star z)\delta(da_1da_2 \star x)) &= \natural\big((x \star z)\delta(a_1a_2 - a_1 \star a_2)\big) \\ &= \natural\big((x \star z)\delta(a_1a_2)\big) - \natural\big((x \star z) \star a_1\delta a_2\big) \\ &\quad - \natural\big(a_2 \star x \star z\delta a_1\big) \\ &= (x \star z)d(a_1a_2) - (x \star z \star a_1)da_2 - a_2 \star x \star z)da_1 \\ &= (x \star z)(da_1)a_2 + (x \star z)a_1da_2 - (x \star z)a_1da_2) \\ &\quad +d(x \star z)da_1da_2 + a_2(x \star z)da_1 + da_2d(x \star z)da_1 \\ &= -b((x \star z)da_1da_2) + (1 + \kappa)d(x \star z)da_1da_2 \\ &= -b(x \star y) + (1 + \kappa)d(x \star y). \end{aligned} \tag{12.3.17}$$

Now we identify the spaces themselves. We have the following array of exact sequences:

$$\begin{array}{ccccccc} \Omega^1 RI + I\Omega^1 R + dI & \longrightarrow & R \otimes I + I \otimes R & \longrightarrow & I & \longrightarrow & 0 \\ \downarrow & & \downarrow & & \downarrow & & \\ \Omega^1 R & \longrightarrow & R \otimes R & \longrightarrow & R & \longrightarrow & 0 \\ \downarrow & & \downarrow & & \downarrow & & \\ \Omega^1(R/I) & \longrightarrow & R/I \otimes R/I & \longrightarrow & R/I & \longrightarrow & 0 \end{array} \tag{12.3.18}$$

Hence

$$\Omega^1(R/I) \cong \Omega^1 R/(\Omega^1 RI + I\Omega^1 R + dI)$$

so

$$\left(\Omega^1(R/I)\right)_\natural \cong \Omega^1 R/\left([R, \Omega^1 R] + I\Omega^1 R + dI\right). \tag{12.3.19}$$

Likewise, one shows that

$$\Omega^1(R/I^n) \cong \Omega^1 R/(\Omega^1 RI^n + I^n\Omega^1 R + dI^n), \tag{12.3.20}$$

hence

$$\left(\Omega^1(R/I^n)\right)_\natural \cong \Omega^1 R/\left([R, \Omega^1 R] + I^n\Omega^1 R + dI^n\right). \tag{12.3.21}$$

□

12.4 The Hodge Approximation to Cyclic Theory

In this section, we show how the homology of $X(RA)$ with δ and β is related to the cyclic homology of A. The approach to periodic cyclic (co)-homology that we have been developing is based upon the idea of representing even and odd classes as traces and cyclic on cocycles on a nilpotent extension of the algebra. Note that $RA/(IA)^n$ is a nilpotent extension of A. We let the kth Hodge approximation to ΩA be

$$F^k = b\Omega^{k+1}A \oplus \bigoplus_{j=k+1}^{\infty} \Omega^j A,$$

so

$$\cdots \subseteq F^{n+1} \subseteq F^n \cdots \subseteq F^1 \subseteq F^0 \subseteq F^{-1} = \Omega A, \tag{12.4.1}$$

and there is a corresponding filtration of vector spaces

$$\cdots \longrightarrow \Omega A/F^{n+1} \longrightarrow \Omega A/F^{n} \longrightarrow \cdots \longrightarrow \Omega A/F^{1} \longrightarrow \Omega A/F^{0} \longrightarrow 0. \tag{12.4.2}$$

We note in particular the exact sequence

$$0 \longrightarrow F^0/F^1 \longrightarrow \Omega A/F^1 \longrightarrow \Omega A/F^0 \longrightarrow 0, \tag{12.4.3}$$

where

$$\Omega A/F^0 = A/b\Omega^1 A = A/[A,A] \tag{12.4.4}$$

and

$$\Omega A/F^1 = A \oplus \Omega^1 A/b\Omega^2 A. \tag{12.4.5}$$

Suppose that $HH_k(A) = 0$ for all $k > n$. Then $\Omega A \to \Omega A/F^n\Omega A$ is a quasi-isomorphism with respect to b which induces an isomorphism of the cyclic theories with respect to the mixed complexes. The nth approximation $\Omega A/F^n\Omega A$ gives the exact value of the cyclic theory of A. In the long exact sequence

$$\xrightarrow{S} HC_{n-1}(A) \xrightarrow{B} HH_n(A) \xrightarrow{I} HC_n(A) \xrightarrow{S} HC_{n-2}(A) \xrightarrow{B} HH_{n-1}(A) \tag{12.4.6}$$

the terms $HH_k(A)$ vanish.

(1) When A is separable, so A is projective as an A-bimodule,

$$\Omega A/F^0\Omega A = A/b\Omega^1 A = A/[A,A] \tag{12.4.7}$$

with $b + B$ gives the cyclic theory of A.

(2) When A is quasi-free, so $\Omega^1 A$ is projective as an A-bimodule, $\Omega A/F^1\Omega A$ gives the cyclic theory of A. We compute the homology of

$$X(A)\colon A \underset{d}{\overset{b}{\leftrightarrows}} (\Omega^1 A)_\natural. \tag{12.4.8}$$

We compare the spaces of the Hodge filtration with the $X(RA)$-complex of the preceding section. Between the complexes $X(R/I^{n+1})$ and $X(R/I^n)$, we introduce

$$\begin{array}{cccc} X^{2n+1}(R,I) & R/I^{n+1} & \rightleftarrows & \Omega^1 R/([R,\Omega^1 R] + I^{n+1}dR + I^n dI) \\ \downarrow & \downarrow & & \downarrow \\ X^{2n}(R,I) & R/I^n + [I^n, R] & \rightleftarrows & \Omega^1 R/([R,\Omega^1 R] + I^n dR) \end{array} \tag{12.4.9}$$

Proposition 12.4.1 *Under the isomorphism $X(RA) \cong \Omega A$ of Lemma 12.3.5, one has an isomorphism of quotients*

$$X^k(RA, IA) \cong \Omega A / F^k \Omega A \tag{12.4.10}$$

and the inverse limits over k satisfy

$$\widehat{X}(RA, IA) \cong \widehat{\Omega} A.$$

Proof For $X^{2n}(RA, IA)$, we have $\natural((IA)^n \delta(RA)) = \oplus_{k=n}^{\infty} \Omega^{2k+1} A$, and we recall from Theorem 12.3.5(ii) that $\beta = b - (1+\kappa)d$. So

$$\begin{aligned}[(IA)^n, RA] &= \beta\big((IA)^n \delta(RA)\big) \\ &= (b - (1+\kappa)d)\left(\bigoplus_{k=n}^{\infty} \Omega^{2k+1} A\right) \\ &= b\Omega^{2n+1} A \quad \text{mod} \quad \oplus_{k=n+1}^{\infty} \Omega^{2k} A. \end{aligned} \tag{12.4.11}$$

It follows that

$$\begin{aligned} RA/\big((IA)^n + [(IA)^n, RA]\big) &\cong \left(\bigoplus_{k=n}^{\infty} \Omega^{2k+1} A\right) \Big/ \left(b\Omega^{2n+1} A + \bigoplus_{k=n+1}^{\infty} \Omega^{2k} A\right) \\ &= \Omega A / F^{2n} \Omega A. \end{aligned} \tag{12.4.12}$$

□

Theorem 12.4.2 (i) *The cyclic homology of A is given by*

$$H_j(X(RA)) = H_j(\Omega A, b + B) = HC_j(A) = \begin{cases} \mathbf{C}, & \text{for } j \text{ even;} \\ 0, & \text{for } j \text{ odd;} \end{cases} \tag{12.4.13}$$

(ii) *the periodic cyclic homology of A is*

$$H_i\big(\widehat{X}(RA, IA)\big) = H_i\big(\widehat{\Omega}, b + B\big) = HP_i(A). \tag{12.4.14}$$

Proof See page 404 of [23]. As in Theorem 7.5.2 which related to ΩA, the proof of Theorem 12.4.2 uses the spectral decomposition for the Karoubi operator

$$\Omega A = P_1 \Omega A \oplus P_{-1} \Omega A \oplus Q \Omega A, \tag{12.4.15}$$

where the generalized eigenspaces are

$$P\Omega A = \mathrm{Ker}((1-\kappa)^2);$$
$$P_{-1}\Omega A = \mathrm{Ker}(1+\kappa);$$
$$Q\Omega A = \bigoplus_{\zeta:\zeta^n=1;\zeta\neq\pm 1} \mathrm{Ker}(\zeta-\kappa). \tag{12.4.16}$$

□

Lemma 12.4.3 *The decomposition is stable with respect to the operators b, d and κ, and is compatible with the Hodge filtration.*

(i) *On $Q\Omega A$, β is invertible, and $\delta = 0$.*

(ii) *On $P_1\Omega A$, $\beta = b$ as an operator $P_1\Omega^{odd} \to P_1\Omega^{ev}$, and $\delta = -nb$ as an operator $P_1\Omega^{2n} \to P_1\Omega^{2n-1}$, where b is exact on $P_{-1}\Omega A$.*

(iii) *On $P_1\Omega A$, $\beta = b - (2/n)B$ and $\delta = -nb + B$.*

Proof (i) We have

$$\beta^2 = (b-(1+\kappa)d)^2 = -(1+\kappa)(bd+db) = \kappa^2 - 1, \tag{12.4.17}$$

which is invertible on $Q\Omega A$, so β is also invertible and $\beta : \Omega^{odd} \cong \Omega^{ev}$. Hence $\delta = 0$.

(ii) *On $P_{-1}\Omega A$, we have $\kappa = -1$, so $\beta = b-(1+\kappa)d = b$, as an operator $P_1\Omega^{odd} \to P_1\Omega^{ev}$. Also on $P_{-1}\Omega^{2n}$, we have $B = \sum_{j=0}^{2n-1}\kappa^j d = 0$, so*

$$\delta = B - \sum_{j=0}^{n-1}\kappa^{2j} b = -nb. \tag{12.4.18}$$

(iii) *On the harmonic space $P_1\Omega A$, it is tempting to jump to the conclusion that $\kappa = 1$; but we have a generalized eigenspace and need to be more cautious. We do have the identities $\kappa d = d$ and $\kappa b = b$, so $B = \sum_{j=0}^{n}\kappa^j d = (n+1)d$ on Ω^n; hence*

$$\beta = b-(1+\kappa)d = d - 2d = b - \frac{2}{2n+2}B \tag{12.4.19}$$

on Ω^{2n+1}; whereas

$$\delta = B - \sum_{j=0}^{n-1}\kappa^{2j} b = B - nb \tag{12.4.20}$$

on Ω^{2n}.

(i) We have $Q = b(Gd)+(Gd)b = (b+B)(Gd)+(Gd)(b+B)$, so $Q\Omega A$ has zero homology, and $H_j(Q\Omega A, b+B) = 0$.

(ii) *We recall that* $bd + db = 1 - \kappa$, *so on* $P_{-1}\Omega$, *we deduce that*

$$b(d/2) + (d/2)b = 1;$$

hence b *is exact and* $P_{-1}\Omega A$ *has zero homology.*

(iii) *On* $P_1\Omega A$ *we can introduce a scaling operator* c *so that*

$$b + B = \begin{cases} c\beta c^{-1}\colon P_1\Omega^{odd} \to P_1\Omega^{ev}; \\ c\delta c^{-1}\colon P_1\Omega^{ev} \to P_1\Omega^{odd}. \end{cases} \tag{12.4.21}$$

Let $c_{2n} = c_{2n+1} = (-1)^n n!$ *and introduce* $c\omega = c_{|\omega|}\omega$. *Then on* Ω^{2n+1},

$$\begin{aligned} c\beta c^{-1} &= c\beta c_{2n+1}^{-1} \\ &= c\left(b + \frac{2}{2n+1}B\right)c_{2n+1}^{-1} \\ &= \frac{c_{2n}}{c_{2n+1}}b - \frac{2}{2n+1}\frac{c_{2n+2}}{c_{2n+1}}B \\ &= b + B. \end{aligned} \tag{12.4.22}$$

There is an isomorphism of complexes that results from this scaling,

$$\begin{aligned} c\colon P_1X(RA) &\overset{\sim}{\longrightarrow} P_1\Omega A; \\ c\colon P_1X^q(RA, IA) &\overset{\sim}{\longrightarrow} P_1(\Omega A/F^q\Omega A); \\ c\colon P_1\widehat{X}(RA, IA) &\overset{\sim}{\longrightarrow} P_1\widehat{\Omega}A, \end{aligned} \tag{12.4.23}$$

where the differentials on the right are $b + B$. *We deduce that* cP_1 *gives a quasi-isomorphism, and since all the homology resides in the component* P_1, *there is an isomorphism at the level of cyclic homology*

$$c\colon H_i(X(RA)) \overset{\sim}{\longrightarrow} H_i(\Omega A, b + B),$$

which passes to each stage in the Hodge approximation to cyclic homology,

$$c\colon H_i(X^q(RA, IA)) \overset{\sim}{\longrightarrow} H_i(\Omega A/F^q\Omega A, b + B))$$

and then passes to the completion

$$c\colon H_i\big(\widehat{X}(RA, IA)\big) \overset{\sim}{\longrightarrow} H_i\big(\widehat{\Omega}A, b + B\big), \tag{12.4.24}$$

which gives the isomorphism of periodic homology asserted in Theorem 12.4.2(ii). □

Lemma 12.4.4 *Let R and R' be quasi-free algebras and $\theta_t\colon R \to R'$ a differentiable family of homomorphisms that induce $(\theta_t)_*\colon X(R) \to X(R')$. Then $(\theta_t)_*$ is independent of t, up to chain homotopy, where*

$$\frac{\partial}{\partial t}(\theta_t)_* = D^* h + h D^*. \tag{12.4.25}$$

Proof of Theorem 12.3.1 Given $A = R/I$ with R quasi-free, we have used the universal case $A \sim RA/IA$ to compute $HP_i(A) = H_i(\widehat{\Omega} A, b + B)$ via $H_i(\widehat{\Omega} A, b + B) = H_i(\widehat{X}(RA, IA))$. Since R is quasi-free, there is by Proposition 8.5.6 a connection D on $\Omega^1 R$, or equivalently, a lifting $\ell\colon \Omega^1 R \to \Omega^1 R \otimes R$ such that $m \circ \ell = Id$. We need to check that $H_i(\widehat{X}(RA, IA)) = H_i(\widehat{X}(R, I))$, so the formula for $HP_i(A)$ is independent of the choice of R in $A = R/I$. Lemma 12.4.4 shows that the complexes $\widehat{X}(RA, IA)$ and $\widehat{X}(R, I)$ are quasi-isomorphic, and concludes the proof of Theorem 12.3.1. □

Theorem 12.3.1 has the following consequence.

Theorem 12.4.5 (Goodwillie) *If $A \to A'$ is a surjective nilpotent extension of algebras, then there is an isomorphism $HP_i(A) \to HP_i(A')$.*

Proof We write $A = R/I$, where R is quasi-free and I is an ideal in R. Next let J be the ideal in R such that $A' = R/J$. Then $I \subseteq J$, and J/I is nilpotent, so $J^N \subseteq I$ for some integer N. Now we have

$$\widehat{X}(R, I) = \lim_{\leftarrow} X(R/I^n) = \lim_{\leftarrow} X(R/J^n) = \widehat{X}(R, J), \tag{12.4.26}$$

and $\widehat{X}$ calculates the periodic cyclic homology by Theorem 12.4.2(ii). □

References

[1] E. Abe (1977) *Hopf Algebras*, Cambridge University Press.

[2] G.R. Allan (2009) *Introduction to Banach Spaces and Algebras*, Oxford University Press.

[3] W. Arveson (1976) *An Invitation to C^*-algebra*, Springer Verlag.

[4] M.F. Atiyah (1967) *K-theory*, W.A. Benjamin.

[5] M.F. Atiyah and R. Bott (1967) A Lefschetz fixed point formula for elliptic complexes I., *Annals of Math.* (2) **86**, 374–407.

[6] M.F. Atiyah, R. Bott and L. Garding (1973) Lacunas for hyperbolic differential operators with constant coefficients II, *Acta Math.* **131**, 145–206.

[7] M. Atiyah, R. Bott and V.K. Patodi (1973) On the heat equation and the index theorem, *Inventiones Math.* **19**, 279–330.

[8] M.F. Atiyah and I.G. MacDonald (1969) *Introduction to Commutative Algebra*, Addison–Wesley.

[9] A. Baker (2002) *Matrix Groups: An Introduction to Lie Groups Theory*, Springer.

[10] B. Blackadar (1986) *K-Theory for Operator Algebras*, Springer-Verlag.

[11] B. Blackadar (2006) Theory of C* algebras and von Neumann algebras, *Encyclopedia of Mathematical Sciences*, p. 122, Springer.

[12] L. Boutet De Monvel and V. Guillemin (1981) *The Spectral Theory of Toeplitz Operators*, Annals of Mathematics Studies, Princeton.

[13] L. Brown, R. Douglas and P. Fillmore (1977) Extensions of C^*-algebras and K-homology, *Annals of Math.* **105**, 265–324.

[14] J.F. Carlson, D.N. Clark, C. Foias and J.P. Williams (1994) Projective $A(\mathbf{D})$ modules, *New York J. Math.* **1**, 26–38.

[15] R. Carter, G. Segal and I. MacDonald (1995) *Lectures on Lie Groups and Lie Algebras*, Cambridge University Press.

[16] J. Cheeger, M. Gromov and M. Taylor (1982) Finite propagation speed, estimates for functions of the Laplace operator and the geometry of complete Riemannian manifolds, *J. Differential Geometry* **17**, 15–53.

[17] P.R. Chernoff (1973) Essential self-adjointness of powers of generators of hyperbolic equations, *J. Funct. Anal.* **12**, 401–414.

[18] A. Connes (1986) Non commutative differential geometry Chapter 1: The Chern character in K-homology, Chapter II: De Rham homology and non commutative algebra, *Publ. Math. IHES* **62**, 257–360.

[19] A. Connes (1994) *Noncommutative Geometry*, Academic Press.

[20] A. Connes (2000) A short survey of noncommutative geometry, *J. Math. Phy.* **41**, 3832.

[21] J. Cuntz (2013) Quillen's work on the foundations of cyclic cohomology, *J. K-Theory* **11**, 559–574.

[22] J. Cuntz and D. Quillen (1995) Algebra extensions and nonsingularity, *J. Amer. Math. Soc.* **8**, 251–286.

[23] J. Cuntz and D. Quillen (1995) Cyclic homology and nonsingularity, *J. Amer. Math. Soc.* **8**, 373–442.

[24] J. Cuntz and D. Quillen (1995) Operators on noncommutative differential forms and cyclic homology, geometry, topology and physics 77–111, *Conference Proceedings and Lecture Notes in Geometry and Topology*, Edited by E.-T. Yau, International Press Cambridge.

[25] E.B. Davies (1989) *Heat Kernels and Spectral Theory*, Cambridge University Press.

[26] P.A. Deift, A.R. Its and X. Zhou (1997) A Riemann–Hilbert approach to asymptotic problems arising in the theory of random matrix models, and also in the theory of integrable statistical mechanics, *Annals of Math. (2)* **146**, 149–235.

[27] P. Deligne (1982) Hodge cycles on abelian varieties, in P. Deligne, J.S. Milne, A. Ogus and K.-y Shih eds, *Hodge Cycles, Motives and Shimura Varieties* pp. 9–100, Springer Lecture Notes in Mathematics **900**, Springer.

[28] S.K. Donaldson and P.B. Kronheimer (1990) *The Geometry of Four-Manifolds*, Oxford Science Publications.

[29] R.G. Douglas (1980) *C^*-algebra Extensions and K-homology*, Princeton University Press.

[30] R.G. Douglas and V.I. Paulsen (1989) *Hilbert Modules over Function Algebras*, Longman Science.

[31] N. Dunford and J.T. Schwartz (1988) *Linear Operators Part I: General Theory*, Wiley Classics.

[32] N. Dunford and J.T. Schwartz (1988) *Linear Operators Part II: Spectral Theory*, Wiley Classics.

[33] D. Eisenbud (2004) *Commutative Algebra with a View toward Algebraic Geometry*, Springer.

[34] H.M. Farkas and I. Kra (1992) *Riemann Surfaces*, second edition, Springer.

[35] B.V. Fedosov (1994) A simple geometrical construction of deformation quantization, *J. Differential Geometry*, **40**, 213–238.

[36] O. Forster (2001) *Lectures on Riemann Surfaces*, Springer.

[37] D.B. Fuks (1986) *Cohomology of Infinite-Dimensional Lie Algebras*, Consultants Bureau.

[38] E. Getzler (1983) Pseudo-differential operators on supermanifolds and the Atiyah–Singer index theorem, *Comm. Math. Phys.* **92**, 163–178.

[39] E. Getzler and A. Szenes (1989) On the Chern character of a theta-summable Fredholm module, *J. Funct. Anal.* **84**, 343–357.

[40] M. Göckeler and T. Schücker (1987) *Differential Geometry, Gauge Theories and Gravity*, Monographs on Mathematical Physics, Cambridge University Press.

[41] J.A. Goldstein (1985) *Semigroups of Linear Operators with Applications*, second edition, Oxford University Press.

[42] M.J. Greenberg and J.R. Harper (1981) *Algebraic Topology: A First Course*, revised edition, Addison-Wesley.

[43] P.A. Griffiths (1985) *Introduction to Algebraic Curves*, American Mathematical Society.

[44] A. Grothendieck and J.A. Dieudonné (1971) *Eléments de Géometrie Algebrique*, Springer.

[45] A. Grothendieck (1964) *Éléments de Geometrie Algébrique IV: Etude locale des schémas et des morphismes de schémas*, Publications Mathématique, Institut des Hautes Etudes Scientifiques.

[46] P. R. Halmos (1982) *A Hilbert Space Problem Book*, second edition, Springer.

[47] B. Hartley and T.O. Hawkes (1970) *Rings, Modules and Linear Algebra*, Chapman & Hall.

[48] R. Hartshorne (1975) On the de Rham cohomology of algebraic varieties, *Publ. IHES* **45**, 5–99.

[49] R. Hartshorne (1977) *Algebraic Geometry*, Springer International Edition.

[50] A. Hatcher (1971) *Algebraic Topology*, Cambridge University Press.

[51] J.W. Helton, P.A. Fillmore and R.E. Howe (1973) Integral operators, commutators, traces, index and homology, in *Proceedings of a Conference on Operator Theory*, ed. P. Fillimore, pp. 141–209, Springer–Verlag.

[52] J.W. Helton and R.E. Howe (1975) Traces of commutators of integral operators, *Acta Math.* **135**, 271–305.

[53] E. Hille (1972) *Methods in Classical and Functional Analysis*, Addison-Wesley.

[54] P. Hilton (1971) *General Cohomology Theory and K-theory*, Cambridge.

[55] G. Hochschild, B. Kostant and A. Rosenberg (1962) Differential forms on regular affine algebra, *Trans. Amer. Math. Soc.* **102**, 383–408.

[56] L. Hörmander (1985) *The Analysis of Linear Partial Differential Operators* III, Springer-Verlag.

[57] N. Jacobson (1974) *Basic Algebra I*, W.H. Freeman and Company.

[58] A. Jaffe, A. Lesniewski and K. Osterwalder (1988) Quantum K-theory 1: The Chern character, *Comm. Math. Phys.* **118**, 1–14.

[59] G. James and M. Liebeck (2001) *Representations and Characters of Groups*, second edition, Cambridge University Press.

[60] M. Kapranov (1998) Noncommutative geometry based on commutator expansions, *J. reine angew. Math.* **505**, 73–118.

[61] M. Khalkhali (2013) *Basic Noncommutative Geometry*, second edition, European Mathematical Society.

[62] F. Kirwan (2012) *Complex Algebraic Curves*, Cambridge University Press.

[63] S. Kobayashi (1987) *Differential Geometry of Complex Vector Bundles*, Princeton University Press.

[64] E. Kunz (1980) *Introduction to Commutative Algebra and Algebraic Geometry*, Birkhauser.

[65] E.C. Lance (1995) *Hilbert C^* Modules: A Toolkit for Operator Algebraists*, Cambridge University Press.

[66] P.D. Lax (2002) *Functional Analysis*, John Wiley.

[67] A. Lichnerowicz (1982) Déformations d'algébres associées a une variété symplectique (les $*_\nu$ produits), *Ann. Inst. Fourier (Grenoble)* **32**, 157–209.

[68] J.-L. Loday (1991) *Cyclic Homology*, second edition, Springer.

[69] P. Malliavin (1997) *Stochastic Analysis*, Springer.

[70] Y.I. Manin (1991) *Topics in Noncommutative Geometry*, Princeton University Press.

[71] J.E. Mardsden and A.J. Tromba (2003) *Vector Calculus*, fifth edition, W.H. Freeman.

[72] V. Mathai and D. Quillen (1986) Superconnections, Thom classes and equivariant differential forms, *Topology* **25**, 85–110.

[73] H. Matsumura (1970) *Commutative Algebra*, W.A. Benjamin, Inc.

[74] H.P. McKean and I.M. Singer (1967) Curvature and eigenvalues of the Laplacian, *J. Differential Geometry* **1**, 43–69

[75] H. McKean and V. Moll (1999) *Elliptic Curves*, Cambridge University Press.

[76] M. Mulase (1984) Cohomological structure in solution equations and Jacobian varieties, *J. Diff. Geom.* **19**, 403–430.

[77] D. Mumford (1971) Appendix to Chapter VII in O. Zariski ed., *Algebraic Surfaces*, second supplemented edition, Springer.

[78] D. Mumford (2000) *Tata Lectures on Theta I*, Birkhauser.

[79] D. Mumford (2000) *Tata Lectures on Theta II: Jacobian Theta Functions and Differential Equations*, Birkhauser.

[80] G.J. Murphy (1990) *Introduction to C^*-algebras and Operator Theory*, Academic Press.

[81] M.A. Naimark (1972) *Normed Algebras*, Wolters-Noordhoff.

[82] V.I. Paulsen (1986) *Completely Bounded Maps and Dilations*, Longman.

[83] P. Petersen (2006) *Riemannian Geometry*, second edition, Springer.

[84] A. Prased (2011) An easy proof of the Stone–von Neumann–Mackey theorem, *Expo. Math.* **29**, 110–118.

[85] A. Pressley and G. Segal (1986) *Loop Groups*, Oxford Science.

[86] D. Quillen (1985) Determinants of Cauchy–Riemann operators over a Riemann surface, *Funct. Anal. Appl.* **19**, 31–34.

[87] D. Quillen (1989) Cyclic cohomology and algebra extensions, *K-Theory* **3**, 205–246.

[88] D. Quillen (1988) Algebra cochains and cyclic cohomology, *Inst. Hautes Etudes Sci. Publ. Math.* **68**, 139–174.

[89] D.G. Quillen, *Notebooks*, edited by G. Luke and G. Segal, Clay Institute, Oxford.

[90] M. Rørdam, F. Larsen and N.J. Laustsen (2000) *An Introduction to K-theory for C^*-algebras*, London Mathematical Society Student Texts, Cambridge University Press.

[91] J. Rosenberg (1994) *Algebraic K-theory and Its Applications*, Springer.

[92] J.J. Rotman (1988) *An Introduction to Algebraic Topology*, Springer-Verlag.

[93] H.H. Schaefer (1974) *Banach Lattices and Positive Operators*, Springer-Verlag.

[94] G. Segal (2011) Daniel Quillen obituary, *The Guardian*, Science, 23 June.

[95] G. Segal (2013) Reflections on Quillen's algebraic K-theory, *Quart. J. Math.* **64**, 645–664.

[96] I.R. Shafarevich (1977) *Basic Algebraic Geometry*, Springer-Verlag.

[97] I.M. Singer and J.A. Thorpe (1967) *Lecture Notes on Elementary Topology and Geometry*, Scott, Foreman and Company.

[98] M.F. Singer (2009) Introduction to the Galois theory of linear differential equations, in M.A.H. McCallum and A.V. Mikhailov eds, *Algebraic Theory of Differential Equations*, pp. 1–82, Cambridge University Press.

[99] M. Singer and M. van der Put (2003) *Galois Theory of Linear Differential Equations*, Springer.

[100] I.N. Sneddon (1972) *The Use of Integral Transforms*, McGraw-Hill.

[101] W.A. Sutherland (1975) *Introduction to Metric and Topological Spaces*, Clarendon Press.

[102] R.G. Swan (1962) Vector bundles and projective modules, *Trans. Amer. Math. Soc.* **105**, 264–277.

[103] J.L. Taylor (1975) Banach algebras and topology, in *Algebras in Analysis*, J.H. Williamson ed., pp. 118–186, Academic Press.

[104] M.E. Taylor (1996) *Partial Differential Equations II: Qualitative Studies of Linear Equations*, Springer.

[105] A. Terras (1985) *Harmonic Analysis on Symmetric Spaces and Applications*, Springer.

[106] N.E. Wegge-Olsen (1993) *K-theory and C^*-algebras: A Friendly Approach*, Oxford University Press.

[107] R.O. Wells (1980) *Differential Analysis on Complex Manifolds*, Springer-Verlag.

[108] H. Kalf, U.-W. Schmincke, J. Walter and R. Wüst (1975), On the spectral theory of Schrödinger and Dirac operators with strongly singular potentials, in *Spectral Theory and Differential Equations*, W.N. Everitt ed., pp. 182–226, Springer.

[109] C. Kassel (1987) Cyclic homology, comodules and mixed complexes, *J. Algebra* **107**, 195–216.

[110] C. Kassel (1988) L'Homologie cyclique des algèbras enveloppantes, *Invent. Math.* **91**, 221–251.

[111] M.F. Atiyah and I.M. Singer (1968) The index of elliptic operators. III Ann. of Math. **87**, 546–604.

List of Symbols

$[\cdot,\cdot]$ Lie bracket;

$\{\cdot,\cdot\}$ Poisson bracket;

A an associative algebra over $\mathbf{k}$, possibly noncommutative;

a_j generic elements of A;

$a_0da_1\dots da_n$ noncommutative differential form in $\Omega^n A$;

$\bar{A} = A/\mathbf{k}$ the unital algebra A reduced by the subspace of multiples of the identity;

$A^+ = A \oplus \mathbf{k}1$ augmented algebra;

$[A, M]$ commutator subspace of A-bimodule M;

b Hochschild differential;

$\mathcal{B}$ bar construction over a non-unital algebra A;

b', B differentials in the sense of Connes's noncommutative differential geometry;

$\mathcal{C}$ Calkin's algebra $\mathcal{L}(H)/\mathcal{K}(H)$;

$\mathbf{C}$ field of complex numbers;

$\mathbf{C}[0]$ complex numbers in degree zero;

$\mathbf{C}[G]$ complex group algebra over group G;

$\mathbf{C}[X, Y]$ algebra of complex polynomials in indeterminates X and Y;

$\mathbf{C}[\mathbf{X}]$ coordinate ring over an algebraic variety $\mathbf{X}$;

c_n Chern class;

ch Chern character;

$C(X; \mathbf{C})$ the space of continuous functions $f : X \to \mathbf{C}$;

$C_0(\mathbf{R}; \mathbf{C})$ the space of continuous functions $f : \mathbf{R} \to \mathbf{C}$ such that $f(x) \to 0$ as $x \to \pm\infty$;

$C^1(S^1; \mathbf{C})$ space of once continuously differentiable functions $f : S^1 \to \mathbf{C}$;

$C^\infty(S^1; \mathbf{C})$ space of infinitely differentiable functions $f : S^1 \to \mathbf{C}$;

C_q Clifford algebra associated with quadratic form q;

Cl_n real Clifford algebra of dimension 2^n;
Cl^0 equivalence classes of divisors of degree zero;
$\cos(t\sqrt{\Delta})$ fundamental solution operator for the wave equation;
$\cup$ or $\cdot$ cup product;
d the Kähler differential on a commutative algebra;
d the formal differential (on bar complex or tensor algebra);
δ operation on cocycles $\delta\colon C^n(A,M) \to C^{n+1}(A,M)$;
D an abstract Dirac operator;
$\mathbf{D}$ the unit disc $\{z \in \mathbf{C}\colon |z| < 1\}$;
Δ co-product;
Δ Laplace operator on Riemannian manifold;
$e, \hat{e}$ an idempotent in A;
E vector bundle over manifold;
$\exp_x$ exponential map of differential geometry at $x \in M$;
$\exp(\mathcal{A})$ image of the exponential map on a unital Banach algebra $\mathcal{A}$;
$\star$ Fedosov's product;
$\star$ free product;
F_θ curvature 2-form;
$\mathcal{F}$ set of Fredholm operators on H;
$\mathcal{F}$ unitary L^2 Fourier transform;
$\mathbf{F}$ set of finite-rank operators;
g^{ij}, g_{ij} Riemannian metric tensor on manifold and its inverse;
G Green's operator of abstract Hodge theory;
$G(\mathcal{A})$ the group of elements of a unital algebra $\mathcal{A}$ that have multiplicative inverses;
$[G,G]$ subgroup generated by the multiplicative commutators $xyx^{-1}y^{-1}$;
Γ Euler's gamma function;
GL_n set of invertible complex $n \times n$ matrices;
Γ^i_{jk} Christoffel symbol;
H complex separable Hilbert space;
H_+ Hardy Hilbert space;
H_- orthogonal complement of H_+ in L^2;
$\mathbf{H}$ quaternions;
H_n Heisenberg group;
$\mathcal{H}_n$ (cover of) Heisenberg group;
He_n the monic Hermite polynomial of degree n;
$\hbar$ positive parameter, suggestive of Planck's constant;
Hess Hessian matrix of second-order derivatives;
$\mathrm{Hom}_A(X,Y)$ A-module homomorphisms from X to Y;

I an ideal in A;
$\sqrt{I}$ radical of ideal;
J an ideal in A;
$\mathbf{k}$ a field characteristic zero;
$\mathcal{K}(H)$ algebra of compact linear operators on H;
κ the Karoubi operator;
$K_0(\mathcal{A})$ group in K theory;
$K_1(\mathcal{A})$ group in K theory;
λ a cyclic permutation operator;
L number operator for the quantum harmonic oscillator, aka the Ornstein–Uhlenbeck operator;
L algebra over $\mathbf{k}$;
L^2 square-integrable complex functions;
LIM Banach limit;
$\lim_{\leftarrow}$ inverse limit;
$\lim^{(1)}_{\leftarrow}$ derived inverse limit;
$\mathcal{L}(H)$ algebra of bounded linear operators on H;
$\mathcal{L}^2(H)$ ideal of Hilbert–Schmidt operators on H;
$\mathcal{L}^1(H)$ ideal of trace-class operators on H;
M Riemannian manifold;
m, μ multiplication on an algebra $A \otimes A \to A$;
$\mathbf{N}$ set of strictly positive integers;
N averaging operator in cyclic theory;
ν grading operator $\nu(\omega) = |\omega|\omega$, where $|\omega|$ is the degree of $\omega \in \Omega$;
∇ a connection or gradient;
$\natural : P \to P_\natural$ the quotient map to the A-bimodule P modulo the commutator subspace;
$O(n)$ group of real orthogonal $n \times n$ matrices;
Ω^n_A Kähler differential forms of order n over a commutative algebra A;
$\Omega^n A = A \otimes (\bar{A} \otimes \ldots \otimes \bar{A}) = A \otimes (\bar{A})^{\otimes n}$ noncommutative differential forms of order n;
ω curvature in cyclic theory;
$\Omega A = \oplus_{n=0}^{\infty} \Omega^n A$ the universal algebra of noncommutative differential forms;
$\Omega A^{ev} = \oplus_{n=0}^{\infty} \Omega^n A$ the universal algebra of even degree noncommutative differential forms;
$\Omega A^{odd} = \oplus_{n=1}^{\infty} \Omega^{2n-1} A$ the universal algebra of odd degree noncommutative differential forms;
$\Omega^1_{\mathbf{X}}$ space of holomorphic differentials on a compact Riemann surface;

$P_\natural = P/[A, P]$ the quotient of an A-bimodule P by the commutator subspace;
P the projection in abstract Hodge theory;
$\check{P} = \mathrm{Hom}_A(P, A)$ the dual of a (projective) module P over a ring A;
P pseudo-differential operator;
$\mathbf{P}^1$ complex projective space, as in the Riemann sphere $\mathbf{C} \cup \{\infty\}$;
$\pi_\hbar : \mathcal{S} \to \Psi DO$ family of maps from symbols in Schwartz's class to pseudo-differential operators;
π_1 fundamental group of a topological space;
$\mathrm{Pic}(\mathcal{A})$ Picard group, the elements of $K_0(\mathcal{A})$ with multiplicative inverses;
$\mathrm{Pic}_0(\mathcal{A})$ connected component of the identity in the Picard group;
$\mathrm{Proj}(A)$ set of projective modules over a ring A;
ΨDO^n set of pseudo-differential operator of order n;
Q parametrix of Fredholm (pseudo-differential) operator;
q quadratic form, used to define Clifford algebra;
$Q(A) = A \star A$ Cuntz algebra over A;
$\mathbf{R}$ field of real numbers;
RA tensor algebra over A, with unital elements identified;
SA suspension of an algebra A;
S^1 unit circle with centre 0;
S^2 sphere in $\mathbf{R}^3$, especially the Riemann sphere $\mathbf{C} \cup \{\infty\}$;
S unilateral shift operator on Hardy space;
S Connes's periodicity operator;
S_m symmetric group of permutations on m symbols;
$\mathcal{S}$ Schwartz class of rapidly decreasing smooth functions;
$\mathcal{S}^m$ Hörmander's class of symbols of pseudo-differential operators;
$sl(n, \mathbf{C})$ set of complex $n \times n$ matrices with trace zero;
$SO(n)$ special real orthogonal group of $n \times n$ real orthogonal matrices with determinant one;
$sp(n)$ symplectic Lie algebra;
$Sp(n)$ quaternionic symplectic group;
$S(V)$ algebra of symmetric tensors on vector space V;
σ cyclic forward shift;
σ spectrum, symbol map;
σ_e essential spectrum;
σ_p point spectrum (the eigenvalues);
$\sigma(X)$ symbol function;
τ a trace;
tr trace on endomorphism bundle over manifold;

trace trace on $\mathcal{L}^1(H)$;
Tr special trace, or space of traces
$\mathcal{T}$ Toeplitz algebra;
T^* cotangent bundle over manifold;
T^* adjoint of an operator $T \in \mathcal{L}(H)$;
$T(A) = \oplus_{n=0}^{\infty} A^{\otimes n}$ the universal tensor algebra over A;
$T_A M$ tensor algebra of a bimodule M over an algebra A;
$T_t = e^{-t\Delta/2}$ heat semigroup;
V vector space;
$\wedge V$ alternating tensors on vector space V;
X complex in cyclic homology theory;
X metric space;
X element of Lie algebra, operator;
$\mathbf{X}$ algebraic variety;
$[X, Y]$ set of homotopy classes of continuous maps $X \to Y$;
$[f, g]$ Lie bracket of f and g;
$\{f, g\}$ Poisson bracket of f and g;
$\mathbf{Z}$ rational integers;

Index of Subjects